QE389.625 .O72 1995 v.1
Origin and mineralogy of
clays

2-1-96

Origin and Mineralogy of Clays

Springer
*Berlin
Heidelberg
New York
Barcelona
Budapest
Hong Kong
London
Milan
Paris
Tokyo*

B. Velde (Ed.)

Origin and Mineralogy of Clays

Clays and the Environment

With 182 Figures and 7 Tables

 Springer

Dr. BRUCE VELDE
Laboratoire de Geologie
Ecole Normale Supérieure
UA 1316 CNRS
24 rue Lhomond
75231 Paris
France

ISBN 3-540-58012-3 Springer-Verlag Berlin Heidelberg New York

Library of Congress Cataloging-in-Publication Data. Origin and mineralogy of clays. p. cm. Includes bibliographical references and indexes. Contents: v. 1. Clay and the environment/ Bruce Velde [editor] ISBN 3-540-58012-3 (v. 1: Berlin: acid-free) 1. Clay minerals. 2. Clay. 3. Clay minerals—Environmental aspects. 4. Clay—Environmental aspects. I. Velde, B. QE389.625.072 1995 552'.5—dc20 95-24195

This work is subject to copyright. All rights are reserved, whether the whole or part of the material is concemed, specifically the rights of translation, reprinting, reuse of illustrations, recitation, broadcasting, reproduction on microfilm or in any other ways, and storage in data banks. Duplication of this publication or parts thereof is permitted only under the provisions of the German Copyright Law of September 9, 1965, in its current version, and permission for use must always be obtained from Springer-Verlag. Violations are liable for prosecution under the German Copyright Law.

© Springer-Verlag Berlin Heidelberg 1995
Printed in Germany

The use of general descriptive names, registered names, trademarks, etc. in this publication does not imply, even in the absence of a specific statement, that such names are exempt from the relevant protective laws and regulations and therefore free for general use.

Typesetting: Best-set Typesetter Ltd., Hong Kong

SPIN: 10467911   32/3136/SPS – 5 4 3 2 1 0 – Printed on acid-free paper

Preface

The need for a solid understanding of the chemistry and physical processes of the solid, silicate-dominated surface environment is at present great and growing. Clay minerals, the most abundant and chemically active parts of the surface mineral world, are the key to understanding the links between nature (life), its substrate (silicates), and a mastery of the total eco-system by man, but unfortunately, the traditions of the study of clays have not always reached in this direction. Past professional interest has most often been bogged down in problems of taxonomy (mineral names), the study of physical and chemical properties without their natural context, and a general imprecision in the definition of interactions that are potentially useful to industrial processes. The growing needs of an industrialized world were not considered. Clay mineralogists took their materials into the laboratory to better study them; agronomists simplified and then basically abandoned the identification of clays and their role in plant growth; and soil engineers abstracted the reality of the chemistry of clays to a neutral, physical use. The different fields of research dealing with the silicate surface environment began to ignore each other.

Today there is an increased awareness that the surface environment is the key to the future of industrial survival, both city and rural. The lessons of the Eastern bloc nations cannot be ignored, nor can we continue to ignore the consequences of human activity on the biosphere; we are part of it, after all. In order to begin to understand the interdependence of the mineral world and the biosphere, it is necessary to understand clay minerals, an agent of exchange in the chemistry of life.

This book, then, attempts to lay the basis of understanding – it is hoped in an understandable way—the why and what of clay minerals. The origins of clays are described in their chemical and geographic context, while the cycle of geological dispersion and concentration is discussed with regard to the type of clays that can be formed and where they can be found. The change and eventual destruction of clay minerals in the cycle of sedimentation, burial, and rock-forming is described, and finally, an explanation is given of the origin of clays which are useful to manufacturing (hydrothermal alteration). In this text, we hope to facilitate the discovery of the origin of clay minerals and their abundance in their various geological habitats. Of course, a brief explanation of the nomenclature and chemical differences between clays is given to start the beginner on his way.

With its five contributors, this book is obviously a group effort. These five people are friends as well as colleagues, and our common interest in writing these chapters was to be as pertinent to present day needs as possible. Hence we have attempted to give readers a sort of introduction for environmentalists to the somewhat confusing world of clay minerals and clay mineralogy. This, the first volume in the series, lays the foundations for a fuller understanding of the real problems of clays in the environment. The second volume then goes on to demonstrate the importance of clays in their different environments as they affect specific problems of interaction between the inanimate and living world.

The volume editor would like to thank his friends and highly competent fellow clay mineralogists for contributing to the endeavor: the organization and conception of the different chapters and the assemblage benefited greatly from their patience and good will. Our publishing editor must also be thanked for his patience. Most of all, our gratitude is due to the home computer.

We hope that the volume will lead young scientists to understand and use this knowledge and to develop a better understanding of clay problems, the use of clays, and the potential they have in limiting the nefarious effects of other actors in the industrial or agricultural cycle. Knowledge is power, and a little prevention based upon some knowledge will certainly go a long way in avoiding much regret in the future.

Paris, July 1995   B. VELDE

Contents

1	**Geology of Clays** B. VELDE	1
1.1	Why Clays Form	1
1.2	Where Clays Form	3
1.3	Clay Formation: The Chemical Necessity	5
2	**Composition and Mineralogy of Clay Minerals** B. VELDE	8
2.1	Introduction	8
2.2	Physical Properties of Clays	9
2.2.1	Particles and Shapes	9
2.2.1.1	Something About Clay Particles	9
2.2.2	Clays and Water	11
2.2.3	Clays in Water	13
2.2.3.1	Mixtures of Water and Clays	13
2.2.3.2	Clays in Water and Transport	14
2.2.3.3	Exchange of Ionic Species (CEC)	15
2.2.4	Summary	17
2.3	Crystallographic Structure of Clay Minerals	17
2.3.1	Tetrahedra	18
2.3.2	Octahedra	19
2.3.3	Layer Structures Through Linkage	19
2.3.3.1	Repeat Distances	20
2.3.4	Crystalline Water	22
2.3.5	Chemical Substitutions in the Structures	23
2.3.5.1	Charge on the Unit Cell for Different Layer Types	23
2.3.5.2	Charge on Ions in a Given Coordination Site	23
2.3.5.3	Substitutional Types	24
2.3.6	Substitutions and Mineral Species	27
2.4	Mineral Families	28
2.4.1	7 Å Minerals (One Octahedral + One Tetrahedral Sheet)	29

2.4.2	10 Å Minerals (Two Tetrahedral and One Octahedral Layer) ...	29
2.4.2.1	Neutral Layer (Charge Almost Zero)	30
2.4.2.2	High Charge Minerals: Mica-Like (Charge near 1, all Dioctahedral)	30
2.4.3	Low Charge (Expanding) Minerals: Smectites	31
2.4.3.1	Dioctahedral Expanding Minerals	31
2.4.3.2	Trioctahedral Expanding Minerals	32
2.4.4	14 Å Chlorites (Two Octahedral + Two Tetrahedral Unit Layers, 2:1 + 1)	33
2.4.4.1	Trioctahedral Chlorite Minerals	33
2.4.4.2	Composition and Substitutions	34
2.4.5	Mixed Layered Minerals	35
2.4.5.1	Mixed Layering Mineral Types	36
2.4.6	Sepiolite–Palygorskite	37
2.4.7	Iron Oxides	38
2.4.8	Zeolites	39
2.4.9	Summary	41
	Suggested Reading	41

3 Origin of Clays by Rock Weathering and Soil Formation
D. Righi and A. Meunier 43

3.1	Introduction	43
3.2	Weathered Rocks and Soils: The Major Factors in Their Development	44
3.2.1	General Organization: Soil and Weathered Rock Domains	44
3.2.2	Basic Factors in Weathering and Soil Formation	46
3.2.2.1	Climate and Water Regime Within the Soil Mantle	47
3.2.2.2	Rock Composition	52
3.2.2.3	Biological Factor: Vegetation and Soil Organic Matter	52
3.2.2.4	Age and Soil History	55
3.2.2.5	Topographic Effects: Translocations and Accumulations	57
3.2.3	Distribution of Major Soil Types at the Surface of the World	60
3.2.4	Structure of Weathered Rocks and Soils	62
3.2.4.1	Weathered Rocks: Inheritance of the Rock Structure	62

3.2.4.2	Soil Structures: Importance of Aggregation	62
3.2.5	Soil Structure-Porosity Relationship	66
3.2.6	Changes in Rock Density and Mechanical Properties	67
3.2.7	Water in Soils: Content and Chemical Potential	69
3.2.7.1	Soil-Moisture Retention Curve	70
3.2.7.2	Flow of Water in Soils and Weathered Rocks	70
3.2.8	Dissolution and Recrystallization Processes	73
3.2.8.1	General Statements	73
3.2.8.2	The Proton-Cation Exchange	76
3.2.9	Basic Factors for Phase Relation Analysis in Rock Weathering	78
3.2.9.1	From Microsites to Microsystems	78
3.2.9.2	Construction of Phase Diagrams	81
3.2.10	Summary and Conclusions	84
3.3	From Rock to Soil: The Granite Example	87
3.3.1	Clay Formation in Weathered Granite Under Atlantic Climatic Conditions	88
3.3.1.1	Weathered Profiles on Granitic Rocks: The First Stages of Weathering	88
3.3.1.2	Construction of Phase Diagrams	93
3.3.1.3	Summary	101
3.3.2	Soil Clays Developed on Granite Saprolite in the Temperate Zone	102
3.3.2.1	Methodology	102
3.3.2.2	Observations	104
3.3.2.3	Clay Genesis in Temperate Acid Soils	107
3.3.2.4	Composition and Properties of the Subfractions	107
3.3.2.5	Conclusion	114
3.3.3	Summary of Weathering Effects	114
3.4	Clays Formed During Rock Weathering	115
3.4.1	Weathering of Basic and Ultrabasic Rocks	115
3.4.1.1	Weathering Profiles	115
3.4.1.2	Weathered Macrocrystalline Basic Rocks	116
3.4.1.3	Weathered Macrocrystalline Ultrabasic Rocks	117
3.4.1.4	Phase Diagrams	119
3.4.1.5	Weathered Serpentinite	124
3.4.2	Weathering of Basaltic Rocks	126
3.4.3	Weathering of Clay-Bearing Rocks	129
3.4.3.1	Weathering of Glauconitic Sandstones	129
3.4.3.2	Weathering of Marls	131
3.4.4	Summary and Conclusions	133
3.5	Clays Found in Soil Environments	134

3.5.1	Clays in Soils from Cold and Temperate Climates	134
3.5.1.1	Nature and Rate of Clay Mineral Formation	135
3.5.1.2	Podzolization and Clay Mineral Evolution in the Temperate Zone: Influence of Organic Matter	138
3.5.1.3	Clay Illuviation in Soils Developed from Glacial Loess Deposits: Movement by Transport of Solids	140
3.5.1.4	Clays in Soils from Heavy Clay Rocks: Selective Transport of Clays	144
3.5.1.5	Summary	144
3.5.2	Clays in soils on Volcanic Rocks: The Short-Range-Ordered Minerals, Allophane and Imogolite	145
3.5.3	Clays in Soils Formed Under Tropical Climate Conditions	147
3.5.3.1	Equatorial Wet Zone: Kaolinite and Al, Fe-Oxyhydroxides in Ferralsols	148
3.5.3.2	Tropical Dry Zone: Smectites in Vertisols	152
3.5.4	Arid and Semi-arid Zones: Palygorskite in Saline Soils and Calcareous Crust	154
3.6	General Conclusions	155
	References	157

4 Erosion, Sedimentation and Sedimentary Origin of Clays
S. HILLIER 162

4.1	Introduction	162
4.2	Origins, Sources and Yields, and Global Fluxes of Clay Minerals	163
4.2.1	Origins of Clay Minerals in Sediments	163
4.2.2	Sources and Yields	165
4.2.3	Global Fluxes	168
4.3	Erosion, Transport and Deposition of Clay Minerals	169
4.3.1	Transport by Rivers	169
4.3.2	Transport in the Sea and Ocean	171
4.3.3	Deposition of Clay Minerals by Settling	177
4.3.3.1	Salt Flocculation	178
4.3.3.2	Differential Flocculation and Settling	180
4.3.3.3	Bio- and Organic Flocculation	180
4.3.3.4	Properties of Aggregates and Flocs	181
4.3.4	Erosion, Transport and Deposition by Wind	182
4.3.5	Erosion, Transport and Deposition by Ice	185

4.3.6	Modifications and Transformations During Transport and Deposition	187
4.3.6.1	Ion Exchange and Fixation	187
4.3.6.2	Pollutant Transport and Regulation	189
4.4	Authigenic (in situ) Formation of Clay Minerals in Sediments	190
4.4.1	Continental Authigenic Smectite	191
4.4.2	Marine Authigenic Smectites	193
4.4.3	Marine Glauconite and Glauconite/Smectite	195
4.4.4	Celadonite and Celadonite/Smectite	198
4.4.5	Non-Marine Glauconite and Ferric Illite	198
4.4.6	Minerals Related to Chlorites and the Verdine Facies	199
4.4.7	Sepiolite and Palygorskite	201
4.5	Mineralogical Patterns in the World Ocean	203
4.6	Environmental Interpretation of Clay Minerals	207
4.6.1	Sedimentary Environments and Provenance	207
4.6.2	Palaeoclimatic Interpretation of Clay Minerals	210
	References	214

5 Compaction and Diagenesis
B. VELDE 220

5.1	The Geologic Structure of Diagenesis	221
5.1.1	Compaction and Porosity	222
5.1.2	Temperature	226
5.1.3	Sedimentation Rate	226
5.1.4	The Kinetics of Clay Transformations	226
5.1.5	Chemically Driven Reaction in Clays	231
5.1.5.1	Solution Transport	231
5.1.5.2	Chemical Equilibrium Among Clay Particle	232
5.2	Major Progressive Clay Mineral Reactions During Burial Diagenesis	234
5.2.1	Mixed Layer Mineral Series	234
5.2.2	Silica Polymorph Change During Diagenesis	235
5.2.3	Zeolite Mineralogy During Diagenesis	236
5.2.4	Changes in Organic Matte	237
5.3	Sequential Mineralogical Changes During Burial Diagenesis	239
5.3.1	The First Kilometer	239
5.3.2	The Stability of Detrital Minerals	240
5.3.3	Clay Mineral Assemblages in the Second Kilometer of Burial	241
5.3.4	The Last Kilometers	242
5.4	Conclusions	244
	Suggested Reading	245

6	**Hydrothermal Alteration by Veins** A. MEUNIER	247
6.1	Introduction	247
6.2	Structure of the Hydrothermal–Wall Rock System	248
6.2.1	Central Deposit and Altered Wall Rocks	248
6.2.2	Fluid Injection Vein Type	249
6.2.3	Fluid Infiltration Vein Type	251
6.2.4	Fluid Drainage Vein Type	251
6.3	Alteration Mechanisms	253
6.3.1	Zone Formation	253
6.3.2	Zone Development	255
6.3.3	Kinetics of Zonation	256
6.3.4	Quantities of Fluids and Flow Regime in Fractures	260
6.3.5	Successive Fluid Circulation	262
6.4	Alteration of Pre-existing Clay Minerals	263
6.4.1	Layer-Charge Control of Clay Hydrothermal Reactions	263
6.4.2	Polyphase Clay Mineral Assemblages	265
6.5	Conclusion	265
	References	266
7	**Formation of Clay Minerals in Hydrothermal Environments** A. INOUE	268
7.1	Initial Statement	268
7.2	Definition of Hydrothermal Alteration	269
7.3	Geologic Settings of Hydrothermal Systems	270
7.4	Physico-chemical Nature of Hydrothermal Systems	271
7.4.1	Temperature and Pressure	271
7.4.2	Fluid Compositions	273
7.5	Formation of Alteration Minerals and Their Zoning	277
7.6	Classification of Hydrothermal Alteration	284
7.7	Distribution and Morphology of Alteration Zones	288
7.8	Case Studies of Hydrothermal Alteration	291
7.8.1	Acid-Type Alteration	291
7.8.2	Intermediate to Alkaline Types of Alteration ...	294
7.8.3	Deep Sea Hydrothermal Alteration	299
7.9	A Brief Summary on the Effect of Rock Type ...	303

7.10	Detailed Mineralogy of Selected Clay Mineral Types	304
7.10.1	Interstratified Illite/Smectite	304
7.10.1.1	Structural Variation	305
7.10.1.2	Chemical Variation	307
7.10.1.3	Morphology Variation	309
7.10.1.4	Stability	311
7.10.2	Rectorite	313
7.10.3	Dioctahedral Smectite	313
7.10.4	Sericite	314
7.10.5	Interstratified Trioctahedral Chlorite/Smectite (C/S)	315
7.10.5.1	Structural Variation	315
7.10.5.2	Chemical Variation	317
7.10.5.3	Stability	318
7.10.6	Trioctahedral Chlorite in Hydrothermal Deposits	318
7.10.7	Aluminous Chlorite and Chlorite/Smectite	320
7.11	Concluding Remarks	320
	References	321

Subject Index 331

List of Contributors

S. HILLIER
Macaulay Land Use Research Institute, Craigiebuckler,
AB92QJ Aberdeen, Scotland

A. INOUE
Department of Earth Sciences, Faculty of Science,
Chiba University, Chiba 263, Japan

A. MEUNIER
Laboratoire Argiles, Sols et Altérations UA 721 CNRS,
Université de Poitiers, 40 Ave Recteur Pineau,
86022 Poitiers, France

D. RIGHI
Laboratoire Argiles, Sols et Altérations UA721 CNRS,
40 Ave Recteur Pineau, 86022 Poitiers, France

B. VELDE
Laboratoire de Géologie UA 1316 CNRS,
Ecloe Normale Supérieure,
24 rue Lhomond, 75231 Paris, France

1 Geology of Clays

B. VELDE

The chapters which follow deal with the occurrence of clays in nature. This concerns the geology of clays: the geological processes which lead to the creation of clays, the transformation of clays and the destruction of clays in different geological environments. Clays, as is the case for most objects on the Earth, are ephemeral. They have a life span which is governed by their geologic history. Clays occur under a limited range of conditions in geological space [time and temperature (essentially depth)]. They are found mainly at the surface of the Earth: their origin is for the most part initiated in the weathering (rock–atmosphere interface) environment. Some clays form at the water–sediment interface (deep sea or lake bottom). A smaller number of clays form as a result of the interaction of aqueous solutions and rocks, either at some depth in the sedimentary pile or in the late stages of magmatic cooling (hydrothermal alteration). Although, this last occurrence is not of great extent it is very important to geologists as they have been called upon to aid human activity. Hydrothermal alteration often leads to the accumulation of useful heavy metals, such as gold, tungsten, and uranium. Therefore, when geologists encounter rocks which have altered to clay, and they determine that this occurred at depth (i.e. not weathering or superficial phenomena), they often take a sample in order to assess the heavy metal content. In addition to the accumulations of interesting metals, the extensive alteration of rocks due to hydrothermal alteration can produce pure clay deposits which are also of economic interest, e.g. clays for ceramics, and silica for industrial uses. Thus, as is often the case, the rare occurrence is of greater interest to humans than the more common, overall changes of geologic materials. However, *in problems of environmental importance, the general, common occurrence is of greatest importance and the rare case is more of an anecdote. Environment is concerned with the everyday, while industry is concerned with the exceptions.*

1.1 Why Clays Form

Most clays are the result of the interaction of aqueous solutions with rocks. The dissolution and recrystallization which occurs at this encounter is the process by which clay minerals are formed and transformed. Clays are not stable in anhydrous environments. The proportion of water, compared to that of the solids (rock), which interacts determines the rate and type of chemical reaction and ultimately the type of clay mineral formed. When large amounts of water are present, the solids in the rock tend to be very unstable and they dissolve for the most part. Dissolution is the first step of most water–rock interactions. The greater the renewal of the water input (rain or fluid circulation) the more dissolution will occur. As the ratio of water to rock approaches one, the reactions are more and more dominated by incongruent dissolution, in which certain elements go into solution and others remain in the solid state in the skeleton of the altered rock. The new solids are generally clay minerals. They are hydrated, having interacted with water, and they have a special physical structure which is very different from that of the pre-existing minerals which originally reacted with the aqueous solution. Because of their hydration, the newly formed minerals have a greater volume than the original minerals. Because the initial stages of alteration, and those that follow, include significant dissolution of rock material, the formation of clays results in an aggregate of lower density than the initial rock. During water-rock interaction voids are usually produced in the alteration or clay-forming process. The proportion of voids produced is a function of the relative amounts of water and rock which interact.

In Fig. 1.1 the relationship of the relative amount of water available compared to the interacting rock are indicated in such a manner that the different types of geological phenomena which give rise to clay minerals are outlined. Each of the geological environments indicated is the subject of a chapter in this book. The geological environments are *weathering*, deposition or *sedimentation*, burial which creates *diagenesis* and *hydrothermal alteration*. In geology, when one thinks of environmental problems one is in fact thinking of the interaction between man-made chemicals, or those chemicals that have been concentrated by the industrial activities of humans. This initial thought is, in fact, that of the soil–atmosphere interface. This is where agricultural chemicals are used, where they are dispersed and, often, where they are eventually

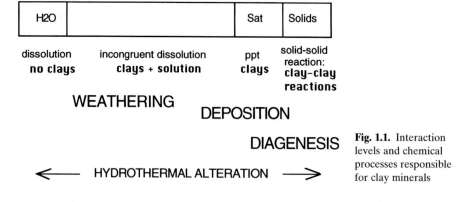

Fig. 1.1. Interaction levels and chemical processes responsible for clay minerals

destroyed by bacterial action. The pathways by which surface waters reach groundwater resources may bring the water into contact with compounds from the use of agrochemicals. Industrial wastes, either released by intent or accident, can be considered in the same framework. In this context attempts to centralize and neutralize man-made chemicals in waste disposal sites come into prominence. The structure and clay mineralogy of soils and surface systems is very important to an understanding of the problems of agro-chemicals and industrial pollution.

However, the environments of sedimentation and the burial of sediments are also very important for the eventual cycling of toxic materials. Environmental problems do not stop at the water table. For example, water-borne contaminants will follow the paths of water flow into rivers and streams, affecting the plant and animal life there. The sedimentation of suspended matter (clays) into sedimentary reservoirs will capture or reject these contaminant materials, either isolating them from the biosphere or concentrating them at certain levels. The sedimentary processes are important for environmental problems. Burial will eventually isolate sediments and pollutants from the biosphere. One must find out how this system works to be able to master its potential.

1.2 Where Clays Form

The different clay mineral environments are, of course, related in space, at or close to the surface of the Earth. The clay environment is limited to a certain range of temperatures and it is also limited in time. For most clays stability is in fact only attained at the very surface of the Earth, in say the upper several hundreds of meters of the Earth's crust. When temperatures exceed 50 to 80 °C the clays are unstable and they begin to change into other minerals, either other clay minerals or different mineral structures such as micas, feldspars, etc. The range of origin and evolution of clay, and its stability in time and temperature coordinates is given in Fig. 1.2. The different geological

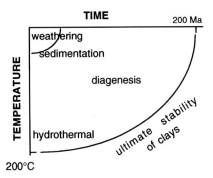

Fig. 1.2. Stability range of clays in T-t space under weathering, sedimentation, burial diagenesis and hydrothermal alteration conditions

environments such as *weathering*, *sedimentation*, burial *diagenesis* and *hydrothermal alteration* form the subjects of major chapters in this book. Long periods of time can cause changes in clay mineralogy. If temperatures are less long lived, i.e. for periods of days or years, the temperatures needed for clay formation can reach several hundreds of degrees centigrade.

The rate of change of clays is dependent upon temperature. This is the common law for chemical reaction; the higher the temperature, the faster the change. However, the rate of change in the temperature acting on the clays is also variable. In geological situations one can have rapid heating, such as the situation in which a magma intrudes a rock or a lava spills out onto the Earth's surface. The rapid increase in temperature (many hundreds of degrees centigrade) can be effected in the space of days or years, this would be a short lived geologic event. Tectonic events can produce fractures and thereby introduce hot hydrothermal fluids (i.e. high temperature, undersaturated aqueous solutions) Possibly associated with magmatic action. These fluids heat the rocks locally at approximately the same rates as the intrusion of magma, but to lower temperatures. Such events are rapid, and they create unstable mineral assemblages due to the high rate of thermal, and induced chemical, change. Therefore, such assemblages are highly localized in the Earth's crust.

By contrast, the normal sequence of sedimentation and burial, formation of sedimentary rocks from clay-rich sediments, is one that can take several millions, or hundreds of millions, of years. One finds that clays are highly transformed in some old, shallow basins (200 million years, 2 km deep, low temperatures of 80 °C) while they are much less affected in young deep basins (2 million years, 5 km deep, temperatures of 180 °C). Thus, the dimension of time can be as important as that of temperature in many instances of clay stability and transformation.

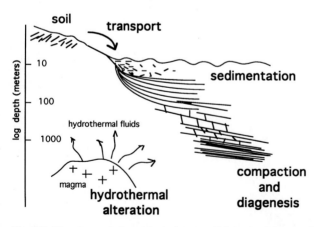

Fig. 1.3. The clay cycle in geological terms. Origin at surface leads to transport and deposition. More deposition leads to burial and transformation, gradually into rocks which in turn can be moved upward to be eroded and altered at the Earth's surface. Vertical scale is log of depth to emphasize the scale of the events. The transformation of clays and their origin is at the surface

In Fig. 1.3 one can see the spatial relations of depth and geological environment where one finds clays. The vertical scale in the figure is logarithmic, which de-exaggerates the importance of depth for the shallower phenomena. In fact, most clays originate at the very surface of the Earth and begin their migration shortly after their formation, by river transport, eventually reaching large bodies of water such as littoral ocean environments. Hydrothermal alteration is revealed at or near the surface of the Earth by the action of erosion, which brings deep-seated materials to the surface. As mountains are eroded their roots are exposed to the eye and the use of mankind.

The contact of rocks and water produces clays, either at or near the surface of the earth.

1.3 Clay Formation: The Chemical Necessity

The origin of clays is found in the interaction of rocks (silicate minerals) and water. This indicates not only that clays are hydrous but also that clays are more hydrous than the minerals in most rocks. The overall reaction of

$$\text{rock} + \text{water} \to \text{clay} \tag{1}$$

is a reasonable starting point. However, things are more complex than that. The mechanism by which water "hydrates" silicate minerals that of hydrogen exchange. Most clay minerals in fact contain (OH) molecules which have a specific role in the mineral structure. The only difference between, say, a potassium ion (K^+) and a hydrogen ion (H^+), contained in water, is that the hydrogen ion can be expelled from the mineral structure at lower temperatures than the potassium ion. In fact the potassium ion will be incorporated into another mineral instead of leaving the solid phase, whereas the hydrogen ion tends to form a gas (combining with oxygen to form water) leaving the system when a high enough temperature is reached (usually between 400 and 600 °C in clay minerals). In a very simple-minded way one can write the stability of clays as

$$\text{clay} + \text{heat} \to \text{rock} + \text{water,} \tag{2}$$

which is roughly the reverse process of clay formation.

The chemistry of the hydration mechanism is one of exchange of cations for hydrogen ions such as:

$$\text{feldspar} + \text{hydrogen ions} \to \text{clay (kaolinite)} + \text{cations, solids, water} \tag{3}$$

$$3KAlSi_3O_8 + 6H^+ = Al_2Si_2O_5(OH)_4 + 2K^+ + 4SiO_2 + H_2O.$$

In most clay-forming reactions, several phases are produced from an initial mineral species and water. The reaction produces clays and other minerals. Silica is a common by-product of hydration reactions; it can go into aqueous solution or it can form a solid phase such as quartz or amorphous silica. Production of clays is an incongruent process, one where solids and give rise to solids and material in solution.

The chemistry of the reverse, dehydration mechanism can be written as

$$\text{clay (kaolinite)} = \text{aluminosilicate} + \text{quartz} + \text{water}, \tag{4}$$

or,

$$Al_2Si_2O_5(OH)_4 = Al_2SiO_5 + SiO_2 + H_2O.$$

The dehydration process creates water but no soluble ions in solution. Thus the reactions are not "mirror images"; some materials is always displaced from solids into solution in clay forming reactions. Reaction (3) is not equivalent to reaction (4). Therefore, the origin of clays is in water and their destruction creates water; however part of the material of the initial anhydrous mineral from which the clay forms is lost to the altering solution.

The altering aqueous solution finds its way, most often, into the ocean where there is a large pool of dissolved ions derived from the alteration of silicate minerals into clay minerals. The ions of greatest abundance are Na, K and Ca. This transfer of material from one geologic environment, crystalline rocks, to another, soils or clay deposits and aqueous solution, is a fundamental process which is very important to an understanding of geological processes at or near the Earth's surface. In this surface environment (aqueous and low temperature) strong chemical segregations occur which effect redistribution of material from one place to several others. Sediments and sedimentary rocks, which are the solids resulting from alteration, tend to have compositions with few elements present. For example, carbonate rocks have high Ca concentrations, whereas sandstones are often over 90% SiO_2, and highly evolved soils tend to concentrate Al and Si.

Although the chemical necessity of exchange of hydrogen for mobile mono- and di-valent ions is the motor of clay formation at the Earth's surface, the process operates at different rates under different climatic conditions, and is very often not fully achieved before erosion strips off the partially reacted material. Sediments are most often a mixture of different phases in different states of chemical equilibrium with each other. The amount of chemical change is governed by the two determinant factors in reaction rate: time and temperature. The higher the temperature, the faster the reaction proceeds and the lower the temperature, the slower the reaction. Transformation of clays and other metastable materials such as amorphous silica on the ocean bottom (4°C) is very slow.

The range of temperatures in which clays form is from 4°C (ocean bottom) to approximately 400 °C (under short thermal pulses during hydrothermal alteration). The time spans can range from hours (laboratory experiments and intrusions) to hundreds of millions of years (burial diagenesis). Since time is a factor, the reaction rate is critical and the relative stability, or instability (distance from thermodynamic equilibrium), are the factors which determine the limits of the clay environmental conditons. Stability, and the rate of reaction or the time necessary for it to be achieved, is important for problems regarding burial containment, e.g. nuclear waste depositories. If clays are used in containment structures, they must maintain their physical and chemical characteristics (remain stable) during periods of time of up to a million years, in order to protect the future of humanity and to a certain extent that of other species present. In regard to soil formation, clays can form from rock fragments in hundreds of years and can form, or be destroyed, in clay-rich samples in a similar time span. Thus the regeneration of clays at the Earth's surface is a process which can take place on the time scale of human life.

The process of hydration of solids, clay formation, occurs at the earth's surface, but it is slow. This slowness means that some material is left in an "unfinished" state, and is not in equilibrium with its fellow particles. This gives the potential for transformation of altered products that is of the greatest importance when dealing with clays. In a soil, one finds clay minerals formed in place, one finds old but partially altered minerals from the parent rock and one finds some original minerals in an un altered state. This aggregate material has a high capacity for change, and it will continue to react if given enough time. Thus sediments have a high reaction potential, both in their initial sites of deposition and as they are buried and their ambient temperature increases.

If one changes the chemistry of the ambient soil or surface materials, for example by introducing an alkaline or basic solution into their milieu, there will be a high potential for mineral change. All the more so if the initial materials are only partially reacted. Thus in pollution problems involving chemical material there will be a high potential for change in the clay mineralogy of the surface material. Chemical spills in soils will certainly cause change to the minerals present and hence their physical behavior. This change can be hastened by changing the temperature of the system. Thus the stability and reaction rate of clay mineral formation is of the greatest importance in environmental problems.

The geologic causes of clay mineral formation are temperature change in chemical change. The different environments of clay mineral origin will be considered in much greater detail in the chapters that follow. However, one should first have a working idea of the mineralogy (chemical composition, crystal structure, and physical nature) of clay minerals, and this is found in the chapter which follows.

2 Composition and Mineralogy of Clay Minerals

B. Velde

2.1 Introduction

Clay minerals were initially defined on the basis of their crystal size. They were determined as the minerals whose particle diameters were less than 2 μm. This limit was imposed by the use of the petrographic microscope where the smallest particle which could be distinguished optically was of this size. Clays were essentially those minerals which could not be dealt with in a conventional nineteenth century manner. Chemical analyses were nevertheless made of fine grain size materials, most often with good results. However, the crystal structure and mineralogical family were only poorly understood. This was mainly due to the impurities present in clay aggregates, either as other phases or in multiphase assemblages. Slow progress was made in the early twentieth century, but the advent of reliable X-ray diffractometers allowed one to distinguish between the different mineral species found in the <2μm grain size fraction. Today we know much more about clay mineral XRD (X-ray diffraction) properties; perhaps too much at times.

In this introductory chapter, we wish to give an overview of the critical properties found in clay minerals in terms of the causes of these properties. This is the most important aspect of clay mineralogy and is crucial to the understanding of clays in nature. The chemical, internal structure of a clay mineral gives very specific characteristics of chemical reactivity. The small size and specific crystal shape give rise to other properties which are more physical. Both contribute to an interaction of the clays with their environment which make them important actors in interactions concerning the biosphere. Clays are, at the same time, physically and chemically active. They combine with water to make pastes, slurries and suspensions, by attracting water molecules to change their effective physical particle size. Clays take various chemical substances (ions or molecules) onto their surfaces or into the inner parts of their structure, so becoming chemical agents of transfer or transformation. Clays can easily exchange these ions or molecules for others given the correct chemical potential of the solution, or they can promote such reactions as hydrogenation. One can see that clays are extremely active and therefore should be studied with care in their role in the Earth's surface environment.

The path to understanding clays is, we believe, through a study of their different properties, chemical or physical. The methods of study (X-ray diffraction, thermal stability, and infrared spectroscopy among others) are almost as complicated as the subject they attempt to study. This is due to the small size of the objects which we wish to investigate which makes them obscure to normal visual investigation. The strong reliance on several different methods of investigation to determine the properties of a clay sample, tends to give us a composite understanding of the object. This is because not all of the properties of a given clay mineral are understood from each of the identification methods. Thus, the diagnostics for one clay mineral group using a particular identification method are different from the diagnostics for the same group using a different method. Such is the situation of science in the late twentieth century.

Much of our knowledge of clay minerals has been obtained using a specific analysis protocol which has not always been correlated thoroughly with results obtained by other methods used to determine other properties of the same types of particles. Hence, in the study of clay minerals, the determined properties and reasons for these properties are not always coherent. Thus the means is not always the message. For this reason, we will look at clays using a step-wise approach: first we will examine their intimate being (atomic structure and chemistry), then we will look at the properties revealed by different analysis methods currently in use.

Studies of clays properties can be divided into three major groups.

1. The *physical properties* of particle shape and the consequent surface areas which give particle "swelling" in aqueous solution.
2. The *crystallographic structure* and the disposition of the constituent atoms in a clay crystal.
3. The *mineral families* determined by chemical substitutions in the structures which gives rise to the chemical interaction of clays and other substances.

2.2 Physical Properties of Clays

2.2.1 Particles and Shapes

2.2.1.1 Something About Clay Particles

As mentioned above, clays are fine-grained minerals with particle diameters of <2 µm. This is the definition of a clay mineral, given in the nineteenth century, which applies to materials beyond the resolution of the optical microscope. This definition was not just one of ignorance; petrographers and mineralogists in those long-forgotten times were aware that there were entites beyond their

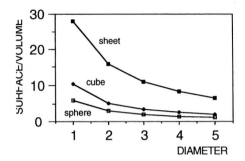

Fig. 2.1. Representation of the change in surface with respect to volume for different grain shapes. Sphere, cube and sheet (with dimensions of $1 \times 1 \times 0.1$ for the sides) are represented by the diameter (the greatest dimension within the volume of the element) and the ratio of surface to volume

power of identification. They saw that individual particles existed in the submicroscopic range, but they could not deal with them in a systematic manner as was possible for minerals of larger size. Thus the name "clay mineral" came into use to describe submicroscopic but crystallized material. As it turns out, most of the silicate minerals of this grain size found in nature have some very special mineralogical characteristics in common, and hence, *a posteriori* the choice of this name for a mineral group is very useful. However, it should be remembered that not all mineral grains in nature below the $<2\,\mu m$ range are of the same mineral type. Non-clay minerals, such as quartz, carbonates and metal oxides, can often form 10–20% or more of a naturally occurring clay-sized assemblage.

The small grain size of the clay crystals automatically gives them a special property; they have a large surface area compared with the volume of the particle. In general, the relative surface area of a grain increases as the diameter decreases (Fig. 2.1).

The properties of clays are in fact dominated by their surfaces. If the clay particles are not chemically active, i.e. charged electrically, they will behave much as other minerals of the same grain size and shape. However, the minerals most commonly called clay minerals have the particularity of being sheet-shaped (hence the name phyllosilicate). This means that they have even more surface area than other minerals of the same grain size but which tend to be cubes or spheres in their fine-grained state. The ratio of thickness to length for sheet-shaped clay particles is normally near 20, which is very large. This makes the surface area of a clay particle nearly three times that of a cube of the same volume. Thus, no matter what its specific surface properties, the surface of a clay mineral crystal is of great importance.

Clays minerals can be divided into three groups on the basis of particle shape:

flakes: sheets of equal dimension in two directions and a thickness of 1/20 in the other.
laths: sheets of a linear aspect where the width is great in one direction and

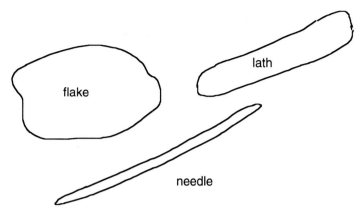

Fig. 2.2. Typical clay particle shapes

much less in the other. The thickness is always much less than the other two directions.

needles: two directions are similar in dimension whereas the last one is much greater. These are rare.

The shapes are shown in Fig. 2.2.

2.2.2 Clays and Water

Small mineral crystals have a very special effect on water molecules. The mineral surface attracts the polar water molecules through weak charge forces, and the crystals are covered by several layers of weakly bonded water molecules. These clay–water units change the physical properties of the aqueous solution. They "thicken" it, changing its viscosity. Thus the combination of minerals and water forms a material with a special state. The action of mixing small mineral particles into an aqueous solution is that of mixing dust and water to make mud. Any silicate mineral, the stuff of surface geology, will attract water molecules. Other mineral species, such as quartz or calcite, also exhibit this behavior. The large surface area compared with the small grain size is the determining factor which makes a plastic material such as mud.

In Fig. 2.1 one can see the difference in the relative surface area for different grain shapes such as spheres, cubes and sheet structures. The relation between particle diameter and the ratio of surface to volume (units squared divided by units cubed) varies greatly for the different particle shapes. The sheet structure, with the same width-to-length ratio but a thickness of only one tenth its length, has a very large surface-area-to-volume ratio which increases greatly as particle size (diameter) decreases.

However, the special sheet-like structures of clays allow them to be compacted more densely than minerals with other shapes since they can be stacked

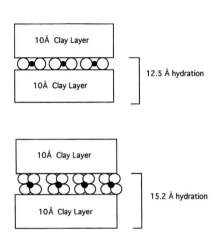

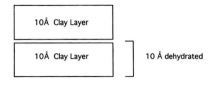

Fig. 2.3. Hydration states of clay minerals indicating a basic 2:1 layer plus a layer of water formed around an exchange cation. The thickness of the combined material is indicated for each state of "hydration". Distances given are in angstroms (Å)

very closely in parallel layers with large surfaces facing one another. This is not the case for spherical particles for example. More isometric shapes leave a significant pore space (up to 40% for spheres), whereas well-stacked sheets leave little pore space. Thus, mudstones (compacted clays), owing to pressure effects, where water is expelled by pressure, have a lower porosity than rocks comprising particles of other shapes, such a sandstones.

Some clay minerals have a special property which allows them to incorporate water molecules into their structure. This water changes the dimension of the clay particles as it goes into or out of the clay structure. These are called *expanding* or *swelling* clays. The other clays are called, by symmetry, non-expanding or non-swelling clays. The incorporation of water molecules into the clay structure is quite reversible under atmospheric conditions, being directly related to the ambient water vapour pressure and temperature. The more humid the air, the more water can be found in between the silicate layers of the clay structures. In the tropics, for example, the expanding clays will tend to be constantly hydrated, whereas those in deserts will only occasionally be hydrated or swelled.

The importance of this swelling is shown in Fig. 2.3. Swelling clays have a basic silicate structural sheet layer of 10 Å. The water introduced around a hydrated cation (usually with a 1^+ or 2^+ charge) forms either a two-layer structure of 5.2 Å thickness or, under less-humid conditions or higher temperatures, a layer 2.5 Å thick. Extreme hydration can produce a more ephemeral 17 Å three-layer structure. In aqueous solution one can form, at times, 19.5 Å structures. All in all, hydration can vary the volume of a clay particle by 95%. If your house is built on expanding clay, you had better be sure that it is either constantly hydrated or constantly dehydrated!

2.2.3 Clays in Water

2.2.3.1 Mixtures of Water and Clays

The importance of particle shape and size is that water molecules are strongly attracted to mineral surfaces. As indicated above, a water layer several molecules thick forms at the clay–water interface. Let us look at this interaction from the two ends of the spectrum: (a) water–clay mixtures and (b) clay–water mixtures.

(a) When clays are added to an aqueous solution, there is a gradual change in the structure of the water solution as the clay particles become more abundant. As more of the water itself is associated with the clays on surface layers, the bulk properties of the solution are modified; a slurry (suspension of clays in water) is formed which becomes viscous in proportion to the amount of clay present. The clay suspension densifies the aqueous solution and increases its viscosity. If other molecules, organic or inorganic, are associated on the clay surfaces, the clay acts as a carrier, keeping the other molecules homogeneously dispersed in the suspension. In modern industry, for example, clays are used to produce paint matricies, in which the pigment is dispersed and held in suspension by the clay particles.

(b) From the other end of the spectrum of physical properties, when one adds water to a clay powder, the clay picks up the water and distributes it around the particles. When relatively little water is present, and the clays are just covered with water layers, the result is a cohesive but plastic mass. The weak cohesive forces of these aggregates allow the particles to slide over one another, giving a certain plasticity, and the mixture is often called mud, at least by children. The properties of mud are well know to any potter and the plasticity of clay–water mixtures formed the basis of the first industries humans developed, those of ceramics and pottery. The easy absorption of the water allows one to model the resulting plastic material. Progressive drying leaves a coherent, solid material of another shape which when heated violently can be transformed into a useful rigid solid, such as a plate or mug.

The greater the surface area of a clay particle compared to its volume (i.e. sheet > lath > needle), the more the surface properties will be apparent in those of the aqueous mixture.

The two extremes of clay–water mixture proportions are useful to us in different manufacturing processes, from making bricks to applying the paints to color their surfaces.

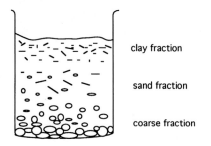

Fig. 2.4. Clay settling illustrated by the effect of different size fraction materials settling in a water-filled beaker. The settling time is a few minutes

PARTICLE SETTLING IN AQUEOUS SOLUTION

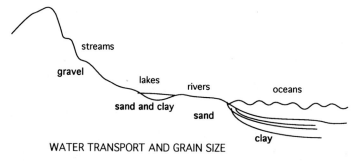

WATER TRANSPORT AND GRAIN SIZE

Fig. 2.5. Clay transport illustrated as a function of topographic relief. Streams in mountainous areas tend to deposit gravel (the coarse fraction) while lakes and streams on the flatter areas deposit sand and clay. The edges of great bodies of water are characterized by sand deposits near shore and clay deposits far shore

2.2.3.2 Clays in Water and Transport

A second important property of small particles is their ability to stay in suspension due to thermal agitation (Brownian motion). Small quantities of clay particles of <2 µm stay in aqueous suspension for many hours due to their small size. The time for which they remain in suspension is augmented by a flat shape which keeps them from falling rapidly, rather as a sheet of paper behaves when taken up in the air on a windy day. Again the large surface area of clays plays a part in their physical properties. The suspension of clay particles in aqueous solution tends to separate them from other minerals of the same grain size which do not have the sheet shape. The effect of particle size is shown diagrammatically in Fig. 2.4, in which particles of different sizes have been thrown into a beaker of water and settling has occurred for several seconds or minutes. This effect causes clays to be transported in aqueous suspension in preference to the other minerals of larger grain size and, as a result, the aqueous suspension preferentially moves clays from one area to another (stream flow, ocean currents) as shown in Fig. 2.5. The sedimentary deposits in mountain streams tend to be dominated by the heaviest particles.

Rivers depositing sediments on hills and plains tend to leave sand-dominated sediments, and oceans and lakes tend to have clay-rich sediments. Therefore clay particle size is important to the properties of an aqueous solution as well as the distribution of clays in a geological landscape. The transportation of clays over great distances tends to concentrate them as sedimentary layers which become shales upon deep burial in sedimentary basins or simply mud layers in lake sediments or along estuarine zones.

In summary, clay minerals are of small particle size. This gives them a large and highly reactive surface when in aqueous suspension. Mixtures of clays and water increase the viscosity of the solution with clays remaining in suspension for long periods of time.

The importance of the particle size, shape and composition is carefully dealt with in Section 4.3.3.2.

2.2.3.3 Exchange of Ionic Species (CEC)

A very important property of clay surfaces is their chemical activity and their interaction with ions in aqueous solution. In these solutions one almost always finds dissolved species. These are normally composed of charged ions or molecular species which can be attracted by a charged surface and *adsorbed* onto the surface. Clays have this charged surface. In some species of clays the activity of the surface (surface charge) is increased by a sort of internal surface into which charged ions or molecules can find their way. These ions are *absorbed* by the clays into internal crystallographic sites. The absorbed ions are normally accompanied by water molecules, expanding the clays, when in aqueous solution. In this way the chemical action of the surface area is increased greatly, by a factor of 25 or more! Figure 2.6 indicates the situations in which cations and water molecules can be attracted to clay mineral particles.

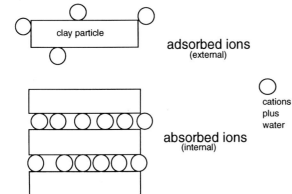

Fig. 2.6. Absorbed-adsorbed ions on clay and the site of their fixation on clay particles. Adsorbed ions are found within the structure (between the layers of the mineral structure) whereas the absorbed ions are found on the surface of the particle

The property of adsorbing and absorbing ionic species in solution is called *cation exchange capacity* (CEC). This capacity is measured in terms of the total number of charged ions which can be fixed onto the surfaces of clays. The measurement is that of the number of moles of ionic charge fixed on 100 g of dry clay. The values are expressed in milli-equivalents of charge (moles)/100 g. The reason for this strange measurement is that such values are often determined for bulk clay or soil samples where one has a mixture of clay and other phases present. If one is dealing with a pure species, it is easier to determine the molecular weight of the mineral and determine the number of moles of charged ions present with respect to a mole of clay. However, in mixed-mineral systems, only a general, bulk value can be determined, and hence milli-equivalents/100 g. Often the CEC is expressed in meq (milli-equivalents) without any reference to the other units of measure.

If the ions or charged molecules in solution can be attracted to the clay surface, a sort of selection process will operate when more than one species is present in the aqueous solution. The more, proportionally, an ion is present in solution, the more of it will be on the clay surface (law of chemical mass action). However, the attraction of the ions onto clay surfaces (internal or external) is not the same for all species. There is also a selection between different species of ions available or present in solution. Some are more strongly attracted to the clay surfaces than others. This selection effect depends upon the species of clay and its chemical constitution, as well as the affinity of the ions to remain in a free hydrated state in the aqueous solution. The composition of the aqueous solution, i.e. the concentration of ions in solution, can affect the attraction for clay sites as well. When an ion is held on a clay (adsorbed at the surface or absorbed within the clay) and displaced by another due to a change in its aqueous concentration, the ion is *desorbed*. If the desorbed ion is replaced by another ionic species introduced into the aqueous solution, it is *exchanged*. The process is known as ion exchange. For simple ionic species in solution, these relations are known as *cation exchange*. The normal laws of mass action are active in the exchange process. The differences or deviations from the ideal one for one exchange (exchange being in direct proportion to the quantity of that ion available for exchange) are of great importance, being the subject of many studies. The selectivity, i.e. preference of the clay for one dissolved species over another, is of great importance to the fate of material as it passes in contact with clays. Some ions are specifically selected onto the clays above their concentration ratio in solution, and hence the clays can be used to capture a specific ion in solution relative to other ions present. This effect is *cation selectivity*.

It should be mentioned here that not only cations in aqueous solution can be fixed on the clay surfaces or in internal sites. Organic molecules are often found to be attracted as absorbed or adsorbed species. Clays can be vehicles for transport of organic molecules. These properties can lead to very interesting situations in problems of waste hazards as one might expect.

The importance of CEC, exchange species and selectivity concerning clays in the sedimentary environments is dealt with in detail in Sections 4.3.6.1 and 4.3.6.2.

2.2.4 Summary

The interaction of clays with aqueous solutions and ions or molecules in solution, due to the surface properties of the clays, is very important. These surface properties play a role in the physical properties of solutions. Also, water modifies the properties of the otherwise dry clays to form a special material with great plasticity. Clays are easily transported by suspension in water and become concentrated in special geologic surface environments. The chemically active clay surfaces allow clays to transport other molecules or ions with them in the aqueous suspension as the clay particles are transferred by water movement. If the clays are fixed in place, as they are for the most part in soils, they play the role of filters as aqueous solutions pass through the soil in contact with the clays. Different ionic or molecular species are retained or released as aqueous solutions pass through the soils.

One can imagine the types of role that clays have in the Earth's surface environment. Absorption of ions or molecules in stream water, for instance, controls the chemical activity of these species as they interact with plants or animals. Such a control could mean life or death for fish or plants in contact with this surface water.

2.3 Crystallographic Structure of Clay Minerals

Clays are called phyllosilicates. This name is given because in most cases their grain shape is that of a sheet, it is much thinner than it is wide or long. This aspect has a fundamental cause. The inner structure, the bonding direction of the constituent atoms, is such that the strong forces are essentially in a two dimensional array. The stronger the bond the more tightly the atoms are held and, conversely, the weaker the bond the more likely it will be broken. Thus, because the bonds are easily broken in only one direction, a sheet structure results. Also, when the crystals are growing, they tend to grow faster in the strongly bonded direction and the result is the same as that for bond breaking, the extension of the crystal is essentially in two dimensions. The thickness compared to width and length in phyllosilicates is often about 1 to 20.

The ionic bonding in clays is highly covalent, roughly half of the ions present are oxygen, and, among the cations, silicon and aluminum are the major constitutents. These ions form highly covalent units which are com-

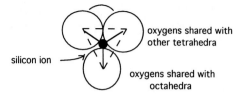

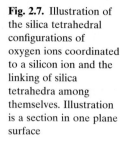

Fig. 2.7. Illustration of the silica tetrahedral configurations of oxygen ions coordinated to a silicon ion and the linking of silica tetrahedra among themselves. Illustration is a section in one plane surface

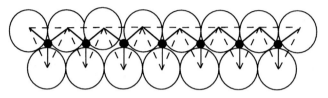

monly interlinked into what is called a *network*. The network is in fact the series of ions (cation and anion) linked by highly covalent bonds.

2.3.1 Tetrahedra

The networks of clay structures are built of interlinked polyhedra composed of oxygen anions and silicon, or frequently aluminum, cations. The majority of cations are silicon and hence the name silicates is given to clays and the other major earth forming minerals. The number of oxygens which can surround a silicon ion, for reasons of bonding orbital geometry and ionic size, is four. The resulting oxygen-defined polyhedron is a *tetrahedron*. The strongest characteristic of these coordination polyhedra is the interlinking through oxygen sharing. Figure 2.7 shows these relations.

In order to maintain electrical neutrality, the charge of the cations in the tetrahedral linkages should be equal to that of the linked oxygens. The positive cationic charge for silicon is four. This value is less than the total negative charge of the four coordinated oxygens, which is eight. This situation is possible due to the fact that each of the oxygens is shared with another cation. The oxygens are shared between tetrahedra in the x-y (*a-b* crystallographic direction) as well as the z direction as we will see later. Hence, only a part (one half) of their charge is compensated by cations within the coordination polyhedron itself.

In sheet silicate layers the linking (oxygen sharing of silicon atoms) for the tetrahedra is restricted to a planar array, i.e. no tetrahedral linking in the z direction, only in the x-y directions. This x-y linkage forms the model for the sheet, phyllosilicate, structure. It is assumed that all of the tetrahedral sites are filled by cations in the clay structures.

In a diagramatic representation of the tetrahedral units a triangle is usually used. This is shown in Fig. 2.7 by a dashed line, with the bonding of the silicon ions with the oxygens shown by a solid line.

2.3.2 Octahedra

Aluminum, magnesium or ferrous iron ions, for reasons of coordination, orbital directions and ion diameter, form polyhedra with 6 oxygens instead of the 4 oxygens found in the silica polyhedral structures. These are octahedrally coordinated polyhedral units (defined by the oxygen anions). This is shown in Fig. 2.8. The larger circles represent anions; either oxygens or hydroxyls.

Unlike the tetrahedra, in the octahedral linkages the number of cations can vary between two and three. For example, one could have 3 Mg^{2+} ions present or 2 Al^{3+} ions in the octahedral sites. The basic requirement is that a total positive charge of six be present for the three possible sites. When three ions are present, the structure is called *trioctahedral*, and when two cations are present, it is called *dioctahedral*. These two types of octahedral occupancy are fundamental to the classification scheme for clay minerals.

The diagramatic figure for the octahedral polygonal unit is usually a diamond (dashed line in Fig. 2.8). This indicates the complex 6-fold coordination of the octahedrally coordinated cations. The anions are not exclusively oxygens but also hydroxyl groups.

2.3.3 Layer Structures Through Linkage

In clay mineral structures, the tetrahedra do not occur alone. Pure silicon–oxygen structures form three-dimensional linkages, not the two-dimensional

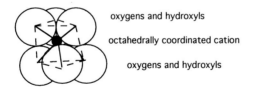

octahedral coordination

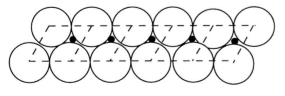

octahedral linkage

Fig. 2.8. Representation of oxygen–cation coordination in an octahedrally coordinated polyhedron and the linkage of these octahedra into a sheet unit or layer

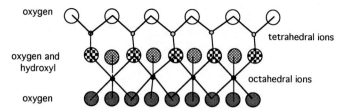

Fig. 2.9. Tetrahedral octahedral linkages through shared oxygens. Checkered ions are those shared between the two layers of different coordination polyhedra; tetrahedral and octahedral. Tetrahedra are represented by *triangles* and octahedra by *diamonds* in Figs. 2.7 and 2.8

sheet structures of phyllosilicates. Further, the octahedrally coordinated layers are rarely found alone. In most clay structures the tetrahedral layers and octahedral layers are linked together in the sheet structure. These sheets have two or more layers of tetrahedra and octahedra. The linkage between polyhedral types, tetrahedra to octahedra, is done through a bridging *apical oxygen*. This oxygen sharing occurs in the z direction of the structure. It is possible to link the layers of octahedra, which are linked between themselves, to layers of silica tetrahedra. When layer linkage occurs, the most common case for clay minerals, upper oxygens of the octahedral units are shared with tetrahedra. The oxygen of a tetrahedron which is shared with an octahedron is obviously one which is not already shared with other silica tetrahedra. The silica tetrahedra have upper oxygen ions which are shared with tetrahedral units and lower oxygen ions shared with octahedral ions. This linkage is shown in Fig. 2.9. The checkered ions are the linking oxygens shared between tetrahedra and octahedra.

2.3.3.1 Repeat Distances

The tetrahedral and octahedral units, when interlinked to form a sheet, have a given and constant thickness which is called thickness of the fundamental sheet structure. The thickness of the tetrahedral layers is considered to be 3.4 Å and the octahedral layers are thinner. Of course, when one layer is interconnected to another through a shared oxygen atom, the combined thickness of the two will be less than the sum of the two individuals. The combinations seen in natural minerals are as follows:

tetrahedral + octahedral layer = 7 Å unit layer, a 1:1 structure,
two tetrahedral + octahedral layer = 10 Å unit layer, a 2:1 structure,
two tetrahedral + two octahedral layers = 14 Å unit layer, a 2:1 + 1
    structure.

Within the structures, the di- and trioctahedral character (occupancy of the octahedral sites by two or three ions per unit cell) can be recognized by observing the *b* crystallographic dimension. This is found commonly by measuring the *060* peak position (using X-ray diffraction). The following results are obtained:

dioctahedral minerals have a 060 reflection near 1.50 Å,
trioctahedral minerals have a 060 reflection near 1.53–1.54 Å.

The repeat distances from one type of unit layer to the next, or layer thicknesses, are the fundamental means of identifying clay mineral species. The structural dimensions in the directions in the plane of the sheet (*a* and *b* crystallographic directions or x-y normal coordinates) are used much less frequently. These layer thickness dimensions can be measured by X-ray diffraction and electron microscopy.

The configuration in the dashed-line portion of Fig. 2.9 is used commonly in texts on clay minerals to show the structure of the linked units. A triangle is used as a symbol for a tetrahedral unit and a diamond for an octahedral unit.

It is also possible to label a composite linked layer using its tetrahedral and octahedral layer content, and represent the structures by parallel lines. In the short hand of clay mineralogists, the 7 Å tetrahedral–octahedral structure is called a 1:1 structure, the two-tetrahedral–octahedral structure a 2:1 structure and the two-tetrahedral–two-octahedral layer structure a 2:1 + 1 structure. This is schematically shown in Fig. 2.10. The reasons for this nomenclature will become a little more clear further on in the chapter.

A further complication, but a minor one, is the possibility of stacking the layer units, 7 or 10 Å layers, one-on-another in different ways giving different crystallographic structures at greater-than-layer distances. This effect is called polymorphism. The units are of the same structure and composition but they give a different internal "morphology" to the aggregate crystallite. Such

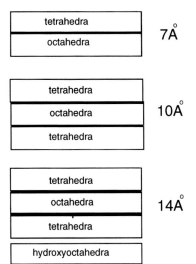

Fig. 2.10. Block diagrams of mineral types according to the combination of octahedral and tetrahedral coordinated sheets

polymorphs can be useful to distinguish different thermal origins for the same mineral phase.

In summary, the basic structural units of clays are silicon ions tetrahedrally coordinated to four oxygen ions, and aluminum or Mg, Fe ions octahedrally coordinated to six oxygen ions. The basic units share oxygen atoms in an array that links them to form sheet structures. Tetrahedra are also linked to octahedra to form a basic sheet structure. The combinations are 1 tetrahedral layer and 1 octahedral layer = 7 Å thick unit layer, 2 tetrahedral and 1 octahedral layers = 10 Å thick unit layer, 2 tetrahedral + 2 octahedral layers = 14 Å. The occupancy of the octahedrally coordinated site by either two or three ions determines another parameter for classification of clay minerals. The identification of the tri- or dioctahedral character is generally made by observing the 060 reflection which gives a peak near 1.50 Å for dioctahedral minerals and near 1.53–1.54 Å for trioctahedral minerals.

2.3.4 Crystalline Water

If only octahedrally coordinated cations are present in a structure, hydrogen ions are associated with the oxygens which form the coordination polyhedra. The hydrated, octahedrally coordinated mineral will be gibbsite, with trivalent aluminum ions and hydroxyls, or brucite with divalent magnesium. If the "top" or "bottom" oxygens are linked to a tetrahedral layer, as described above, through a shared oxygen anion, the number of hydrogen ions decreases. The octahedra bring hydrogen ions to the phyllosilicates; however only a portion of the upper layer ions are hydroxyl, the others being linked oxygens to the silica tetrahedra. If none of the oxygens are linked to another polyhedron type through apical ions, the oxygens are bonded to hydrogen ions forming a hydroxyl (OH) ion complex, and the whole octahedral unit is hydrated, i.e.

$(Al)_2(OH)_6$ = the one-layer mineral *gibbsite*.

If the upper oxygen anions of the octahedra are linked with silica tetrahedra to form a two layer clay structure (7 Å), the number of hydroxyl ions compared to cations decreases to give

$Si_2(Al)_2O_5(OH)_4$, as in the two-layer mineral *kaolinite*.

There are still hydroxyl units in the upper portion of the octahedral unit, which are not linked to the tetrahedra. If both the upper and lower cations of the octahedra are linked to silica tetrahedra, the resulting structure (10 Å) has even less hydrogen ions per cation, i.e.

$Si_4(Al)_2O_{10}(OH)_2$, as in the three-layer mineral *pyrophyllite*.

The decrease in hydrogen for a constant number of octahedrally coordinated ions (two aluminum ions in the cases cited here) shows that with an increase in layer linkage, i.e. more tetrahedral layers, the water content in the crystalline structure of the clay mineral decreases.

2.3.5 Chemical Substitutions in the Structures

2.3.5.1 Charge on the Unit Cell for Different Layer Types

It is possible to look at the ions and charge balance in a clay mineral either by considering the cations or the anions in a given portion of the structure. For example, one can consider the tetrahedral unit either as a silicon cation surrounded by oxygens or as four oxygen anions enclosing a silicon ion. In most chemical structural formulae for minerals, both the cation and anion content are given. The charge on both must match. Normally the cations are listed first (positive charges) and then the anions. For example kaolinite, a 7 Å dioctahedral mineral is given as $Al_2Si_2O_5(OH)_4$. The cations (two Si and two Al cations for a total charge of 14) are balanced by five oxygens and four hydroxyls of the same total negative charge. In fact the negative, cationic, charge is used to determine the total cationic charge when one calculates a structural formula. Anionic charge is a given, stochiometric constant in dealing with chemical determinations of mineral formulae.

The standard structurally linked phyllosilicate units of 7, 10 and 14 Å are given as follows (unit = oxygens/unit cell layer):

$7 \text{ Å} = O_5(OH)_4 = 14^{e-}$

$10 \text{ Å} = O_{10}(OH)_2 = 22^{e-}$

$14 \text{ Å} = O_{10}(OH)_8 = 28^{e-}$

2.3.5.2 Charge on Ions in a Given Coordination Site

Tetrahedra

Each cationic layer in a layer structure has a given charge per unit cell, e.g. the tetrahedral unit has a charge of +4 per site. In the silica tetrahedral unit each oxygen is shared with two cationic polyhedra, due to linking. This diminishes the average charge on each oxygen for a given cation bonding by a factor of two. Instead of an oxygen anion with a charge of two, each oxygen has a charge of one for a given silicon ion. This gives a total charge of four for the four oxygens of a potential negative charge of eight. Each tetrahedra is shared on average twice. It is assumed by convention that all of the silica tetrahedra are occupied in all clay structures, and that there are no hydrogen ions associated with the linked oxygen ions of the silica tetrahedra. Thus the layer charge is zero when all sites are occupied by silica.

Octahedra

The octahedral cation sites have a total charge of +2 per site. In the octahedra there are six anions, either oxygens or hydroxyls. These anions are arrayed in two layers encompassing the octahedral cations. A significant proportion of the anions in the octahedra are hydroxyl units.

The combination of octahedral and tetrahedral linkages into sheets of strongly linked cations gives rise to sheet structures characteristic by their thickness, i.e. the number and type of tetrahedral or octahedral layers linked into a sheet. Figure 2.10 shows the resulting sheet structures as they are formed by the linking of different combinations of tetrahedral and octahedral layers. For example, one octahedral layer and one tetrahedral layer form a sheet structure of 7 Å thickness. One octahedral layer between two tetrahedral layers gives a 10 Å thick structure and a tetrahedral–octahedral–tetrahedral plus hydroxy-oxide layer gives a 14 Å structure. The latter structure is often called a 2:1 + 1 structure. These are the basic units of clay mineralogy, the thickness of the sheet structures; 7, 10 and 14 Å.

2.3.5.3 Substitutional Types

Constant Charge (Solid Solution)

As in most natural minerals, different elements can be found in the two types of sites described above. Such a continuous array of compositions is called *solid solution*, referring to the gradual change in composition possible without abrupt gaps. For example, aluminum can substitute for silicon in the tetrahedral site and for iron or magnesium in the octahedral site. The major consideration in such substitutions is that of ionic charge balance. If an ion of the same charge is substituted into a site, no other compensation is necessary. For example if Mg^{2+} substitutes for Fe^{2+}, because they are both divalent no other compensation is necessary in the structure:

$$Mg^{2+} + Fe^{2+} + Fe^{2+} = Fe^{2+} + Fe^{2+} + Fe^{2+}$$
$$(6+) \qquad\qquad\qquad (6+)$$

Figure 2.11 shows substitutions of ions having the same charge in the same structural site (homo-ionic).

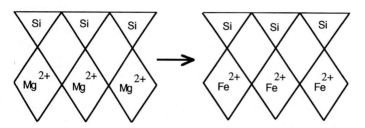

Fig. 2.11. Substitutions of ions having the same charge in the same structural site (homo-ionic)

Charge Imbalance Substitutions

Ionic substitution of cations with non-equivalent charges into one co-ordination layers of the clay structure (tetrahedral or octahedral) often necessitates a compensating substitution in another coordination layer of the structure. Charge imbalance is often compensated between the coordination layers. It can be compensated within the octahedral layer by changing the number of ions present.

There are four types of charge substitution which can create charge imbalance.

Constant Ionic Site Occupation

If Al^{3+} substitutes for Si^{4+} in the tetrahedral site, a charge imbalance is created in the tetrahedral site and so another substitution of charge compensation must be made elsewhere in the structure in order to bring the oxygen network forming the coordination polyhedra into electrostatic balance.

The ion pairs have a total of six charges thus the substitution conserves electrostatic charge balance on the clay structure. Subscript roman numerals are often used to indicate the type of coordination site. The value IV indicates tetrahedral coordination (four coordinated oxygens) and VI indicates octahedra (six coordinated oxygen ions).

In such a substitution scheme, the tetrahedral and octahedral cations are brought into electrostatic equilibrium by means of having a substitution of lower charge in one site and a substitution of an equivalent higher charge in the other. It should be noted here that the substitution of one ion for another in the tetrahedral site is accompanied by the substitution of an equivalent number (one) in the octahedral site of the structure. The total number of ions in the structure is constant as in the first charge-equivalent substitution scheme. Figure 2.12 shows the tetrahedral-octahedral substitution scheme in a 1:1 structure:

$$Si^{4+} + R^{2+} = Al^{3+}oct + Al^{3+}tet.$$

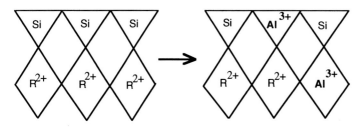

Fig. 2.12. Substitution in tetrahedral and octahedral layers, of 1:1 structure, of ions of different charge bringing the structure into electrostatic balance

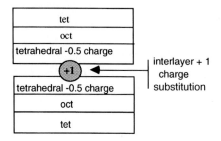

Fig. 2.13. Substitutions bringing about an interlayer substitution. Each unit layer (2:1 in these examples) contributes to the charge imbalance which necessitates the insertion of a compensating interlayer ion

Interlayer Substitutions

It is possible to change the charge on a given cationic layer (tetrahedral or octahedral) and compensate the charge imbalance created by inserting an ion between the unit layers of the structure. This occurs only in the case of 2:1 layer minerals, those having two tetrahedral layers and an intermediate octahedral layer. The basic interlayer distance is 10 Å (see Fig. 2.10).

If the substitution giving rise to the charge imbalance is in the tetrahedral site (for instance a substitution of Al^{3+} for Si^{4+}) the resulting charge deficiency (−1 in this case) will be compensated by the insertion of an ion of balancing (+1) charge between the layers of outer tetrahedral oxygen sheets facing one another. Both tetrahedral layers (oxygens and cations) on either side of the interlayer site will have a charge deficiency. Figure 2.13 illustrates this situation.

Non-Stochiometric, Di- and Trioctahedral Substitution

Another type of substitution which changes the number of ions in the structure is one exclusively found in the octahedral site. This is a change of two to three, or vice versa in the number of ions present. It is possible to fill the three sites in the octahedral layer, which are included in a layer silicate unit cell, with either three divalent ions ($3R^{2+}$) or with two trivalent ions ($2R^{3+}$). In the first case we speak of a *trioctahedral* structure, in the second of a *dioctahedral* structure, referring to the number of ions found in the three octahedral sites. This is a major subdivision of the clay minerals, defining their composition and hence their origin. The trivalent ions (usually Al^{3+} and Fe^{3+}) correspond to specific conditions of chemical potential, and hence form minerals under specific geological and environmental conditions. The same is of course true of trioctahedral minerals. The substitution scheme is shown in Fig. 2.14.

In clay minerals there is a lack of continuous substitution between the di- and trioctahedral minerals. This substitutional gap, or solid-solution gap, cre-

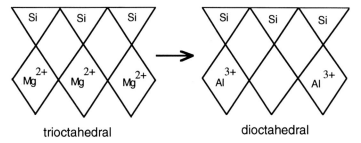

Fig. 2.14. Dioctahedral–trioctahedral substitution. Total charge is the same, but the number of ions varies thus giving rise to a non-stoichiometric substitution

ates the fundamental mineralogical differences in most clays. There is almost no substitution of the trioctahedral structure into the dioctahedral minerals. The dioctahedral minerals have almost exclusively two cations in their octahedral site. However, there is significant substitution of dioctahedral ions (two per three sites) into the trioctahedral structure. The number of octahedral ions ranges from 3 to about 2.5 and about 2.0 in these structures. Thus the trioctahedral end of the series shows substitution but the dioctahedral pole shows very little if any.

The di- and tri-octahedral nature of a structure can be detected using the crystallographic *b dimension*. This is most often observed as the 060 reflection, near 1.5 Å.

2.3.6 Substitutions and Mineral Species

The substitution of different ions, one for the other, of the same or different charge, gives rise to what is called solid-solution-type chemical variation in clay minerals. This means that a continuous range of compositions is possible for different crystals having the same structure. The idea that a single structure can have a certain variation or range in composition due to continuous substitution of ions, one for the other, gives an insight into the fundamental reality of mineral chemistry. Most natural minerals have not one, fixed and constant composition, but show variations in composition which respond to, or reflect, the changes in chemistry of their environment, as well as the limits which are determined by pressure and, above all, temperature. The mineral compositions found at low temperature, say 30 °C, will not be those found for the same mineral at 220 °C. These ranges in composition can be represented as compositional fields on appropriate composition plots. Figure 2.15 shows the ranges in composition of smectites as a function of their octahedral-site ionic composition.

As a result, a large proportion of clay minerals do not have a single, fixed chemical composition. Instead, their compositions are determined by the

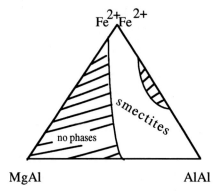

Fig. 2.15. Smectite solid solution as a function of octahedral ions in dioctahedral minerals. Representation is made in a triangular system in which all of the ion pairs present are summed to 100% and the relative positions of the compositions are shown as a function of the octahedral ions present in the mineral

chemical environment under which the mineral forms and is at equilibrium. As it turns out, clays have a large range of possible compositions at low temperatures which is greatly reduced as the ambient temperature is increased. The variability of clay compositions is strongly restricted when they occur at higher temperatures (e.g. at greater depth in the earth's crust). This is the reverse of what is known for silicates which have formed at higher temperatures.

At this point we can recapitulate the characteristics of clay by the following:

1. Clays are made up of sheets of tetrahedral and octahedrally coordinated cations bound to oxygens.
2. Clays never have only a tetrahedral or octahedral unit-layer structure, but have a multilayer structure in which the tetrahedra are linked to an octahedral sheet.
3. Clay structures frequently show ionic substitutions in the tetrahedral and octahedral sites which involve ions of the same charge or different charges. Overall electrostatic charge balance can be maintained when non-constant charge substitutions occur by substitutions in both tetrahedral and octahedral sites, in the octahedral layer alone, or by substituting ions in the interlayer site.
4. Gaps or discontinuities in substitutions often create mineral groups which are based upon structure and chemical compositional range.

2.4 Mineral Families

There are two different properties of clays which are used to describe and hence to classify clay minerals: their swelling properties (expanding minerals and non-expanding minerals) and the basic crystallographic repeat unit of the layer structures. These classifications encompass most of the varieties of clay minerals. There are a few strays, sepiolite and palygorskite, which are needle shaped and do not fit easily into the sheet silicate – clay mineral categories. Non-sheet structures of materials found in the clay-size fraction, oxides and zeolites, can be found in clay assemblages formed under conditions of clay mineral genesis. These minerals will be dealt with, briefly, at the end of this chapter.

The layer structure distance method gives the following categories when octahedral–tetrahedral linkage occurs: 7, 10 and 14 Å. The 7 and 14 Å (2:1 + 1) minerals never swell or change dimension, and thus only the 10 Å category will be subdivided upon its swelling or expandability properties.

2.4.1 7Å Minerals (One Octahedral + One Tetrahedral Sheet)

These mineral types can be divided into two clearly defined groups based upon the octahedral ion species present.

Kaolinite has only two Al^{3+} and very minor amounts of Fe^{3+} ions; they are, therefore, dioctahedral minerals. *Dickite* is a high temperature polytype (minor structural variation) of kaolinite.

$Al_2Si_2O_5(OH)_4$

Halloysite has the same composition and structure as kaolinite, except for an additional water layer between the layers which distorts the sheet structure into a tubular form. This water is easily removed at temperatures near 110 °C.

Berthiérine–Serpentines have between 3 and 2.5 ions present with the majority of the ions being divalent (2+). They are hence trioctahedral minerals. Substitution of Al or possibly Fe^{3+} into the tetrahedral site compensates for a proportion of the trivalent ions in the octahedral site. The substitution of from 3 to 2.5 ions is called the dioctahedral substitution.

$((R_2,R_3)_{3-2.5})$octahedral $(Si,Al)_2$tetrahedral $O_5(OH)_4$

Electrostatic balance is maintained by substitution such that one in every two divalent ions is replaced by a trivalent ion and a half vacancy, for an average occupancy of 2.5 ions

$(3R^{2+}) = 1.5R^{2+}R^{3+}$

2.4.2 10 Å Minerals
(Two Tetrahedral and One Octahedral Layer)

The various 10 Å mineral types are presented in Table 2.1.

Table 2.1. The various 10 Å mineral types arranged according to their charge and octahedral occupancy

Neutral	Low charge	High charge
Dioctahedral		
Pyrophyllite (Al)	Montmorillonite (Al,Mg)	Illite (Al,Mg)
	Beidellite (Al)	
	Nontronite (Fe^{3+})	Glauconite–Celadonite
Trioctahedral		
Talc (Mg)	Vermiculite (Mg,Al)	
	Saponite (Mg,Al)	

2.4.2.1 Neutral Layer (Charge Almost Zero)

These minerals are either dioctahedral or trioctahedral. Their neutral charge structure means that the basic layers are only weakly bound together. Since there is no charge imbalance no other ions are found between the layers, but because of their weak binding they can be displaced or pried apart easily. This gives the minerals a slippery texture due to layer sheets sliding one upon the other, and they are frequently used as lubricants. The basic 10 Å layers slide over one another.

Pyrophyllite, $Al_2Si_4O_{10}(OH)_2$. This mineral has Al in the octahedral site and Si in the tetrahedral site. The unit layer structure has two tetrahedral sheets and one octahedral sheet forming a 9.6 Å thick (almost 10 Å) unit layer. The cationic composition is very nearly exclusively Al and Si.

Talc, $(R^{2+})_3Si_4O_{10}(OH)_2$, always has nearly three divalent ions in the octahedral site with a small number of trivalent ions in the octahedral and tetrahedral sites. Two tetrahedral layers and one octahedral sheet give a 9.6 Å thick unit-layer structure. Significant Fe for Mg substitution occurs.

2.4.2.2 High Charge Minerals: Mica-Like (Charge near 1, all Dioctahedral)

In these minerals, the charge imbalance is between 0.8 and 1.0 and, as a result, there is an interlayer ion between the layer units which strongly binds the mineral into a coherent unit of from several to many 10 Å layers. These minerals are micas or mica-like minerals and they are exclusively dioctahedral in low temperature environments. The interlayer ion is almost exclusively potassium. The charge on the structures is always slightly less than 1.0 per unit cell, that of a mica, and hence these minerals should be called mica-like. They are not true mica structures. True micas have a different composition as far as charge is concerned and also they are generally found in rocks which have experienced higher temperatures. Micas are usually of greater grain size than clay minerals, i.e. >2μm.

Illite, $K_{0.8-0.9}(Al,Fe,Mg)_2(Si,Al)_4O_{10}(OH)_2$, is an aluminous, 10 Å mineral with some substitution of Fe_{3+}, Mg, and Fe_{2+} in the octahedral site, and some Al in the tetrahedral site, which gives rise to the greatest part of the layer charge imbalance. Si content is usually less than 3.50 ions. The mineral has two tetrahedral and one octahedral layer, with an interlayer ion population (potassium) holding the layers firmly together give a near 10 Å unit layer.

Glauconite–Celadonite, $K_{0.8}(Fe,Mg,Al)_2(Si,Al)_4O_{10}(OH)_2$, are the iron-bearing micaceous minerals with di- or trivalent iron greatly exceeding aluminum, and with Si content greater than 3.20 ions. The high iron content gives the minerals a green color and, when they are concentrated, this color is evident in hand specimen. These mica structures are very frequently interlayered with iron-rich smectites.

2.4.3 Low Charge (Expanding) Minerals: Smectites

Low charge on a 10 Å structure allows hydrated ions or polar ions to be insterted between the layers (absorbed); the average interlayer distance changes and the mineral is called *swelling* or expanding. This property gives these minerals a very special character. Essentially, a charge of 0.7 to 0.2 allows the layers to absorb hydrated cations and polar molecules between the 10 Å sheets. These minerals are given the group name of *smectites* which does not distinguish the specific type (di- or trioctahedral) of expanding mineral. The usual interlayer distance for swelling minerals is greater than the 10 Å layer unit due to the presence of hydrated cations. The normal basal spacings are 12.5 Å for a monohydration state (one water layer) and 15.2 Å for two water layers (Fig. 2.3). Commonly, ehtylene glycol is used to stabilize a swelling state which is independent of the humidity of the air. In this state the structure has a basal spacing of 17 Å. It is the most common reference state for swelling clays in laboratory mineralogical determinations.

2.4.3.1 Dioctahedral Expanding Minerals

Beidellite, $M^{0.n}(Al,Mg)_2(Si,Al)_4O_{10}(OH)_2 \cdot xH_2O$, where n < 0.5. This is an aluminous mineral, with two tetrahedral layers dominated by Si ions, in which the Al substitution is the major source of charge imbalance. The octahedral layer is minaly aluminous.

Montmorillonite, $M^{0.n}(Al,Mg,Fe^{2+})_2Si_4O_{10}(OH)_2 \cdot xH_2O$, is an aluminous mineral in which the two tetrahedral layers are almost exclusively occupied by silicon. The charge imbalance comes from divalent ion substitutions, Fe or Mg, for the trivalent aluminum ions in the octahedral site.

Nontronite, $M^{0.n}(Fe^{3+},Fe^{2+},Al)_2(Si,Fe^{3+})_4O_{10}(OH)_2 \cdot xH_2O$, is a ferric mineral with minor substitution of Al and occasionally Mg ions in the octahedral site. These substitutions, as well as some substitution of ferric iron in the tetrahedral site for Si^{4+}, give rise to the interlayer charge.

The different minerals given above, for the dioctahedral, expanding minerals, have some solid solution (i.e. intermediate compositions between the different types). The exact extent is not well known but it seems that there is a continuous sequence between the different types mentioned above.

2.4.3.2 Trioctahedral Expanding Minerals

The trioctahedral 10 Å structures do not form any high charge, micaceous phases in the clay environments, however, they do form expanding minerals, i.e. low charge phases. Unfortunately, not a great deal of information is available on exact ranges of composition for these minerals at present. There is no reason to believe that the substitution of Fe_{2+} for Mg is not extensive, yet for the most part, we know only of the magnesian end members. Hence, the chemical description is given schematically as being exclusively Mg-bearing, but it should be understood that such a description is a simplification.

These expanding, trioctahedral minerals are certainly common in soils, for instance, but they cannot be identified easily at present because they are rarely abundant enough to be identified with any precision concerning their chemical composition. Most often they go unidentified as specific species when accompanied by other expanding phases which are dioctahedral. Mineral mixtures of expanding clays tend to be called by the general term smectites and are left at that.

For the minerals which are sufficiently pure to be correctly identified, the following types have been described.

Saponite, $M^{0.n}(Mg,Al)_{3.0-2.5}(Si,Al)_4O_{10}(OH)_2 nH_2O$. These minerals have a chare imbalance largely dominated by substitutions in the octahedral site by the introduction of divalent ions and by the presence of vacant sites which lower the positive charge balance necessitating a compensation in the interlayer site.

Stevensite, $M^{0.n}(Mg)_{<3}(Si)_4O_{10}(OH)_2 nH_2O$. The interlayer charge imbalance is derived from a low octahedral site occupancy, giving rise to a low positive ion total. Little or no aluminum or iron has been found in these minerals.

Vermiculite, $M^{0.x}(Mg,Fe,Al)_{<3}(Si,Al)_4O_{10}(OH)_2 nH_2O$. These minerals are chemically known but the material most commonly used for their characterization comes from rather special environments, the hydrothermal alteration of biotites of igneous origin. The clay minerals from other environments, especially soils, which have the physical characteristics of vermiculites are not well identified. However, vermiculites can be described as being high charge smectites (i.e. minerals with charges between 0.7 and 0.5, compared to a mica with 1.0 charge) which do not expand nor contract fully under conditions of hydration and heating, seemingly holding much of the absorbed interlayer polar ions between the layers of the structures without interacting with their chemical environment. This is ascribed, for many soil clay minerals, to the existence of hydroxyl-ion structures in the interlayer site which do not behave as fully exchangeable ions in the way that those in normal smectites do. Possibly, the high charge induces a certain organization of a structured

Table 2.2. Two criteria to show the classifications of clay minerals

Layer type	Tetrahedral		Octahedral		Interlayer
	Majority	Minority	Majority	Minority	
10 Å					
Micas					
illite	Si	Al	Al	Mg,Fe	K
glauconite	Si		Fe	Mg,Al	K
celadonite	Si		Mg,Al	Fe	K
Smectites					
dioctahedral					
montmorillonite	Si		Al	Mg,Fe	Ca,Na
beidellite	Si	Al	Al		Na,Ca
nontronite	Si	F^{3+}	$Fe^{3e}+Mg$		Ca,K
trioctahedral					
saponite	Si	Al	Mg	Al	Ca,Na
7 Å					
Kaolinite	Si		Al		
Chamosite	Si	Al	Fe,Al,Mg		
14 Å					
Chlorite	Si	Al	Fe,Al,Mg		Mg,FeAl
Mixed layer					
Compound name	Si	Al	Fe,Al,Mg		Na,Mg,Fe Ca,K
Chain					
Sepiolite	Si		Mg		
Palygorskite	Si	Al	Mg,Al		

interlayer material tending towards that found in the chlorites (Table 2.2). A detailed description of the occurrence and identification of such minerals in soil environments is given in Sections 3.3.2.4 and 3.5.1.

2.4.4 14 Å Chlorites
(Two Octahedral + Two Tetrahedral Unit Layers, 2:1 + 1)

2.4.4.1 Trioctahedral Chlorite Minerals

In low temperature environments these minerals are strictly trioctahedral, with dioctahedral-type substitutions of trivalent ions (Al and Fe^{3+}) in up to half of the octahedral sites for the normal divalent ion. This substitution is also found in the berthiérine–serpentine (7 Å) minerals. Some substitution of trivalent ions (Al^{3+}) in the tetrahedral site occurs which compensates for a proportion of the trivalent ion substitution in the octahedral site. Thus, the

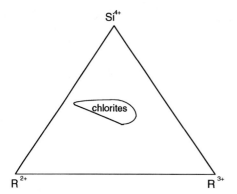

Fig. 2.16. Chlorite compositional range in Si–R^{2+}–R^{3+} coordinates where R^{2+} = Mg, Fe^{2+} and R^3 = Al, Fe^{3+} ions in the mineral structure. Here both the tetrahedral ions (Si and Al) and the octahedral ions (Mg, Fe, and Al) are represented. R^{2+} = Fe and Mg, R^{3+} = Al for the most part

chlorite compositions are the result of complex, simultaneous substitutions, presenting several types of ionic substitution at the same time. The basic structure of chlorites is in fact similar to that of a mica, or micaceous mineral – one where two layers of silica tetrahedra are linked by an octahedral layer, and where there is an interlayer substitution between the resulting 10 Å layers of the basic structure (see Fig. 2.13). However, the substitution of the interlayer ions, between the 10 Å layers, is not that found in micas, nor in expanding minerals, in that the interlayer ions in chlorites are in fact well organized hyroxy-complexes of divalent and divalent–trivalent ions. These interlayer substitutional units are similar to the structures of *brucite* $(Mg(OH)_2)$, for the divalent hydroxy-complex, and *gibbsite*, for the trivalent end-member $(Al(OH)_3)$. The charge balance is generally maintained both by substitutions in the octahedral–tetrahedral 2:1, 10 Å layer, structure and by compensating substitutions between this structure and the brucite–gibbsite layer. The decrease in layer charge on the 10 Å unit is compensated for by an increase in the charge on the interlayer unit.

2.4.4.2 Composition and Substitutions

Complete homo-ionic substitutions are known to occur between Fe^{2+} and Mg^{2+} in chlorites. This substitution changes the cell dimensions considerably, the Fe^{2+} ion being larger than the magnesium ion. Some Fe^{3+} is found in 14 Å chlorites, usually in minor quantities of less than 15% of the octahedral sites available. However Al^{3+} is common if not necessary to the chlorite structure. The amount of Al in the octahedral site ranges from 10% to around 33%. Substitutions of Al in the tetrahedral site range from 0 to 50%. Dioctahedral–trioctahedral substitutions can occur up to around 50%, that is the octahedral site occupancy can range from 3.00 to 2.50 ions.

These substitutions give a compositional range in coordinates of the major composition elements Si–R^{3+}–R^{2+} as shown in Fig. 2.16. Chlorite compositions as plotted in the compositional variables of Si, R^{3+} and R^{2+} ions where R^{3+} can be Al or Fe^{3+} and R^{2+} can be Mg or Fe^{2+}.

Minor substitution of other ions such as Ni, Zn etc. can also occur, and under certain circumstances they become major components. In general, the chlorites, from the highest temperature clay environments, hydrothermal, are more magnesian than those from surface environments. However, the chemistry of the environment of formation of the chlorites is certainly very important to their composition. Thus chlorites from diagenetic conditions of crystallisation (50–180 °C) can have different compositions, depending upon either temperature of formation or the bulk composition of the chemical system active in their formation.

2.4.5 Mixed Layered Minerals

The mineral types and structures described above are relatively simple, being composed of either two layer or three layer units to form either 7 or 10 Å minerals; chlorites can be considered to be a derivative of the 10 Å structure. There are, however, a relatively large number of cases where a single clay crystal is made up of a composite of different basic structures. Since the clays are phyllosilicates, the mixed layering occurs in the layer plane. For example, a layer of mica can be substituted for a smectite layer in a smecitite mineral. The result is a sheet by sheet chemical mixture on the scale of the crystallite. These minerals are called interlayered or mixed-layered minerals referring to their composite structure, which is a series of different layers of compositions corresponding to mineral species. They are generally considered to be a more or less stable assemblage (at least persistent) of different layers in crystallographic continuity. This being the case, definite crystallographic and thermodynamic properties should be found for them. However, this is not always so, as will now be discussed.

Mixed layer minerals seem to be, very often, an expression of change in mineral stability, one phase becoming unstable and another becoming stable. The change in phase is apparently effected by the production of a series of intermediate-composition crystallites which are actively changing their composition during the time over which the conditions which provoke the instability of the initial mineral are maintained. This observation seriously puts in doubt the idea of a stable phase status for the mixed-layer minerals. If the mixed-layer minerals are in fact an expression of a transition state, even though they are regularly and systematically present in nature, their thermodynamic status is difficult to treat because they are composed of, on the one hand, a stable phase and, on the other, an unstable phase.

Not all mixed-layer phases are in the category of intermediate minerals; some are formed in specific conditions with neither precursor nor apparent successor minerals. They do not show a gradual transition in bulk composition. However, whether or not mixed-layer minerals are stable phases, they do exist and can be characterised by X-ray diffraction and other methods. Therefore, mixed-layered minerals must be dealt with in a systematic manner. We will describe them here according to their crystallographic characteristics; those "seen" by X-ray diffraction.

2.4.5.1 Mixed Layering Mineral Types

Regular Mixed Layering

If the elements in a mixed-layer mineral are repeated with regularity the mineral is called a regular mixed-layer mineral. We will take the most prevalent case of two-layer types, here called A and B. The structure will look like that in Fig. 2.17.

The cases of regular mixed-layer minerals with equal proportions of two components are relatively limited, and the most abundant examples are of the following minerals:

mica:smectite

1. dioctahedral = illite/smectite, (a) *rectorite* (sodic mica) or
    (b) *allevardite* (potassic mica)
2. trioctahedral = biotite/smectite, *hydrobiotite*

chlorite:smectite

1. trioctahedral = *corrensite*
2. dioctahedral = *sudoite*
3. dioctahedral – trioctahedral = *tosudite*

These minerals are almost always found in high energy, or high temperature, situations such as hydrothermal alteration or the upper limits of diagenesis. In these instances the regular mixed-layer minerals seem to be proper phases, as they are not related to a precursor nor successor phase.

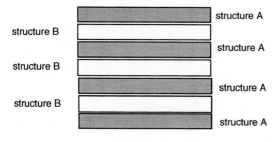

Fig. 2.17. Regular mixed layering with unit layer types A and B

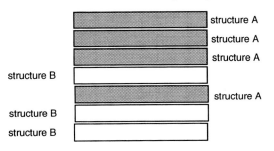

Fig. 2.18. Example of random mixed layering with two types of layer unit, *A* and *B*

Random Mixed-Layer Minerals

These minerals seem to be the most closely related to the transition between two different minerals, usually the two components in the complex mineral crystallites. Most often there are only two components but occasionally there has been evidence for three component minerals and there might be four-component crystallites, but as they would be difficult to identify, they have not as yet been detected.

A schematic diagram of such crystallites is given in Fig. 2.18.

The types of random mixed-layer minerals commonly encountered are as follows:

mica:smectite

1. illite/smectite
2. biotite/smectite
3. celadonite/smectite
4. glauconite/nontronite

chlorite:smectite

kaolinite:smectite

mica:chlorite

These minerals are more commonly assoicated with the phase change from one of the components (smectite usually) to the other, 10 or 14 Å. This change can be effected by a change in a chemical parameter, such as weathering, or by change in temperature, such as in burial diagenesis.

2.4.6 Sepiolite–Palygorskite

These clay-sized minerals are not sheet silicates in the strict sense. They have a needle-like morphology. However, they are hydrous with crystalline water

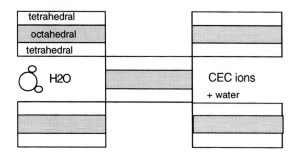

Fig. 2.19. Block diagram of sepiolite–playgorskite structures. The diagram is a cross section of a portion of a needle or lath-shaped crystal

Sepiolite - palygorskite structures

(OH units) and they have chemically absorbed water within their structures similar to that of smectites. They occur at the surface of the Earth, the low temperatures of clay environments. Hence they are classified as clays. Their internal crystal structure is based upon shortened layers. For all intents and purposes they are clay minerals, but with a special morphologic structure.

The sepiolite–palygorskite structure is largely two dimensional. The needles are formed by the structural linkage of tetrahedral–octahedral strips. These linkages are indicated in Fig. 2.19 in a schematic view of the linkage of the tetrahedral–octahedral unit chains in the a-b crystallographic direction. The structure extends in a semi-infinite direction into the plane of the page in the c direction. Water molecules and exchange ions are found in the inner sites inside the linked chains of tetrahedra–octahedra. In the inner channels, along the interlinked chains, exchange ions and water are found. The cation exchange capacity of these minerals is less than that of smectites, about one-third as much. Nevertheless, these clay minerals have much in common with smectites because they can absorb organic molecules and are reversibly hydrated.

The difference between palygorskite and sepiolite is in the composition of the unit cell. Both have similar a and b crystallographic dimensions, but palygorskite has a 12 Å repeat distance while sepiolite has a repeat distance of 10.5 Å. Palygorskite has a larger unit cell, and also contains much aluminum, whereas sepiolite is nearly aluminum-free being composed almost exclusively of magnesium and silicon. In most environments both minerals are almost exclusively magnesian, excluding the exchange cations of course. The aluminum substitution is similar to that in chlorites, extending from a trioctahedral pole towards a dioctahedral end member.

2.4.7 Iron Oxides

The oxides of iron are not generally considered as clay minerals although they are of small grain size. They are not silicates and, hence, are often neglected in the discussions of clay minerals. This is a great injustice to iron oxides because they are very apparent, despite their general low abundance in soils, sediments and sedimentary rocks. The simple fact of the presence of iron oxides is that they are very strong coloring agents. These colors are quite remarkable and they also tell us something about the chemical conditions under which the materials containing them formed. The colors are basically as follows:

Color	Mineral	Composition
red	hematite	Fe_2O_3
yellow to brown	goethite	α-FeOOH
orange	lepidocrocite	γ-FeOOH
black	maghemite	FeO

Under oxidizing conditions goethite is favored at low pH and hematite at neutral to high pH. Arid conditions and a lack of water also favor hematite. At low pH and under reducing conditions lepidocrocite is favored. Maghemite is favored by very low pH and very reducing conditions.

Hence the color of a soil, often given by very low quantities of iron oxide, as little at 0.5 weight percent, will reveal the chemical conditions (pH and eH) which are active in the soil environment.

2.4.8 Zeolites

Zeolites are common in some clay mineral environments. They generally indicate the existence of a high silica activity in the aqueous solutions affecting silicate crystallization. Zeolites are not phyllosilicates and for the most part they have crystal sizes above the 2μm limit given as a definition of clay minerals. However, their frequent association with clays in their finer fractions leads us to give a very brief treatment of these minerals.

Zeolite composition is dominated by the substitution of Al for Si accompanied by either K, Na or Ca ions. Charge balance is maintained of course, and hence a Ca substitution is the equivalent of two K or Na substitutions. As Al content increases so does alkali and alkaline-earth content. Water content (H_2O) increases with increasing silica content. The structural relations (crystalline space groups) are very complicated in zeolite mineralogy, and in fact the same chemical composition can be expressed in two different, and coexisting, minerals. Overlaps in composition and slight differences in crystal structure lead to a great complexity in zeolite mineralogy. This problem is greatly increased by the very active field of laboratory zeolite synthesis, which in many cases simplifies the problems of structure and composition but does not help in unravelling the problems of old classifications used by mineralogists for complex natural samples. The difficulties often arise from the differences in temperature of formation, the synthetic minerals being formed at higher temperatures and more rapidly than those in nature. Are the gaps in composition and structural continuity due to true structural differences or only to incomplete sampling or natural occurence? These problems are of course

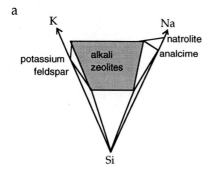

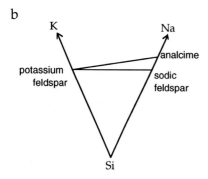

Fig. 2.20. The alkali zeolite mineral assemblages as a function of the variables K, Na and Si. Alkali zeolites form a more or less continuous series of compositions in the center of the chemical space shown (*shaded*). The pure sodic phases are natrolite, analcime and sodic feldspar. The potassic phases are potassium feldspar. **a** The phase associations under sedimentary and early diagenesis. **b** The phase relations of the highest temperature zeolite, analcime and the feldspars which replace the zeolites in later diagenetic stages

interesting, but they lead us some distance from clay minerals and we will not deal with them here. In fact we will give the rudiments of composition and nomenclature and leave it at that.

The zeolites can be, approximately, divided into alkaline earth dominated types and calcic ones (see Chap. 7 for a more exhaustive description of zeolite minerals). In general, the calcic zeolites are found in higher temperature environments than the alkaline earth forms. The Na, K zeolites have both higher and lower silica contents than alkali feldspars while the calcic forms have higher silica contents than the calcic feldspar anorthite. As temperature increases in geologic environments, the zeolites are replaced by feldspars or the mica equivalents. The temperature range of zeolite occureence is between 4 °C (ocean bottom) and towards 90 °C for alkali zeolites. There seems to be a certain kinetic relation which gives a lower overall temperature stability to alkali zeolites in older rocks. By contrast, calcic zeolites are found in low-grade metamorphic rocks where they are often major components in materials of basaltic or basic composition.

The sequence of common alkali zeolites in sedimentary and diagenetic environments is (as a function of increasing silica content) natrolite, analcime, phillipsite, erionite, (feldspar), stilbite, heulandite, clinoptolite, mordenite. Natrolite and analcime are almost exclusively sodic, whereas the others are Na–K dominated with various amounts of Ca present. The more common calcic zeolites are (in increasing silica content) (feldspar), gismondine, lawsonite, scolectite, chabazite, epistilbite (see Chap. 7).

Each of these minerals has a range of chemical compositions (alkali or alkaline earth) for Al substitution, and often compositions will overlap with those of another, neighbouring structure. Figure 2.20 shows the general relations of zeolites as a function of their Na, K and Si content. The central region indicates the area where the complex series of minerals of varying compositions lies. There are overlaps between one structure and another in chemical space. Most have some capacity to accept Ca ions into their structure. The most important observations regarding zeolites in clay environments is that they are cation exchangers, as are some clays, and the alkali zeolites are not stable at temperatures above 100 °C. They are highly siliceous and probably cannot coexist with stable quartz. They will be replaced by an assemblage of feldspar and quartz when true chemical equilibrium is attained.

These minerals are found in hydrothermal alteration of eruptive rocks and will be further discussed in Chapter 7. They are less common, but do occur, in the transformation of volcano-clastic material and, as such, are discussed in Chapters 5 and 6, which discuss sedimentation and diagenesis. Zeolites are extremely rare in weathering environments and hence are not discussed in Chapter 3.

2.4.9 Summary

The major groups of minerals are simple to define using simple structural criteria; types of minerals are defined using chemical criteria. As could be expected, the chemistry of the clay will reflect the chemistry of the environment in which it formed. Hence, the clay mineral species reflects the chemistry of its origin and can be used as a tracer of provenance.

There are of course many other names for specific minerals which contain small or large amounts of less common elements such as Ni, Cr, etc. These minerals are interesting but are not of vital interest to a useful comprehension of clay mineralogy.

This chapter has been written to give a general framework of understanding of the mineralogy of the different clay minerals most common in nature. The use of specific mineral names has been kept at a strict minimum, hoping to give a workable understanding of the problem of classification without boring the student with useless details.

Suggested Reading

General works

Barrer RM, Tinker PB (1984) Clay minerals: their structure behaviour and use. Proceeding of the Royal Society, London, 432 pp
Grim RE (1968) Clay mineralogy. Harper and Row, New York, 254 pp
Newman ACD (ed) (1987) Chemistry of clays and clay minerals. Mineralogical Society of Great Britain, Monograph 6, Longmans, London, 480 pp
Velde B (1985) Clay minerals: a physico-chemical explanation of their occurrence. Elsevier, Amsterdam, 423 pp
Velde B (1992) Introduction to clay minerals. Chapman Hall, London 193 pp
Weaver CE (1989) Clays, muds and shales. Elsevier, Amsterdam, 819 pp
Jasmund K, Lagaly G (eds) (1993) Tonminerale und Tone. Steinkopff Verlag, Darmstadt, 490 pp

Journals specializing in Clay Mineralogy

Clays and Clay Minerals. Jounral of the Clay Minerals Society (USA) P.O. Box 4416, Boulder, CO, USA
Clay Minerals. Journal of the European Clay Group, published by The Mineralogical Society, 41 Queen's Gate, London, SW7 5HR

Applications of clay minerals to specific fields of interest

American Mineralogist
Contributions to Mineralogy and Petrology
Journal of Sedimentary Petrology
Economic Geology
Geoderma
Soil Science
Journal of Soil Science
Journal of the Soil Science Society of America
Journal of Colloidal Science

3 Origin of Clays by Rock Weathering and Soil Formation

D. Righi and A. Meunier

3.1 Introduction

It is a fact that mankind's domain of influence at the surface of the planet is roughly that of clay mineral formation: soils, weathered rocks, diagenetic series, continental and marine sediments, geothermal fields. These clay resources have been exploited since the discovery of fire. It is now important, for environmental studies, to know as well as possible, how and where these minerals form. Curiously, among the numerous works published until now, only a few are devoted to the mechanisms of clay formation at the scale of a soil profile, i.e. the metric scale in temperate zones. Indeed, more is known at the scale of a country (km) or the mineral–fluid interface (nm). For example, at the scale of a country, weathering can be considered as a homogeneous process. As a consequence, it is possible to model chemical transfers and clay-mineral stability fields using calculation codes. On the other extreme, the intimate dissolution–recrystallization mechanisms at the fluid–mineral interface scale are studied on isolated pure crystals in order to simplify the chemical system.

The major difficulty of studying clay formation at the scale of eye observation (m to µm) lies in the heterogeneity of natural materials which results in an enormous variability of chemical and mineralogical properties. Nevertheless, it is only at this scale of observation that important problems in environment conservation can be approached. The aim of this chapter is to show that the apparent variability is not an effect of chance, but that it is in fact governed by a few general laws. Consequently, a methodology will be proposed to researchers involved in clay formation problems.

Any student concerned with soils and weathered rocks must be aware of the difficulty of drawing a boundary between these two formations. Besides the intrinsic complexities of soils and weathered rocks, one reason for this difficulty is that they were most often studied by different categories of researcher; i.e. pedologists and geologists. As no definition is completely satisfying, a distinction will be proposed here in order to enhance what is important for us and to avoid unnecessary details. It is hoped that the reader will be informed of the current knowledge regarding the processes controlling the formation of clay minerals in surface conditions. The primary objective is to document how natural processes work. The second objective is to set out a method of practicing petrology of altered rocks and soils, since such a method is the most useful tool to characterize the physical and chemical properties of materials subjected to human activities.

This chapter is composed of four parts which it is hoped will give fundamentals as well as examples of profile studies. The first part is devoted to the description of the influence of the main parameters in terms of soil formation, water–rock interactions and dissolution–crystallization processes. It includes an introduction to a general method of reasoning with chemical and mineralogical data using different types of phase diagrams. The second part is devoted to a detailed presentation of the formation of soils and weathered profiles on a chosen parent rock: the granite. The aim is to give an example of the changes in mineralogy, chemical composition and structures which effect a hard rock at the surface horizon under vegetation in temperate conditions. The third part presents the effects of the mineralogical and chemical compositions of rocks on the formation of a weathered profile. Several examples of profiles developed on magmatic, metamorphic and sedimentary rocks are presented. Mineral reactions are interpreted using phase diagrams. The fourth part is devoted to the description of the major categories of soils which can be observed in different climatic conditions.

3.2 Weathered Rocks and Soils: The Major Factors in Their Development

3.2.1 General Organization: Soil and Weathered Rock Domains

Weathering profiles and soils often show a polarity between the organic horizons at the surface and the rock underneath (Fig. 3.1). Although it is totally artificial to draw a line between the weathered rock and the soil overlaying it, such a separation must be made for practical reasons. As a limit must be put somewhere it will be considered that the lower limit of soil is reached where the original structure of the rock is still preserved. In other words, the soil is considered as the unconsolidated, restructured material built with materials derived from the weathered rock. Moreover, the soil structure may change seasonally through physical processes such as drying and wetting cycles or biological activity. In contrast, the weathered formation is generally characterized by a permanent structure inherited from the parent rock.

Other features make soil different from weathered rock and allow one to consider it as an original body:

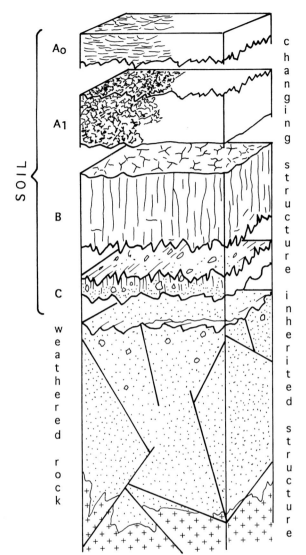

Fig. 3.1. Representation of the different horizons in a soil and weathered rock sequence

1. Soil is the place where the transformation of rock minerals into clays is most important. In some soils, all the original minerals have been transformed.
2. The upper soil layer is where plant roots and animals are living. After their death, organic matter from plants and animals in incorporated into soil material and intimately mixed with the mineral part of soil. After a complex course of transformations, including biological and chemical reactions, specific soil organics, the humic compounds, are formed. These humic compounds may contract close and strong associations with the mineral phase, especially with clay minerals. For most soils in the temperate zone, the

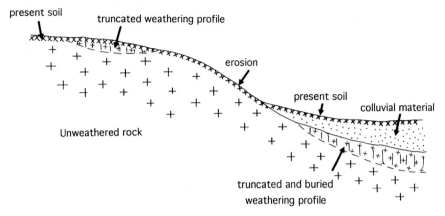

Fig. 3.2. Effects of erosion and colluviation on the ideal soil–rock weathering sequence

evolution of soil clays is related to that of organic compounds and the associations that they are able to form (Macias and Chesworth 1992).

In nature, the theoretical sequence from fresh to weathered rock and soils can be observed only at a few particular sites. These are in very stable landscapes where erosion is minimised. Usually, the sequence is disturbed by erosion and colluviation events; consequently the soil may develop from allochtonous materials, in regards to the rock underneath (Fig. 3.2). The present soil may also develop from a truncated weathering profile of which the upper part has been removed by erosion. This is typical for weathering profiles and soils developed in temperate and cold areas that were affected by glacial and periglacial erosion during the Quaternary period. Where ice fields were present, the whole soil and weathered rock would have been disturbed and removed, leaving either the fresh hard rock or thick glacial tills as surface materials. The present-day soils developed directly from the rock or glacial material. Periglacial disturbance was less drastic, generally preserving the lower layers of the weathering profiles. The present soil develops from this previously weathered material; however, in this case, the soil and the weathered rock underneath are of different ages and have not developed under the same climatic conditions.

3.2.2 Basic Factors in Weathering and Soil Formation

Weathering and soil formation do not lead to the formation of a homogeneous mantle over the Earth's surface. Several factors determine the transformation of rocks into soils of different types. These factors are climate, rock type (or parent material), vegetation, age and topography (FitzPatrick 1980; Duchaufour 1991).

3.2.2.1 Climate and Water Regime Within the Soil Mantle

Transformation of rock minerals into clays occurs through chemical reactions between the minerals and water which comes from rainfall. Therefore, rainfall is essential for weathering and soil formation. The amount of rainfall and its distribution over the year, determine the type of weathering and the type of soil that is formed. Temperature increases the kinetics of chemical reactions. Moreover, hot and humid conditions favour biological activity and, thereby, the transformation of organic matter. Temperature and rainfall define a climate; therefore it must be expected for soils to be of different types within the different climatic zones of the world.

Atmospheric and soil-temperature variations are the most important manifestations of the solar energy reaching the surface of the Earth. The main effect of temperature on soils is to influence the rates of reactions; for every 10 °C rise in temperature the rate of a chemical reaction increases by a factor of 2. The rate of biological breakdown of organic matter and the amount of moisture evaporating from the soil are also increased by a rise in temperature. Since most climates are seasonal, the rate of chemical and biological activity will vary during the year. In the warm season chemical weathering and biological activity are greater, providing there is an adequate supply of water. In the cool or dry season the speed of reactions is reduced. Thus, the rate of soil formation may vary seasonally. Soils have well-marked daily and annual temperate cycles. The daily variations extend to a depth of about 50 cm whereas the annual cycles extend to a depth of 2 m, below which the temperature is more or less uniform. During the daily cycle heat moves downwards during the day from the surface, which is warmed by solar radiation, and upwards during the night as the surface cools. During the annual cycle, in countries with contrasting seasonal climates, the soil is warmer during the summer and cooler during the winter.

Heat moves very slowly down through the soil so that the temperature fluctuations within the soil are greater at the surface than in the lower horizons. Daily fluctuations are strongly reduced at a depth of 25 cm as shown in Fig. 3.3 and, furthermore, the maximum temperature at depth occurs many hours later. The annual fluctuations of the temperature penetrate much deeper than the daily fluctuations. During summer, the diurnal mean surface temperature is higher than that in underlying layers, but in winter the reverse is true as shown in Fig. 3.4. The mean temperature at 2 m depth is approximately the same as the annual mean of the temperature of the air above the surface.

Soil temperature is mainly under the influence of two factors; latitude and slope orientation. Land surfaces normal to the rays of the sun are warmer than those at smaller angles. This is the case in tropical areas where the distance traveled by the sun's rays through the atmosphere is also shorter. To the north and south of the tropics, surfaces normal to the sun's rays are also warmer but, as the distance from the equator increases, the mean annual temperature of

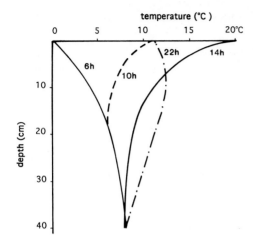

Fig. 3.3. Soil temperature at different depths at 6:00, 10:00, 14:00 and 22:00 h for a temperate climate

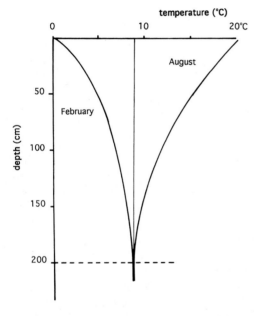

Fig. 3.4. Soil temperature at different depths in February and August under temperate climatic conditions

such surfaces steadily decreases. Differences in temperature between surfaces are particularly important in the middle latitudes where sloping land surfaces facing the equator generally have a warmer and drier climate than lands facing towards the poles. Altitude also influences climate. In the middle latitudes, mean annual temperature decreases at about 0.8 °C for each 100 m rise in elevation. Fluctuations of the soil temperature are strongly buffered by the vegetation cover. During the day, a large proportion of the radiant energy is adsorbed or reflected by vegetation. At night, vegetation reflects the heat radiated from the warm soil to the cooler atmosphere. The buffering effect of

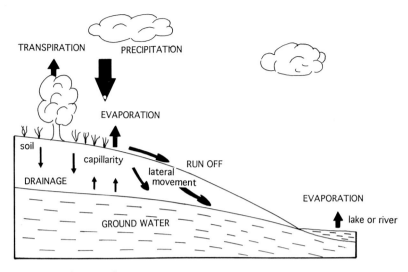

Fig. 3.5. Moisture cycle

vegetation is of particular importance in the middle latitudes where it prevents rapid heat loss during the winter and reduces the penetration of frost into the ground.

Not all of the water from rainfall enters the soil (Fig. 3.5). Depending on the slope and the state of the soil surface, a more or less important part of the rainfall water goes as runoff to streams and rivers. The other part enters the soil porosity. Depending on climatic parameters such as temperature, wind, and the density of the vegetation cover, part of the soil water evaporates and returns to the atmosphere: this process is called evapotranspiration. Soil water which is not evaporated moves downwards and finally joins the water table: this volume of water is accounted for as drainage.

For a particular site the annual balance of water may be written as:

$P = R + E + D$ (see Fig. 3.6),

where: P is rainfall (precipitation), R is runoff, E is evapotranspiration and D is drainage.

The amount of drainage is a very important parameter in rock weathering and soil formation. Large volumes of drainage water mean large volumes of water in contact with rock which results in large amounts of matter being lost through dissolution in rather diluted solutions. Conversely, a small volume of drainage water indicates that most of the soil water is evaporated. Losses of matter are reduced and soil solutions are concentrated through evaporation at some period of the year.

Such a balance may be computed for a year or for shorter periods of time, such as 1 month or 10 days. For the last two cases it is necessary to introduce the concept of field water capacity of soils. Soils are able to hold water in their

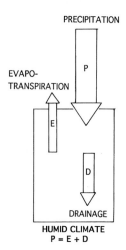 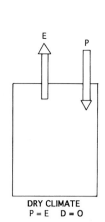

Fig. 3.6. Annual mass balance of water for humid and dry conditions. *P* Precipitation; *D* drainage; *E* evapotranspiration

pores, which does not immediately move downwards: this amount of water is called the field capacity and it is available for plant growth or evaporation after the rainfall period. A soil water regime is therefore defined by the variations of the volume of water held at the field capacity and the amount of drainage.

Two major soil water regimes can be distinguished. In one regime the amount of precipitation exceeds evapotranspiration, and some water moves down through the soil to the ground water at some time during the year. In the other regime water moves into the soil but is withdrawn by evapotranspiration, leaving behind precipitated carbonates and more soluble salts. For general considerations regarding losses of soluble materials or their accumulation in part of the weathering profile, this broad distinction is adequate. For the understanding of chemical and biological processes in soils and weathered rocks much more detail is needed. For example, a soil can be saturated with water and subjected to leaching in the winter when it is too cold for optimum chemical and biological activity. In the summer it can be too dry for any significant soil–solution reaction, even if the temperature is high.

Some typical water regimes are given as examples in Fig. 3.7. The first diagram illustrates the water regime in a soil in an *arid climate*. Evapotranspiration exceeds precipitation in all months of most years. The soil is dry for more than half of the year, and there is no leaching. In contrast, the second diagram illustrates the water balance in a soil in a *humid climate*. Precipitation exceeds evapotranspiration in all months of the year. The soil is permanently moist and water moves through the soil throughout the year. The third example typifies the soil water regime in a *temperate humid climate* with the rainy season in the spring and summer. There is enough rain during the summer such that the amount of stored water in the soil plus the rainfall is equal to the amount of evapotranspiration. The soil never dries for a long period and limited leaching occurs at the end of the winter and in early spring. The last

Origin of Clays by Rock Weathering and Soil Formation 51

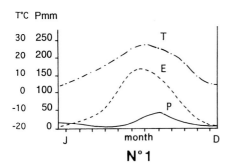
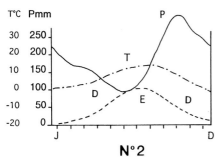
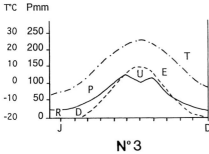
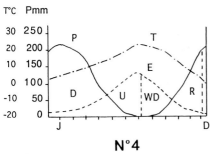

Fig. 3.7. Climatic data and soil water balance in some typical regions. *1* Arid; *2* humid; *3* temperate humid; *4* cool moist winter, warm dry summer. *T* Temperature; *P* precipitation; *E* evapotranspiration; *D* drainage; *R* soil water reserve; *U* utilization of water reserve; *WD* water deficit

example illustrates a water regime in a climate where *winters are cool and moist* and *summers warm and dry*. The moisture input in winter, when evapotranspiration is at minimum, is particularly effective for leaching. In summer, the stored moisture is used but is not enough to prevent the soil from drying out for two or three months. Because atmospheric climatic data do not always properly reflect the climatic conditions within the soil, soil scientists have produced classifications of soil climates.

The classification of soil moisture regimes used in the US Soil Taxonomy (USDA 1975) is given below as an example:

Aquic moisture regime. The soil is generally saturated with water and more or less free of oxygen so that reduction can take place.

Aridic moisture regime. The soil is dry for more than 50 percent of the growing season or never moist for more than 90 consecutive days during the growing season. There is little leaching in this regime and soluble salts usually accumulate.

Udic moisture regime. The soil is not dry for as long as 90 consecutive days. These soils occurs in humid climates with well distributed rainfall with enough during the growing season and leaching in most years.

Ustic moisture regime. The soil has a limited amount of moisture but it is present in sufficient quantity during the growing season. In the tropics this moisture regime occurs in monsoon climates.

Xeric moisture regime. The soil is dry for 45 consecutive days in summer and moist during winter. This soil moisture regime is found in Mediterranean climates.

3.2.2.2 Rock Composition

Soil formation is strongly dependent of rock composition. Soils with contrasting properties develop in granitic areas and areas of calcareous sediments. In spite of these very different soil features and properties, the clay minerals in the soils are not so different. Indeed, all the soils in the temperate zone contain a mixture of kaolinite and vermiculite as their dominant clay minerals. This situation is enhanced in equatorial climates where the rock composition does not control the clay mineral type that forms: basalt, as well as granitic rocks, produce soils which contain the same kaolinite + oxides assemblage. So long as a sufficient time has elapsed, the mineralogical composition of soils depends more on climatic conditions than on the chemical composition of the parent rock. Conversely, the rock composition is a determinative parameter for clay chemistry in the early stages of weathering, whatever the climate may be. This will be extensively discussed in Section 3.5.

3.2.2.3 Biological Factor: Vegetation and Soil Organic Matter

Organic matter is added to the soil system is by vegetation. The amounts of organic matter entering the soil is dependent on the vegetation cover. The

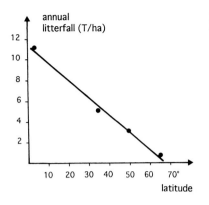

Fig. 3.8. Distribution of annual litter fall as a function of latitude

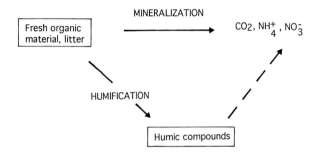

Fig. 3.9. Biological and chemical transformations of organic material in soils

organic compounds reaching the soils differ according to their origin and their resistance to biodegradation. Their persistence in soil and reactivity are highly variable. Therefore the type of vegetation may influence soil formation.

Most soil organic matter comes from the remains of dead plants falling on the soil. The quantity of vegetable material produced, and subsequently returned to the soil, can range from a trace, in arid and Arctic regions, to several tons per hectare in warm and humid climates where plant growth occurs throughout the year (Fig. 3.8).

Organic compounds from plant remains may be involved in two major processes: mineralization and humification (Fig. 3.9; Stevenson 1982). Mineralization is the transformation of organic molecules by microbes into CO_2, NH_4 and H_2O. Humification leads to more complex organic polymers, typical of the soil medium, so-called humic compounds. Humification is a slow process in which both biological and physico-chemical reactions are involved. The rate of humification versus mineralization determines, for an equal input of organic residues, the organic-matter content of the soil and, therefore, some important physico-chemical properties of the soil medium in which the clay minerals are formed and react. In response to factors such as climate, type of vegetation, and soil pH, the amount and types of humic substances that are formed may be very different. It is well known for instance that soils of the warmer climatic zones generally have low organic-matter content, despite the large quantity of plant material produced and returned to the soil. This is attributed to the high activity of microorganisms in the warmer temperatures. In contrast, in areas with cold climates, poorly transformed plant residues accumulate at the soil surface as a consequence of restricted biological activity. The highest levels of soil-organic matter are found in grassland soils of continental climates. The frequent wetting and drying cycles occurring in these soils are thought to be the best conditions for the synthesis and preservation of soil-organic matter.

The chemical composition of the mineral part of the soil may also strongly influence the processes of decomposition of plant remains and incorporation of stable organic compounds into the mineral soil. Parent materials which provide sufficient amounts of nutrients (calcium, phosphorus) through weathering favor biological activity and consequently a rapid and complete transformation of organic residues. Association with the clay fraction of the soil may also protect organic molecules against attack by micro-organisms and thereby contribute to the preservation of soil-organic matter.

Because humic substances are very complex and include many components, there is not yet a single and clear theory of humification. However, there is general agreement for two major processes. The first process gives great importance to the degradation of lignin (a component of vegetal cell wall) as a source for humic substances. The second process promotes microbial synthesis of aromatic substances which polymerize to form humic compounds. There is evidence that both processes are active, even in the same soil; however, certain soil conditions like acidity, excess of water, favor one or the other of these mechanisms (Fig. 3.10).

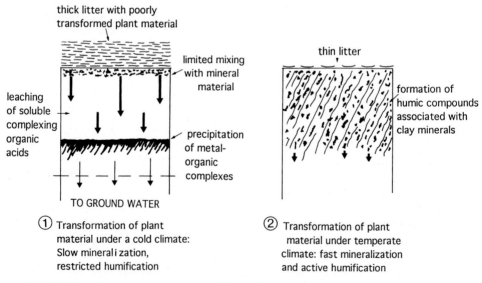

Fig. 3.10. Transformation of organic material in soils

When slow decomposition of plant remains is induced by cold climates or cation deficient parent materials, low-molecular-weight organic compounds are formed. These are complexing acids able to dissolve minerals and form organic complexes with metals such as Al and Fe. These organic complexes are soluble and promote exportation of the complexed metals out of the weathering system. In such conditions, transformation of initial minerals into clays is strongly disturbed as less Al or Fe is available for the formation of new clay minerals. Conversely, in a mild climate, an abundant clay fraction and a neutral pH promote the stabilization of humic compounds which contract close associations with the surfaces of clay minerals. As a consequence, no water soluble organic components are present in soil solutions. In this case contribution of organic matter to weathering is achieved through production of CO_2 which reacts with water to give carbonic acid.

3.2.2.4 Age and Soil History

Weathering of rocks, and the formation of soils and clay are slow processes that require hundreds, thousands and even millions of years. Moreover, periodic changes of climate and vegetation have often occurred which have changed the original pathways of soil development, and not all soils have been developing for the same length of time. Even in a single soil, some layers differentiate more rapidly than others. Incorporation of organic residues into the surface layer of a soil developed from unconsolidated rocks takes only a few decades, but translocation of a noticeable amount of clay from the top to the medium part of the soil profile may require 5000 years. The complete weathering of rocks to form the several meter thick kaolinitic layer in tropical soils is thought to take more than one million years (Tardy and Roquin 1992).

In most cases the interpretation of soil features, such as the type of clay present, as being the result of interactions under present-day environmental conditions would be erroneous. Due to climatic changes at a scale below that of soil formation, it is evident that most soils have experienced a succession of different climates which have induced changes in pedogenesis. Therefore, soils are not developed by a single set of processes, but undergo successive waves of formation. Furthermore, each wave imparts certain features that are inherited by the succeeding phases. In some cases their properties are so strongly expressed that they remain observable thousands and even million of years after their formation. Consequently, soils should be regarded as the result of a developmental sequence which exhibits not only the present-day factors and processes of soil formation, but also many preceding phases.

The oldest land surfaces, such as those in parts of west Africa, developed in the mid-Tertiary period. The soils on these surfaces are very old. At that time tropical conditions existed in many of the same areas where they are found today, as well as in many of the present subtropical and arid areas. These warm conditions caused the rock to undergo profound weathering.

Considerable erosion also took place during this prolonged period, to form the characteristic flat surfaces associated with old landscapes, but great thicknesses of soil and weathered rocks were maintained. The processes of weathering were so complete in many places that rocks of all types were transformed into kaolinite, and hydroxides and oxides of iron and aluminium. The warm, humid conditions were maintained in tropical areas but became cooler in higher latitude areas. This cooling of the climate culminated in the Pleistocene period leading to repeated glaciations in Eurasia and North America (Fig. 3.11). Glaciations and associated periglacial conditions effectively removed most of the deep soils and weathered rock that formed during the Tertiary period. Consequently, the formation of soils in Western Europe

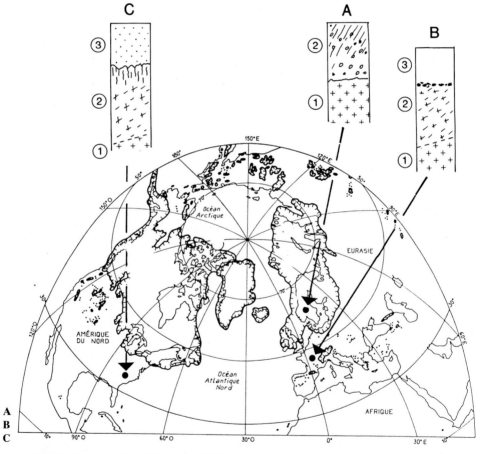

Fig. 3.11. The extension of the ice field during the last Quaternary glaciation *Heavy line* shows ice cap. Soil profiles. **A** Recent soil (<5000 years) from *2* glacial till; *1* eroded fresh rock; **B** recent soil (<10000 years) from *3* reworked material, overlaying *2* an old saprolite, *1* eroded fresh rock; **C** strongly developed soil *3* produced by a long weathering period

and North America has been achieved in the relatively short period of 10 000 years. Thus, it must be expected that the transformation into clay of the primary minerals is far less advanced in these soils than in soils from the tropical zone.

3.2.2.5 Topographic Effects: Translocations and Accumulations

Climate and vegetation are relevant to the distribution of soil at the world scale, but within a smaller area topography strongly disturbs spatial distribution of soils. Topography (slope) changes the direction of the water flow which may be not only vertical, but also lateral along slope. Lateral transfers of matter through solution or suspension may occur from the higher to the lower part of the slope: soils at the higher parts become impoverished in clays whereas soils lower down are enriched.

Nearly all material in soils may be moved by one process or another. In fact, many of the processes of soil formation are concerned primarily with the reorganization and redistribution of material within the upper layers of the weathering profile. Translocation may be performed in solution or suspension.

In a humid environment, part of the water which migrates through the soil is lost as drainage water. Water–soil interactions lead to dissolution of materials which are lost from the soil as the solution moves downward. The highly soluble simple salts, including nitrates and chlorides, are completely removed, whereas carbonates usually precipitate in lower layers of the soil where the solution becomes concentrated. If lateral movement of water occurs along a slope, soluble products of weathering can migrate laterally downslope (Fig. 3.12). Enrichment in basic cations is generally observed in the lower

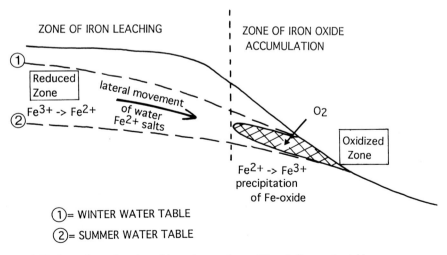

Fig. 3.12. Lateral translocation of ions along a slope with a shallow water table

positions of the landscape. Such a high-base environment (high pH) favours the formation of smectite.

The progressive leaching of divalent cations by drainage water, induces a partial desaturation of the exchange complex of clay minerals with respect to these cations. This allows colloidal clays to disperse as a stable suspension. Through this process clays are transported from the upper layers to the medium part of the soil profile. This is one of the most important process that affects clay minerals in soils. This process is called *clay illuviation*. Clays produced by weathering and soil formation do not stay in place, but frequently move to another place, even over long distances (hundreds of meters). Impoverishment in clays, observed in surface layers of soil developed from poorly permeable materials, is attributed to translocation of clays moving laterally with seepage water over poorly permeable subsurface soil layers.

Another important material that can migrate in soils is organic matter. Usually, organic compounds produced during humification are fixed on clay surfaces, and therefore are unable to move except with the clay suspension. However, in acid soils with low clay content, a more or less important fraction of the organic compounds is translocated in solution or as colloidal suspension. These mobile organics contain complexing organic acids which fix Al, Fe or other metals. In this way, elements such as Al and Fe can move, even under pH and Eh conditions where they are not expected to be soluble. Dissolved organic compounds and organo-Al and Fe complexes, are generally precipitated at depth in the soil profile, forming cemented layers. Some of the dissolved organic compounds may escape precipitation and enter rivers and lakes, giving typical brown waters such as those of the Rio Negro in the Amazon.

Considering now the "normal" evolution of a soil in a humid climate, the first process one may expect is the leaching of cations. Consequently, the ionic concentration of the soil solution decreases and clays are able to disperse and move from the upper soil layers. When a definite stage of clay impoverishment is reached, organic compounds can no longer be totally fixed and some move. To summarize, the succession of mobile materials in soils under humid climate is: cations (Mg, Ca), clays and finally organic matter (Fig. 3.13).

By contrast, in an arid climate, where evapotranspiration largely exceeds precipitation, there is only a short period following rainfall during which some of the most soluble materials are dissolved and are able to move. Since the rainfall under these conditions is small, any downward movement is quickly reversed by intense evaporation which induces upward movement of the soil solution and a deposition of salts at, or near, the surface.

A major consequence of translocation of soil material from one place to another is the change of soil structure. As soil structure greatly determines the pattern of water flow within soils, any translocation of material will lead to the disturbance of water flow. As water flow strongly influences weathering of initial minerals and clay formation, any translocation will change the local physico-chemical parameters of weathering. In other words, even with stable

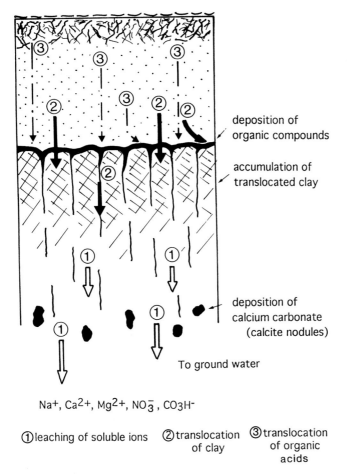

Fig. 3.13. Theoretical soil profile showing the successive translocations of cations, clay minerals and organic compounds in a wet climate

conditions of climate, rock chemistry etc., soil forming conditions are permanently changed by the process of soil formation itself.

One of the best examples that illustrates this process are Luvisols, i.e. soils where clay translocation occurs. It will be described in detail later on (Sect. 3.4), but can be summarized as follows. Clays move from the upper to the medium soil layer (the B horizon) where they accumulate. The increase in the clay content of the B horizon progressively reduces its permeability in such a way that an intermittent perched water table occurs above it during the rainy season. Such a waterlogging induces low redox conditions. Iron (from oxyhydroxides) is reduced to the ferrous state, becomes more soluble and moves, leaving iron-free materials. The clay assemblage shifts from a clay-oxide complex to essentially clay minerals.

3.2.3 Distribution of Major Soil Types at the Surface of the World

Soil classification has been a matter of debate between pedologists for years. The idea now is that soils cannot be classified as has been done for plants or animals. However, large groups of soils with common features and properties may be defined with general agreement among specialists. These major groups are used for the legend of a soil map at the world scale (Pédro 1985; Fig. 3.14).

The first observation is that the major soil units roughly reproduce the extension of major climatic zones: equatorial, tropical, arid, temperate, cold

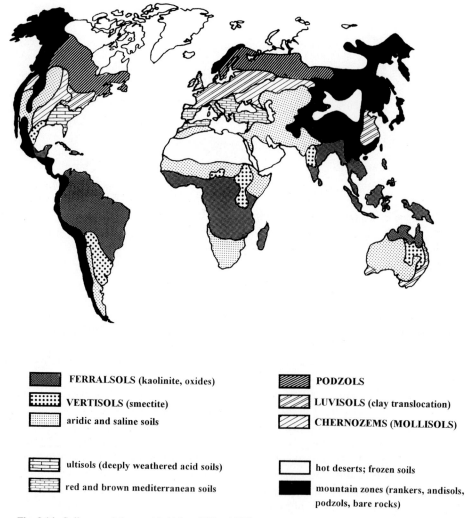

- FERRALSOLS (kaolinite, oxides)
- VERTISOLS (smectite)
- aridic and saline soils
- ultisols (deeply weathered acid soils)
- red and brown mediterranean soils
- PODZOLS
- LUVISOLS (clay translocation)
- CHERNOZEMS (MOLLISOLS)
- hot deserts; frozen soils
- mountain zones (rankers, andisols, podzols, bare rocks)

Fig. 3.14. Soil map of the world. (After Pédro 1985)

etc.. The large unit centered on the equator is that of *Ferralsols* which are thick soils (15 m) with a yellow to red colour. The mineral part of Ferralsols is almost exclusively clays. These clays are dominantly kaolinite associated with a more, or less, important fraction of Fe and Al oxides and hydroxides (goethite, hematite, gibbsite). CEC of the whole soil is very low (10 to 15 meq/ 100 g soil) due to the dominant fraction of kaolinite clay. The pH is about 5 throughout the soil profile. Organic-matter content is very low because organic components are quickly biodegraded by an active microflora under a wet and hot climate. The area covered by ferralsols is about 30% of the Earth's surface.

At the northern and southern edges of the Ferralsol unit one finds large areas of arid soils and deserts. Within these areas soils are not more than sand or an accumulation of stones. At some places accumulation of soluble salts occurs (Na-chloride, Na-carbonate). Between the Ferralsol and arid soil units one can see a soil unit of lesser extension. It is the area of *Vertisols* which are clay-rich soils, as are ferralsols. These clays are of a different type, typically smectites. Vertisols are far less thick than ferralsols, 1 to 2 m, with a very dark colour (black). The pH is in the range of 6.5 to 8.5. CEC (cation exchange capacity) is rather high; up to 80 meq/100 g dry soil (Chap. 2; Sect. 2.2.3.4). The best-known property of Vertisol is the large clay fraction dominated by smectites, which causes these soils to shrink greatly and crack upon drying.

In the Eurasian northern hemisphere, a latitudinal zonation of soils can be seen from the North to the South of the continents. In the southern part of the northern areas, where permanently frozen soils occurs, one finds the area of *Podzols*. Podzols are characterized by a slow and incomplete transformation of organic residues, and the production of large amounts of complexing organic acids. Initial minerals are strongly attacked by these acidic organic compounds, and the formation of clay minerals is weak. The fine mineral fraction of a Podzol is dominated by amorphous organo-mineral compounds that move in the soil profile and accumulate in a specific horizon. Podzols are acid soils with a low CEC. Going further south, the next area is that of *Luvisols*. The climate is milder and allows a rather good decomposition of organic residues. However, weathering of initial minerals is far from complete and most of the clay fraction is derived from the transformation of pre-existing phyllosilicates. Clays are typically vermiculite and clays with mixed layers of illite, chlorite and smectite. Luvisols are also characterized by the illuviation of clays. The pH of Luvisols is usually close to neutral.

South of the Luvisol area, where climate is continental, Luvisols are replaced by *Chernozems* which are characterized by an accumulation of humic compounds over a great depth (1 m). Calcium carbonate accumulations are frequent at the lower part of the soil. As in luvisols, clays are mainly derived from pre-existing phyllosilicates. They are dominated by smectite and illitic mixed layers. The pH is about 7 and increases up to 8.5 when calcium carbonate is present. Further to the south, arid soils are found.

The distribution of the great groups of soils also gives the distribution of some great groups of clay minerals: kaolinite is the dominant clay in the wet tropics, a dry tropical climate produces smectite, and temperate zones have mixtures of vermiculite and complex mixed layers (Pédro 1968). However, relating clay type too closely to the current climate is dangerous. Even at the world scale, the time of evolution and rock chemistry also control the soils and clays that form. Ferralsols are generally old soils developed from various rock types, but vertisols are restricted to calcium-rich rocks or occur in depressions where calcium-rich solutions concentrate. Podzols are young soils that develop, most often, in sandy or gravely materials such glacial tills. Luvisols are typically developed from loess, a fine-grained, well-sorted, recent material, in which clay movement appears to have been favoured. Clays and soils are dependent on climate but they are also conditioned by the their source material, the bed rock.

3.2.4 Structure of Weathered Rocks and Soils

3.2.4.1 Weathered Rocks: Inheritance of the Rock Structure

Under temperate climates, because of erosion, creep on slopes, and subsequent redeposition, the weathered rock and the soil are generally not in continuity, the boundary being marked by a stone line. Weathered and unweathered rock structures are typically in continuity. Most often, the macroscopic features of the unweathered rock (fractures, diaclases, dikes, veins, etc.) are still recognizable even in the highly weathered zones.

The structural differences between rock and its weathered products are mainly observable at the microscopic scale. Primary minerals are replaced by a porous fine-grained material which is composed of clays and tiny debris of the parent crystals. The initial fabric is conserved as long as the volume of altered zones in each primary crystal, do not reach a yield value beyond which the mechanical resistance of the rock collapses. In more intensely altered zones, the framework of parent minerals is destroyed and the original rock fabric locally disappears. Moreover, in some cases, the rock structure totally vanishes, giving place to a clay-rich horizon. This is observed in Fe, Mg-rich crystalline rocks (gabbros, amphibolites, serpentinites) where a clayey restructured horizon represents a transition zone between the weathered rock and the soil. In summary, the weathered rock domain is considered here, to be limited to the presence of observable macroscopic parent rock features.

3.2.4.2 Soil Structures: Importance of Aggregation

In most soils the individual particles do not exist as discrete entities, but are grouped into aggregates with fairly distinctive shape and size. The degree and type of aggregation determine aeration and permeability, and therefore, the infiltration capacity and extent of moisture movement. Structure also influ-

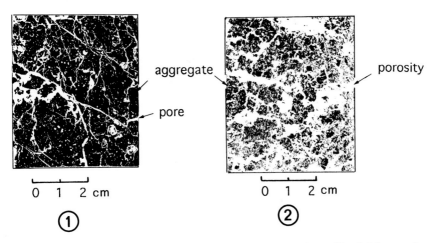

Fig. 3.15. Two typical soil structures. *1* Angular blocky structure formed by shrinkage and cracking of clay-rich soil materials; *2* crumb structure produced by burrowing animals

ences the erosive potential, since the presence of surface horizons with massive structure may reduce infiltration, which increases runoff thereby increasing the erosion hazard. Figure 3.15 shows two typical soil structures.

The genesis of soil structure involves, at least, three main processes. One of these is the expansion and contraction of clay-rich materials in response to wetting and drying. As the soil dries, its volume decreases; high tensions are created which are released by cracks opening in the soil material. Polygonal bodies are formed of different shapes and sizes, according to the importance of the retraction of the soil material. The fine granular structure of the upper layers of Vertisols is a typical result of this process.

A second origin for soil structure is physico-chemical. During the course of weathering of Fe–Mg phyllosilicates, Fe is exuded from the mineral, is oxidized, and precipitates as poorly crystalline, strongly hydrated oxy-hydroxides, which form bridges between, and covering, the remnants of weathered minerals (Fig. 3.16). As a result, small (100 μm) stable aggregates are formed, which leads to high soil porosity. Figure 3.17 shows how such a structure develops in a soil formed from a weathered micaschist (Righi and Lorphelin 1987).

A third origin for soil structure is the biological one. Animals, especially earthworms, strongly disturb the soil material and produce rounded aggregates as dejections. The structure of Chernozem soils is made essentially by earthworm galleries and dejections.

A major property of the soil structures is their possible evolution with time. For instance, clay-rich soil materials change their structure according to wetting and drying cycles. When dry, these materials are strongly aggregated, which leads to the development of numerous large pores in which water moves rapidly. As the soil is wetted and absorbs water, aggregates expand and

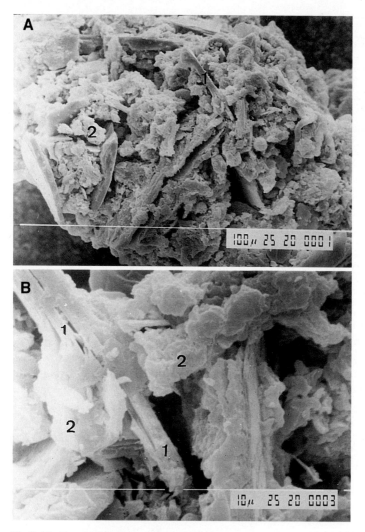

Fig. 3.16. Scanning electron microscope observation of a soil aggregate. **A** General view; **B** detail showing the edge–face association of the weathered inherited phyllosilicates and bridges of amorphous weathering products. *1* Inherited phyllosilicates; *2* iron-rich amorphous material

porosity is progressively reduced to very fine pores; water transfer becomes very slow. In such soils, moisture movement may follow two contrasting regimes. If rain falls on the dry soil, water flows directly to depth, by-passing interactions with soil material (Fig. 3.18). Conversely, if the soil has been previously wetted, water enters and moves slowly, allowing a close contact of solutions with the soil material for long periods. Another important point to be stressed is that the soil structure can be very different from one layer to another. Thus water circulation in soils changes greatly with time and space.

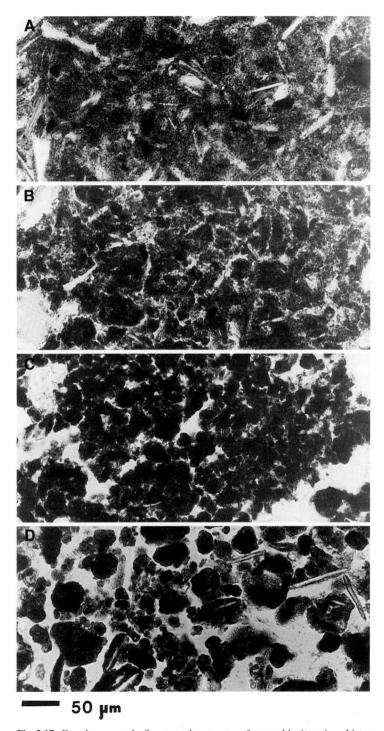

50 µm

Fig. 3.17. Development of a fine granular structure from a chlorite-micaschist saprolite. $A \to B \to C \to D$ from the bottom to the top of the soil profile

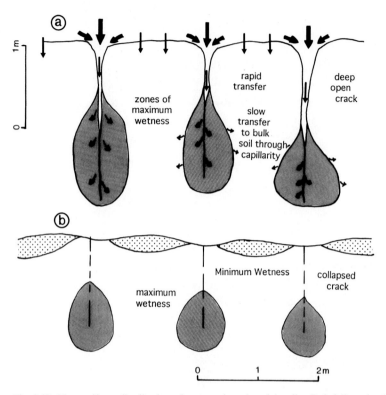

Fig. 3.18. Non-uniform distribution of wetness in a clay-rich soil. **a** Rainfall on the dry soil; **b** after a certain time of desiccation after saturation

This may explain why different clay minerals may be found in different soil layers.

Why put such an emphasis on the structural aspect rather than the chemical or mineralogical properties of these complex materials? The basic reason is that the circulation of water is the decisive parameter which controls mineral reactions. Dissolution of pre-existing minerals, as well as precipitation of new ones, depends on the local chemical composition of the solutions which, in turn, depends on the flow rate of water. Usually, the solutions are diluted in the larger pores, where water circulation is fast, and they are more concentrated in dissolved elements in narrow pores in which water moves slowly. In summary, the way that water penetrates into the soil depends on the structure, but the soil structure changes with wetness as the seasons change.

3.2.5 Soil Structure–Porosity Relationship

Soils, as well as weathered rocks, are discontinuous materials in which the solids (minerals, aggregates) are separated by voids. The spatial distribution of

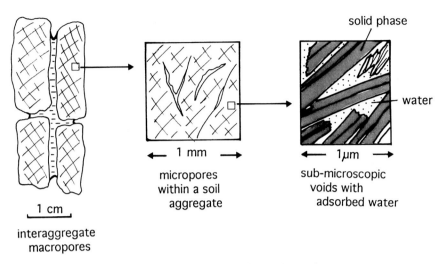

Fig. 3.19. Soil structure and porosity at different scales of observation

these voids depends on the structure of the rock or the soil. At a given time, in spite of structural discontinuities, the soil horizons and the weathered rock are interconnected by a complex void network. The largest interconnected voids form a fast-water pathway from the soil surface, through the weathered rocks, to the groundwater table and springs. Such interconnections may change with the seasons, because the soil structure is controlled by wetness and biological processes, wheras the weathered structure remains stable through the year.

In the smaller pores (Fig. 3.19) on the microscopic scale, capillary forces become important. Soil scientists consider that the competition between gravity and capillary forces arises in voids with diameter less than 200 µm. Below 0.2 µm in diameter, voids are not accessible to root hairs. Thus, the total volume of the voids with a diameter between 200 and 0.2 µm, represents the available (for plant growth) water capacity of the soil. Below the 0.2 µm boudary, voids are considered to belong to a sub-microscopic class. In such pores the fluids are retained on surfaces of solid particles by electrostatic forces. The chemical composition of these waters depends on the dissolution–precipitation phenomenon occurring at the interface.

3.2.6 Changes in Rock Density and Mechanical Properties

One of the major effects of the weathering processes is the transformation of a massive rock into a porous friable material. The increase of porosity and its consequence, the decrease of density, were discovered very early in the nineteenth century (Ebelmen 1847 for example). Considering the case where the structure of the parent rock is conserved, the volume balance of a given alteration reaction can be calculated as follows:

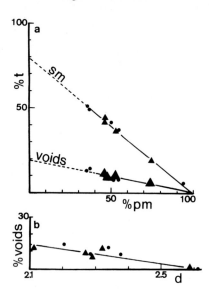

Fig. 3.20a,b. Variation of the relative volume of **a** secondary minerals and **b** dissolution voids (% t) and density (d) as a function of % unaltered primary minerals in the zones of two weathered profiles in which the initial structures of the granitic rocks are conserved. *Dots* La Pagerie profile; *triangles* La Rayrie profile; *sm* secondary minerals; *voids* dissolution voids. (Data from meunier 1980)

x% vol primary mineral → y% vol secondary minerals + z% vol voids

where: x = y + z.

The majority of documented physical analyses are found in studies of granite weathering. This is because granitic rocks are the dominant outcrop at continental surfaces. Modal analysis (microscope observation) of samples from two altered granite profiles in western France, confirm the above equation: y and z increase linearly as x decreases (Fig. 3.20a). As a consequence, density is inversely proportional to porosity (Fig. 3.20b). This relation is not applicable in zones where the initial structure of the rock has disappeared. The question therefore is, why is the conservative alteration process limited to a maximum porosity of 12–15% in granite weathering?

A possible answer to this question can be found in studying the change in mechanical properties of rock during the alteration process. Baudracco et al. (1982) showed that the resistance of an altered granite to uniaxial strain, and the Young modulus, are strongly reduced in the early stages of weathering (Fig. 3.21). This means that the initial rock structure collapses when an alteration threshold is reached. The decrease in the Young modulus indicates that the rock progressively changes from a brittle to a ductile material.

In summary, the mechanical resistance markedly decreases at the beginning of the weathering process, giving rise to modification of the parent rock by different processes near the surface: collapse under gravity, penetration by roots. In this way, the parent rock is transformed into a new material whose physical and chemical properties are different from those of the saprock. In some cases, a clay-rich horizon (30–50% in weight) is formed; the density of the horizon increases while the porosity decreases. This is frequently observed in the weathered profiles developed on basic rocks (Ildefonse 1980; Proust

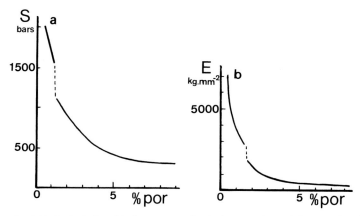

Fig. 3.21a,b. Relation ship between mechanical properties and the porosity of an altered granite from Sidobre, France (data from Baudracco et al. 1982). **a** Variation of the uniaxial stress (S); **b** variation of the Young modulus (E)

1983). The structure of these clay-rich materials is governed by the swelling power of expandable clays. The seasonal drying–wetting cycle induces local stresses which provoke the formation of new fractures and gives rise to the transfer of solids.

3.2.7 Water in Soils: Content and Chemical Potential

The variable amount of water contained in a volume of soil is an important factor affecting the chemical state and reactivity of the soil. In addition to this amount, which characterizes the soil wetness, soil water must be characterized in terms of physico-chemical condition or state. The energy state of soil water, i.e. the *potential*, is characterized by its free energy per unit mass. The most important components of the total potential of soil water are described below (Hillel 1980; Chamayou and Legros 1989).

Gravitational potential (physical): the gravitational potential of soil water at each point is determined by the elevation of that point relative to some arbitrary reference level (for instance the ground water level or the soil surface). All of the categories of soil water are affected by gravity, independently of the chemical and pressure conditions.

Osmotic potential (chemical): the presence of solutes in soil water affects its thermodynamic properties and lowers its potential energy.

Pressure potential (physical): in a soil saturated with water, the pressure potential originates from the pressure caused by the water column above the considered point.

Matrix potential (chemical): this potential results from the capillary and adsorptive forces due to the soil matrix. These forces attract and bind water in the soil and lower its potential energy below that of free water. The matrix potential is developed only in unsaturated soil (three-phase system: solid, liquid, gas). Capillarity results from the surface tension of water and its contact angle with the solid particles. The matrix potential can be described as a negative pressure potential, the so-called water suction or tension: consequently matrix potential and water suction have opposite signs.

The various potentials can be expressed in terms of an equivalent head of water, which is the height of a liquid column corresponding to the given pressure. The state of soil water is characterized in terms of a hydraulic head (H) which is the sum of (omitting the osmotic potential) the gravitational (z) and the pressure (Hp), or the matrix (Hm), potential heads: $H = z + Hp$ (saturated soil) or $H = z + Hm$ (unsaturated soil).

The energy state of water in soils can also be expressed by the following relations:

$$\Delta\mu_w = \mu_{sw} - \mu_{fw} = RT \log(P_{sv}/P_{fv}) = -2\sigma V/r = -\rho g V h_c$$

where:

$\Delta\mu_w$: energy of water in voids,
μ_{sw}: chemical potential of water in voids,
μ_{fw}: chemical potential of free water,
P_{sv}: vapor pressure in the void
P_{fv}: vapor pressure of free water in the standard conditions
σ = surface tension, r = radius of capillary meniscus,
V = molar volume of water,
ρ = water density, g = gravitational field, h_c = capillary height.

These equations indicate that the energy state of water, or its potential, decreases with the radius of the capillary pore (Table 3.1).

3.2.7.1 Soil-Moisture Retention Curve

If a slight suction is applied to a saturated soil, no outflow will occur until this suction reaches a value exceeding that applied by the soil matrix on water in the largest pores. Remembering that water is more strongly retained by soil as pore radius decreases, a gradual increase of the suction applied to the soil leads to a progressive removal of water from pores of decreasing sizes. Increasing suction is thus associated with decreasing soil wetness. This function is measured experimentally and it is represented by a curve, known as the soil-moisture retention curve (Fig. 3.22). The shape of this curve is controlled by the pore-size distribution and thus, strongly affected by the soil texture and structure.

3.2.7.2 Flow of Water in Soils and Weathered Rocks

Before discussing flow of water in soils and weathered rocks, it is necessary to distinguish between saturated and unsaturated soils. In saturated soils all the

Table 3.1. Energy state of water in soils. The relationship between suction pressure (bars) and pore diameter

Water suction (bars)	Chemical potential (J/Kg)	Water activity	Maximum size of pores (μm)
0.01	1	0.999993	150
0.10	10	0.999927	15
1	100	0.99927	1.5
10	1 000	0.9927	0.15
15.8	1 580	0.9888	
100	10 000	0.927	0.015
500	50 000	0.695	
1000	100 000	0.485	0.0015

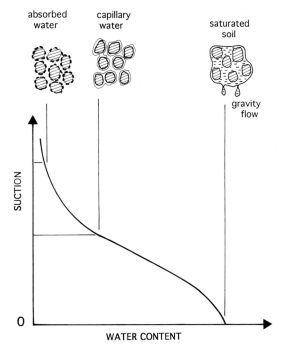

Fig. 3.22. Soil-water retention curve

pores are filled with water whereas in unsaturated soils some pores are filled with water and the other ones (generally the largest) are filled with air. Laws that govern flow of water are more complex in unsaturated soils.

Flow of Water in Saturated Soils

Figure 3.23 shows a horizontal, uniform, saturated soil body through which a steady of water flows from an upper reservoir to a lower one; each reservoir

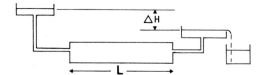

Fig. 3.23. Water flow in a horizontal soil column. L Length, ΔH hydraulic head drop

has a constant water level. The volume of water V, flowing through the column per unit time is proportional to the area of the cross section of the column and to the dropin hydraulic head ΔH, and inversely proportional to the length of the column L. Obviously, no flow occurs in the absence of a hydraulic head difference, i.e. when $\Delta H = 0$. The head drop by unit of distance in the direction of flow is the hydraulic gradient ($\Delta H/L$), which is, in fact, the driving force. The specific discharge (volume of water flowing through a unit cross-sectional area per unit time) is called the flux and is indicated by q. Thus, the flux is proportional to the hydraulic gradient: $q = K\Delta H/L$. In this equation, known as Darcy's law, K is designated as the *hydraulic conductivity*. A more exact and generalized expression of Darcy's law is in differential form: in a one-dimensional system it is

$$q = -K\,dH/dx.$$

Water flow in a horizontal column occurs in response to a hydrostatic-pressure gradient. Flow in a vertical column is caused by gravitation as well as pressure. At any point, a gravitational head, designated by z, can be determined as the elevation of the point above an arbitrary reference plane, in which case the pressure head (Hp) is the height of the water column resting on that point. The total hydraulic head is the sum of these two:

$$H = z + Hp.$$

K, the hydraulic conductivity, is obviously affected by structure and texture, being greater if a soil is highly porous, fractured, or aggregated than if it is compact and dense. Hydraulic conductivity depends not only on total porosity, but also, and primarily, on the size of the conducting pores. For example, a gravely or sandy soil with large pores can have a conductivity much greater than that of a clay soil with many narrow pores, even though the total porosity of a clay is generally greater than that of a sandy soil. In a saturated soil with a rigid structure the hydraulic conductivity is characteristically constant. Its value ranges from $10^{-2}\,\text{cm}\,\text{sec}^{-1}$ in a sandy soil to $10^{-7}\,\text{cm}\,\text{sec}^{-1}$ in a clayey soil.

Flow of Water in Unsaturated Soil

Most of the processes involving soil-water interactions in the field occur while the soil is in an unsaturated condition. As we have seen, the moving force in saturated soil is the gradient of hydraulic potential which is the sum of the pressure potential and gravitational potential. When the same reasoning is applied to an unsaturated soil, flow takes place in the direction of decreasing potential, the potential gradient being the driving force. However, in an unsaturated soil, the pressure potential is replaced by the matrix potential. Matrix potential is often changed by water suction or water tension which is a negative pressure in unsaturated soils. Suction is due to the physical affinity of water for soil-particle surfaces and capillary pores. Water tends to be drawn from a zone where the hydration envelopes surrounding the soil particles are thicker, to zones where they are thinner, and from a zone where the capillary menisci are less curved (larger radius) to zones where they are more highly curved (smaller radius).

Water flows spontaneously from zones where suction is lower to where it is higher. This is especially true when the moving force is large, such as at a wetting front zone where water enters into an originally dry soil. In this zone, the suction gradient can be many bars greater than the gravitational force, and water can move "uphill".

The most important difference between unsaturated and saturated flow is in the hydraulic conductivity. When the soil is saturated all the pores are water filled and conducting, so that hydraulic conductivity is at a maximum. When the soil becomes de-saturated, some of the pores (the larger ones) become air-filled and the conductive portion of the soil decreases. As suction develops, the first pores to be emptied are the largest ones, which are the most conductive, thus leaving water to flow only in the smaller pores. For these reasons, the transition from saturation to unsaturation is generally characterized by a large reduction in hydraulic conductivity which may be lowered by several orders of magnitude (Fig. 3.24). At saturation, the most conductive soils are those in which the overall pore volume is consists of large and continuous pores, wheras the least conductive are those soils with numerous micropores. For example, a saturated sandy soil conducts water more rapidly than a clayey one. The opposite may be true when the soils are unsaturated. Large pores are quickly empty and become non-conductive as suction, even weak suction, is applied, and the conductivity rapidly decreases. Small pores retain water, even at rather high suction, thus the conductivity does not decrease as rapidly and it may be greater than that of a soil with large pores subjected to the same suction (Fig. 3.25). Figure 3.26 illustrates how conductivity (K) increases and suction decreases as soil wetness increases.

3.2.8 Dissolution and Recrystallization Processes

3.2.8.1 General Statements

The energy state of water varies from place to place in a soil or a weathered rock, between a minimum value, in the absorbed layer on the solid surfaces, to

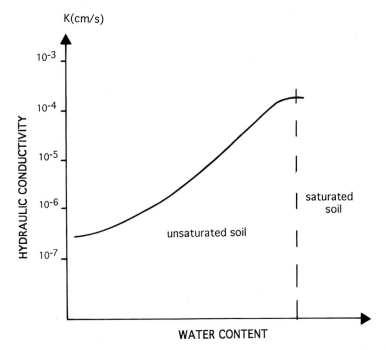

Fig. 3.24. Dependence of hydraulic conductivity on soil wetness

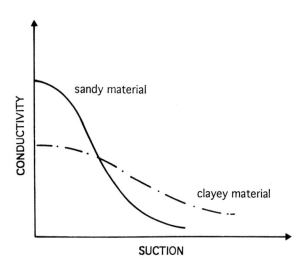

Fig. 3.25. Decrease in hydraulic conductivity as suction pressure increaces, for two materials with different textures

a maximum value which is that of free water in large voids. This is of importance for chemical reactions at the solid–fluid interface. These reactions could be regarded as two opposing processes: the growth and the dissolution of crystalline phases. During dissolution of silicates, cations will (1) detach from the oxygen bonds, (2) diffuse away from their site in the crystalline lattice, (3)

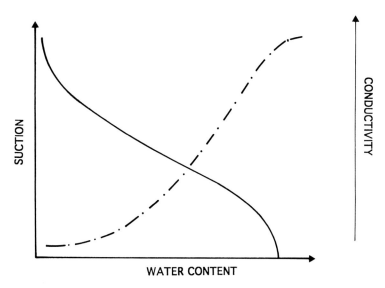

Fig. 3.26. Variation of wetness, suction pressure and hydraulic conductivity in an unsaturated soil

desorb from the crystal surface and (4) diffuse into the bulk solution (Hochella and White 1990). From a chemical point of view, growth as well as dissolution processes, could be reduced by the effects of ion concentration gradients at the crystal surface.

Dissolution of pre-existing mineral phases as well as nucleation and growth of secondary ones are controlled by the differences in chemical concentrations of the solutions far away and adjacent to the surface of the solids (C_α and C_{eq} respectively). A crystal is dissolved if the solution at the interface is maintained in an undersaturation state (i.e. $C_\alpha - C_{eq}$ is negative); on the other hand, a crystal will grow if the solution is oversaturated ($C_\alpha - C_{eq}$ is positive). Both phenomena occur in weathering processes in which the preexisting minerals are seen to be replaced by new ones.

The rate at which these processes occur is controlled either by transport of ions, reactions at the surface, or by a combination of both. These phenomena were formalized by Berner (1980) and are summarized in Fig. 3.27. In spite of the fact that the mechanisms of growth or dissolution are complex, an approximate value of the rate can be calculated using the following equation:

$$dr_c/dt = vD_s(C_\alpha - C_{eq})/r_c,$$

where:

r_c = average radius of crystals,
V = molar volume of the crystalline substance,
D_s = coefficient of molecular diffusion in aqueous solution,
C_α = concentration in solution away from the crystal surface,
C_{eq} = equilibrium concentration adjacent to the crystal surface,
t = time.

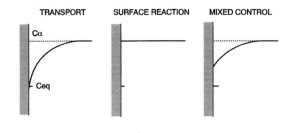

Fig. 3.27. Schematic representation of the concentration gradients of a given chemical component at the fluid-solid interface during growth or dissolution stages. (Berner 1980)

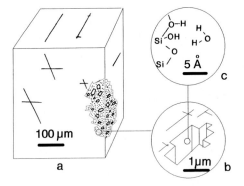

Fig. 3.28. Alteration features of a silicate at different scales. **a** Formation of the "alteroplasma" in place of the parent crystal. Most often, the alteration process is volume conservative. The alteroplasma is composed of parent mineral debris mixed with neogenic clay and/or oxide minerals; **b** view of an etch pit at the corroded surface. Etch pits frequently have geometrical shapes due to preferential dissolution along some crystallographic directions; **c** the Si–O or Al–O bonds are broken at the solid–fluid interface; cations are replaced by H^+

3.2.8.2 The Proton–Cation Exchange

The dissolution of primary minerals in aqueous solutions implies the rupture of inter-atomic bonds in the crystalline lattice (Fig. 3.28). Most often, the first step of the reaction is the exchange of alkali and alkaline earth cations in the crystal with proton in the solution. For example, the attack of feldspars by H^+

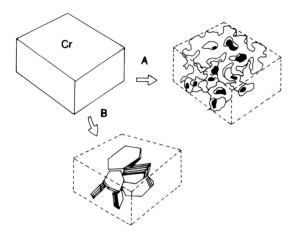

Fig. 3.29. The two alteration pathways for a crystal *Cr*. **A** Incongruent dissolution: the parent crystal becomes porous but no secondary phases are observed. The original crystalline framework is conserved; **B** dissolution–precipitation: the parent crystal is replaced by a microporous assemblage of secondary minerals

will lead to a surface which has lost the alkali cations; it is composed of SiOH or $SiOH_2^+$ groups. Further attack by H^+ or by H_2O will hydrolyze the Si–O–Al and Si–O–Si bridging bonds as shown in the reaction

$$Si–O–Si + H_2O \rightarrow Si–OH + OH–Si.$$

This process is affected by the diffusion of H^+, H_3O^+ and molecular water into the crystalline surface from the mineral–solution interface (Petit et al. 1987). The diffusion layer is several tens of nanometers thick. Mogk and Locke (1988) showed that cation diffusion occurs at the same scale changing the mineral composition near the interface. The size and properties of this intermediate layer at the interface are still a controversial subject. Nevertheless, except for phyllosilicates (micas, chlorites), the dissolution of silicates in natural rocks appears to be congruent at the micrometer scale. In other words there is no solid residue with a different chemical composition, and the solids which replace the pre-existing crystal are neo-formed secondary phases (Fig. 3.29). Phyllosilicates must be considered separately because their crystalline structure is similar to that of the clay minerals which replace them. The question is how to distinguish between those which are strictly neo-formed and those which could be considered as resulting from an incongruent dissolution. This will be discussed in more details in the chapter "Clays in soils from granite saprolite in the temperate zone" (Sect. 3.3.2).

Dissolution and growth are locally activated at excess energy sites such as lattice defects emerging on crystalline faces, edges or corners of crystals, twin boundaries, and micro cracks. Dissolution smoothes the angles and produces etch pits which become deeper and finally coalesce with time. This process results in a very irregular shaped interface showing corrosion gulfs. Usually the corrosion gulfs created by the dissolution of primary silicates are partially filled by the remains of the primary minerals and by secondary minerals. In other words, the continuous solid is replaced by a porous polyphase mineral assemblage (Fig. 3.28).

Secondary minerals formed in the corroded zones of primary ones can originate from two different processes: precipitation from solutions enriched in dissolved components (nucleation and growth), and transformation of the parent crystal lattice by diffusion. The latter is typical of phyllosilicate alteration (mica, chlorite) but it is also observed in non-phyllosilicate alteration. Indeed, olivine, pyroxene and amphiboles, weather in a biopyribole-sequence type (Veblen and Busek 1980; Eggleton and Boland 1982). In this case, the secondary phases grow in structural coherence with the parent crystal lattice; the coalescence of silica chains produces 2:1 layer silicates with minimum disruption of the pre-existing mineral structure. Nevertheless, the nucleation-growth process is the most common process involved in the crystallization of clay and oxide minerals in weathering or soil conditions. Nucleation is activated by the catalytic effect of defects on the dissolved surface of primary minerals (heterogeneous nucleation). In summary, the dissolution–recrystallization processes which occur at the mineral–aqueous solution interface can be reduced to an exchange of protons for soluble cations (hydrolysis). For example:

K-feldspar + H^+ → kaolinite + K^+ + Si^{4+}

The alkali bearing mineral (feldspar) is replaced by a hydrogen bearing one. These hydrolysis reactions can be used to determine the relative stabilities of mineral species.

3.2.9 Basic Factors for Phase Relation Analysis in Rock Weathering

3.2.9.1 From Microsites to Microsystems

The striking fact when observing altered rock at different scales is the great variation in physical properties (color, hardness, porosity, etc.) from place to place. At the thin-section scale, alteration reactions can be observed in numerous microsites where the pre-existing minerals are corroded. However, in spite of their apparent diversity, these microsites belong to just a few different types according to their location (intergranular joints, intracrystalline microcracks, transcrystalline fractures, porous secondary assemblages).

The dissolution–recrystallization processes occurring in each microsite are controlled by local chemical equilibria. Reactions are made possible by the presence of aqueous solutions which, unfortunately, are not analyzable at this scale. Nevertheless, these fluids must be taken into account as they exchange

Origin of Clays by Rock Weathering and Soil Formation 79

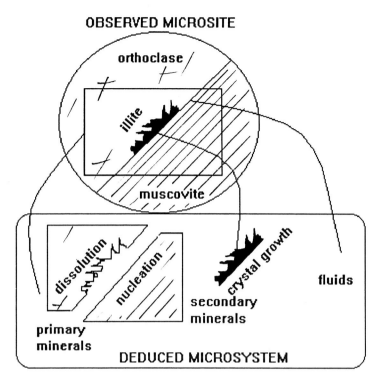

Fig. 3.30. Schematic representation of the different components of a microsystem. The example chosen is from a weathered granite where the observed mineral reaction takes place in the orthoclase–muscovite intergranular joint

energy and chemical components with the surrounding solids. Each microsite can be considered as a microsystem composed of three parts: primary minerals, secondary minerals and fluids (Thompson 1955; Korzhinskii 1959) (Fig. 3.30).

The microsystem concept is a useful tool to express mineralogical observations and chemical analyses in terms of phase equilibrium. It allows one to apply easily the Gibbs' phase rule and the classical laws governing the construction of phase diagrams. At this point, it is necessary to remember some basic definitions. A *phase* in thermodynamics is defined as a "pure substance, or mixture of pure substances, wherein the intensive properties do not vary with position" (Sposito 1981, p. 4). In fact, natural minerals are rarely homogeneous (zonations, inclusions etc.); nevertheless, they could be considered as phases in the Gibb's definition until their reacting interface with fluids becomes homogeneous. An *intensive variable* (V_{int}) does not depend upon the size of the system: for example temperature (T), pressure (P), density (ρ) and chemical potential (μ_x). *Extensive variables* (V_{ext}) depend on the size of the system: weight, enthalpy etc. The Gibbs' phase rule was extended to a more general expression by Korzhinskii (1959): $F = V_{ext} + V_{int} - P$ where F is the variability of the system (or *degree of freedom*) and P the number of phases. This expression is used here.

For example, in the mineral reaction: K-feldspar + H^+ → kaolinite + K^+ + Si^{4+} described above, the composition of the solution depends on the concentration of three mobile components: H^+, K^+ and Si^{4+} while Al remains fixed in solid phases (inert component). According to Korzhinskii (1959), the chemical potential of the three "mobile" components may act as intensive variables, if they are controlled by the external environment of the system. This is the case in water dominated systems. In contrast, Al whose concentration determines the quantities of solid phases involved in the reaction, is typically an extensive variable.

The quantities and the energy state of fluids involved in each microsystem vary with the dimensions and the interconnection of voids. Three types of microsystems are active in rock weathering (Meunier and Velde 1979):

Contact Microsystem. The earliest weathering reactions in macro-crystalline rocks, whatever their composition, have been observed in intergranular joints. Generally, the secondary phases nucleate on the surface of one of the parent crystals and grow into the other one. The microsystem is dominated by the chemical compositions of pre-existing phases; it is roughly closed.

Plasmic Microsystems. "Plasma" is the term describing the porous material composed of a primary mineral debris mixed with secondary minerals. In the saprock (in which the primary rock structure is conserved), the plasmic microsystems appear inside the primary minerals along internal microcracks, cleavages or twin boundaries (alteroplasma). In the saprolite, the primary rock structure disappears. Here another type of plasmic microsystem is observed in the restructured zones in which clay minerals, and the remaining debris of preexisting minerals, are physically intermixed (pedoplasma). New clay minerals form from old ones due to local changes in chemical composition, caused by physical mixing. As the pores are in the microscopic class, the fluids cannot circulate quickly. Fluids control the chemical potential of the more soluble components of the rock in all the microsystems through which they percolates. Thus, in a given plasmic microsystem, the new phases will depend on the local chemical composition and the value of the chemical potential of the more soluble species.

Fissural Microsystems. Coatings on the walls of fractures and large pores are composed of identical secondary mineral assemblages, whatever the size and the position of the open fractures in the weathering profile. The sedimentary features observed in the coatings are interpreted as microstratigraphic formations in which each deposition stage corresponds to seasonal variations in fluid dynamics. The size of these voids is large enough to reduce the capillary effect to negligible values. Consequently, the fluids circulate under gravity and are quickly renewed. In this situation, the stability of the mineral phases which constitute the coatings is controlled by the chemical composition of the fluids; the microsystem can be considered as totally open.

3.2.9.2 Construction of Phase Diagrams

The Different Types of Phase Diagrams

Several types of phase diagrams can be used in rock weathering studies. Considering that the effects of P, pressure, and T, temperature, variations are negligible for the formation of mineral phases at Earth surface conditions, it is possible to reduce the number of intensive variables to the chemical potentials of elements which are active in the different microsystems. According to Prigogine and Defay (1954), the chemical potential μ_i of a component i, can be defined by the following equation:

$$\mu_i = \mu\theta_i(P,T) + RT\ln\gamma_{ixi} + g^E_i - \mu^\theta_i.$$

The terms are defined as follows:

μ^θ_i: This term refers to a pure substance. In a given system, the variation in chemical potential is a function of the intensive variables, pressure (P) and temperature (T).

$RT\ln\gamma_{ixi}$: This term refers to the effect of variable composition of a phase upon the chemical potential of component i; γ_{ixi} is the activity of i; R is the gas constant $(1.987 \times 10^{-1} \text{kcal}\,°C^{-1}\text{mol}^{-1})$ and T is the temperature on the Kelvin scale. This is the function of mixing.

g^E_i: This term refers to the effect of structural configuration of a phase upon the chemical potential of component i which gives an excess function of mixing.

When two or more phases coexist at equilibrium, the chemical potential of all the components in the system must be identical in each phase.

Phase diagrams can be used to describe the changes of mineral assemblages induced by weathering processes at the scale of microsystems. If the microsystem considered is strictly closed (water is strongly retained on the mineral surfaces and the submicroscopic pores), there is no intensive variable acting in the formation of new phases. The components of primary mineral chemicals are recombined inside the microsystem without any exchange with its environment. Generally, the secondary minerals produced are polyphase minerals. If the microsystem is partially open (water can move in microscopic pores), the chemical potential of the more soluble component becomes an intensive variable. The dissolution–precipitation reactions modify the chemical composition of the microsystem; the number of secondary phases is reduced compared to the closed microsystem. Finally, if the microsystem is totally open (free water circulating under gravity in macroscopic pores and fractures), several chemical components become totally soluble; their chemical potentials act as distinct intensive variables. Figure 3.31 summarizes the different possibilities in a three-component system.

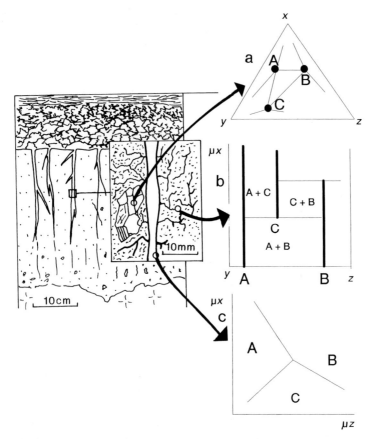

Fig. 3.31. Sketch showing the positions of the different microsystems in a soil. The adapted phase diagrams are related by the *arrows*; phase relations are depicted in a three–component chemical system (x-y-z). **a** Closed system (crystal microfissures or intergranular joints), the chemical components are inert; phases A, B and C are determined by the relative concentrations of x, y and z (*triangle*); **b** semi-open microsystem (altero or pedoplasma); one component (x) becomes an intensive variable; phase relations are controlled by the chemical potential of this component μx-y-z system); **c** open microsystem (mineral coatings in fissures where fluids circulate under gravity); phase relations are controlled by the chemical potential of two mobile components, only one component remains inert (μx-μz-y system)

In summary, phase relationships in contact, plasmic and fissural microsystems can be graphically studied using triangular composition, potential–composition and potential–potential diagrams respectively.

How can one construct the correct phase diagram?

First Step. It is necessary to define a simplified chemical space in which the composition fields of the mineral phases which have been identified, are represented in the different microsystems (identification controlled by X-ray diffraction; composition given by microprobe analysis). Usually, in spite of the fact that natural rocks are composed of ten major elements, it is possible to

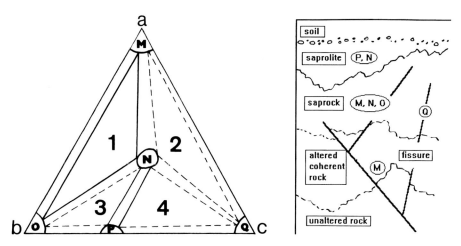

Fig. 3.32. Theoretical representation of a weathered profile in which the secondary phases are the following: M; $M + N + O$; $P + N$ and Q, in the altered coherent rock, the saprock, the saprolite and the fissures respectively. Composition fields of these phases are projected in the *a-b-c* chemical system. Observed and deduced mineral assemblages are represented by *heavy* and *dotted* tie-lines respectively

reduce the chemical system to a few groups of elements which react, chemically, in the same way. Indeed, some elements can be neglected because they are not decisive in the crystallization of clays and oxides or hydroxides phases (titanium for example); others can be grouped because of two reasons:

1. They are not independent in crystalline structure of these phases (for example, in the silicate group, the number of O^{2-} and $(OH)^-$ anions is fixed in a given crystalline structure);
2. They exhibit quite similar chemical behavior in the crystallization processes of the secondary phases (for example, Mg^{2+}, Mn^{2+} and Fe^{2+} cations are interchangeable in the octahedral layer of phyllosilicates). Thus, one can attempt to simplify the representative chemical systems in order to select three elements, or groups of elements, which determine the presence of phases. For example, the composition fields of phases M, N, O, P and Q identified in a weathering profile are projected in the triangular chemical system *a-b-c* as shown in Fig. 3.32.

Second Step. The mineral assemblages are determined by microscope observations, and are represented in the *a-b-c* triangle by tie-lines joining the phases which coexist. Coexistence means co-precipitation or replacement. The Gibbs' phase rule indicates that the maximum number of coexisting phases is only 3 in a three component system:

$$F = V_{ext} + V_{int} - P$$

as $V_{int} = 0$ in a closed system in the surface conditions, the equation becomes

$$F = 3 - P$$

if F = 0, the maximum number of coexisting phases is P = 3; three phase assemblages are represented by sub-triangles in the a-b-c system. Of course, not all the assemblages depicted in the diagram are observed in the weathered profile, because the chemical composition of the unaltered rock is represented by a limited field in the three-component system; nevertheless, when a sufficient number of tie-lines are known, there is only one solution to draw the missing ones. For example, in Fig. 3.32, the identified assemblages M + N + O; P + N; Q and M are represented by heavy lines. The presence of the N–O phase tie-line excludes M–P from coexisting (tie-lines must not cross, Niggli 1938). Consequently, the O–P, P–Q, N–Q and M–Q tie-lines (stippled lines) are necessary in order to respect the phase-rule relations (subtriangles).

Third Step. The secondary mineral assemblages change throughout the weathering profile from the nearly closed microsystems, which are numerous in the altered coherent rock, to the totally open systems in fractures (Fig. 3.32). These changes are due to variations of the chemical potential of one or two components. For example, if component a becomes an intensive variable, the succession of mineral assemblages in the profile is controlled by the variation of μ_a from high values in the closed microsystems to low values in the open ones. Thus, a sequence of μ_a equipotentials can be determined which depicts the mineral succession in the profile. The total number of equipotentials (N) necessary to describe all the mineral assemblages represented in the a-b-c system is given the following relation: $N = N_{tri} + 2$ where N_{tri} is the number of three-phase assemblages represented in the triangular diagram (Meunier and Velde 1979). A μ_a-b-c phase diagram can be derived from the triangular one (Fig. 3.32).

Fourth Step. In totally open microsystems such as fractures in the rock, mineral reactions are controlled by the chemical potentials of two elements. For example, if component b remains inert while components a and c become active, the best adapted diagram will be the μ_a-μ_b-c one. Such a diagram can be constructed in combining the μ_a-b-c and μ_c-a-b diagrams derived from the a-b-c one (Figs. 3.33 and 3.34) as shown in Fig. 3.35.

3.2.10 Summary and Conclusions

Soils and weathered rocks appear to be complex materials in which different physical, chemical and biological processes continuously transform organic and non-organic substances. The most important facts which must be kept in mind through out this chapter are the following:

1. The soil structure may change seasonally, whereas the weathered rock structure is permanent and inherited from the parent rock.
2. Although many parameters act in the formation of soils and weathered rocks (climate, age, topography, vegetation etc.), it seems that the most important parameter is the quantity of water which flows through.

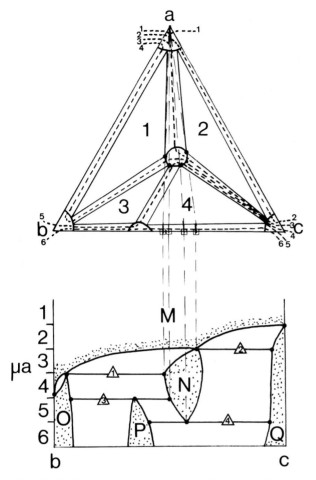

Fig. 3.33. Derivation of a potential-composition diagram from the composition diagram; assuming that component *a* becomes an intensive variable, all the mineral assemblages defined in Fig. 3.31 are depicted by a sequence of 6 equipotentials. A μ_a–b–c diagram is derived as shown. The triphased assemblages are formed for some fixed values of μ_a; they are indexed by the same numbers as in Fig. 3.31. The composition fields of O, P, N, Q are shown by areas; the stability field of phase M is represented by a *stippled area* (see text for details)

3. In spite of the fact that soils as well as weathered rocks seems to be dramatically complex at the microscopic scale, mineral reactions proceed in a few types of chemical microsystems.

Whatever the apparent complexity, an accurate observation of minerals and structures at different scales is the best tool we have, to determine the mechanisms of such continuous transformations. Thus, a general method for the study of soils or weathered rocks can be proposed:

1. As far as it is possible, observations must be made at different scales in order to determine the geometrical properties of the structure.

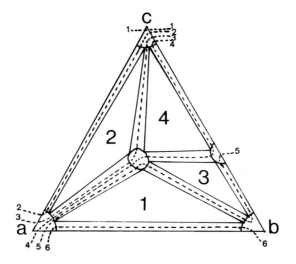

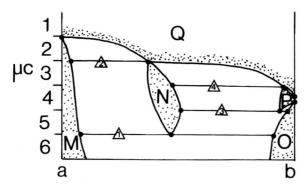

Fig. 3.34. Derivation of a μ_c–a–b diagram. The procedure is identical to that given in Fig. 3.32; the mobile component is c

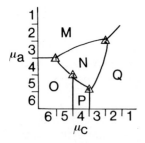

Fig. 3.35. Construction of a μ_a–μ_c–b diagram depicting the phase relations when two components (a and c) behave as intensive variables. Triphased assemblages are formed only for fixed values of μ_a and μ_c (no degrees of freedom); they are represented by *small triangles*; *numbers* are identical as in Figs. 3.31, 3.32, and 3.33

2. Then, mineral identification (X-ray diffraction) must be accurate enough to recognize the location of the phases or assemblages in the structure at the microscopic scale.
3. The chemical compositions of the different primary and secondary minerals involved in the reactions must be measured at the scale of the microsystems (electron microprobe).
4. Finally, petrographic observations, chemical data (microprobe) and mineral identification (XRD analyses) must be linked together by means of phase diagrams, in order to determine the mechanisms of the mineral reactions.

This four-step method is essentially applied to the study of weathered profile. It needs to be modified for soil study because microsystems are more complex and more difficult to observe directly under optical microscope. This will be explained later on.

3.3 From Rock to Soil: The Granite Example

The guide lines for phase analysis and the determination of the influences of the physical parameters on rock weathering and soil formation have been outlined above. However, a demonstration of their applicability has not been made, yet. We would like to demonstrate the analysis method described above as applied to a real case of the alteration of a rock type under a given climate at the Earth's surface. In doing this, a careful analysis is made to illustrate the use of the methods. The conclusions are drawn and the influences of structure, chemistry and the physical parameters are assessed as a function of their relative importance. In doing this it is hoped that the analysis method can be understood and that the subsequent analyses of soils and rock weathering which are presented will become clearer to the reader. The method proposed is that shown above applied to soil formation and the weathering processes developed on a granite. Why granites? They are among the most widespread rocks found at the surface of continents; thus, soils derived from granites, as well as weathered rocks, are well represented in large areas of many countries. The detailed study which is proposed in this section aims to describe the mechanisms which govern the transformation of a hard rock into a friable sandy material and then into a soil. It appears that soil properties depend more on the water regime than on the mineralogical and chemical properties of the parent rock. The formation of more or less horizontal layers (horizons) indicates that the governing mechanisms are chemical transfers and particle translocation from the atmosphere interface. In contrast, weathering produces a friable rock in which the structural, mineralogical and chemical characteristics are strongly dependent on that of the parent rock. Usually, the transition between soils and weathered rocks is sharp. Both can be considered at first glance as independent environments. Soils and weathered rocks will be studied separately. As the first secondary minerals form from the primary granitic minerals, it seems more logical to begin with the weathering reactions.

3.3.1 Clay Formation in Weathered Granite Under Atlantic Climatic Conditions

3.3.1.1 Weathered Profiles on Granitic Rocks: The First Stages of Weathering

The weathering intensity is classically considered to regularly increase from the freshest rock at the bottom of the profile to the most altered zones at the top. Thus, the profile is composed of a vertical sequence of horizons: fresh rock, coherent slightly altered rock, saprock and saprolite. This is usually an oversimplification, since such a vertical sequence is rarely observed. In particular, weathering does not produce a regular and continuous coat above the granitic massifs. Erosion, as well as preferential alteration in the fractured zones, greatly modify the theoretical vertical distribution of the different horizons. It is common to observe weakly altered zones above deeply altered ones. Besides, heterogeneities are commonly observed in a given zone at the centimeter scale: for example, small volumes of coherent rocks are often seen in a friable zone. Figure 3.36 shows the most common occurrence of weathered profiles developed under an Atlantic climate in the Armorican Massif (Meunier and Velde 1979). It is composed of the following zones: fresh rock, coherent slightly altered rock, saprock (structure conservative friable material) and saprolite (restructured clayey zone).

The Coherent Slightly Altered Rock

This zone is not systematically found at the bottom of the profile: for example, it is seen in the right part of the La Rayrie profile as large blocks outcropping through the soil cover (Fig. 3.36). The granite remains coherent, representing the initial stages of alteration via the weathering process. New phyllosilicate phases are found along the microcracks and discontinuities of the rock structure. These are grain boundaries, internal microcracks in grains and polycrystalline microfissures. Consequently, the energy state of water is high; the microsystems can be considered as closed. Two stages are noted:

1. The first new mineral to be formed is illite which grows on the (001) faces of muscovite or biotite found in grain to grain contact with potassium feldspar (Fig. 3.37a). At a slightly later point in stage 1, dioctahedral and trioctahedral vermiculite are found along the muscovite–biotite interfaces

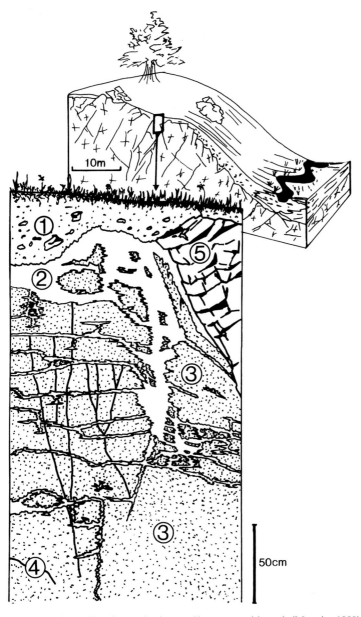

Fig. 3.36. The La Rayrie weathering profile on a granitic rock (Meunier 1980). *1* Colluvial soil; *2* white zone in which the initial granitic structure is modified; *3* saprock; *4* fissures filled by a reddish deposit; *5* altered coherent rock

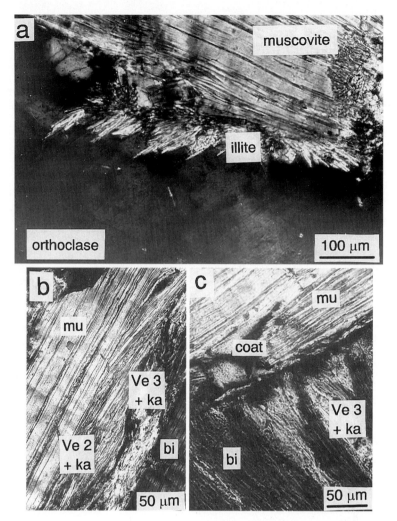

Fig. 3.37. Microphotographs of some weathering microsites in the La Rayrie profile. **a** Illite growth on a muscovite–orthoclase intergranular joint. This microsite can be considered as a closed microsystem; **b** alteration zones at a muscovite–biotite intergranular joint (parallel to the 001 crystal faces). Muscovite *mu* and biotite *bi* are replaced by a dioctahedral vermiculite *Ve 2* + kaolinite *ka* and a trioctahedral vermicilite *Ve 3* + kaolinite secondary mineral assemblage; **c** microfissure outlined by a clay coating *coat* at the muscovite–biotite intergranular joint (orthogonal to the 001 crystal faces); the coating is composed of kaolinite + Fe-hydroxides

(Fig. 3.37b). These minerals are the only ones to be found at grain contacts where two preexisting phases participate in the formation of a weathering mineral (contact microsystems).
2. The primary minerals destabilize internally. Kaolinite is found in microcracks in plagioclases. Beidellite–montmorillonite (smectite) zones are

apparent in potassium feldspars in more altered grains of the slightly altered rock. Kaolinite is at times also present. It should be noted that there is no evidence of new mineral growth along quartz grain boundaries, nor is there evidence of dissolution even though these zones are quite permeable (Parneix and Meunier 1983). The accessibility of water to minerals is obviously the more important factor in the early stages of alteration. This accessibility is strongly dependent on the initial microfissuration as shown by coloration techniques (Rassineux et al. 1987).

The Saprock

The altered rock becomes friable (i.e., individual grains are loosened with one's fingers) in hand specimen but the structure of the crystalline granite is still visible, i.e. initial feldspars can be distinguished with respect to quartz and mica grains, even though their color and general aspect is greatly modified. In this zone, all primary minerals except quartz are internally destabilized (alteroplasma) showing single or multi-mineral assemblages of phyllosilicates. Biotite produces a trioctahedral vermiculite + kaolinite + iron oxide assemblage. Muscovite forms a dioctahedral vermiculite + kaolinite assemblage (Fig. 3.37c). Plagioclase recrystallizes into single-phase kaolinite zones. Orthoclase is seen to recrystallize into alteroplasma zones sometimes with sericite (largely due to deuteric alteration in the late stages of the magmatic evolution) but more often with vermiculite or beidellite and kaolinite. In all cases, the initial magmatic minerals form a new alteroplasma which wholly or partially replaces the initial grains without changing the dimension of the grain zone. Thus, the mineral transformations observed depend upon the original chemistry of the minerals which were destabilized by the altering solutions. In other words, several chemical components are extensive variables of the system (inert components: Fe, Al) while others are intensive (mobile: alkalis, Ca and Mg) variables. The rock structure becomes one of the mosaic of microsystems (Korzhinskii 1959) which are in internal equilibrium with an altering fluid.

The Saprolite

The altered rock becomes very friable and porous. The granitic structure is destroyed. In temperate climates, the saprolite zones do not form continuous horizons, as is the case in tropical climates (Nahon 1991). Instead, white patches or bands are found which cut the saprock mass. Parental minerals are largely destroyed, except quartz, and one finds a general plasma (pedoplasma) structure in thin section. The reorganization of old and new phyllosilicate phases in the pedoplasma creates new chemical potentials which crystallize new minerals. In this zone, the assemblage illite + kaolinite has been identified, it replaces such minerals as dioctahedral vermiculites and smectites, mixed with remaining debris of orthoclase.

Fissures

Micro- and macrofissures (from micrometer to centimeter widths) are found throughout the alteration profile. They are always coated with clay deposits which have been transported by the aqueous flow in the system. Thus, at any point in the profile (slightly altered rock, saprock, saprolite) the only minerals found in fissures or on fissure walls are kaolinite and iron oxides (Fig. 3.37c). If the altering fluid in a profile is collected, i.e. that fluid which flows easily in the profile through the large fissures, it will be in equilibrium with kaolinite (Feth et al. 1964). This indicates that the other clay minerals which are also transported (vermiculite, smectite) are not stable in such diluted solutions. Their absence must not be taken as an effect of a selective translocation but rather as a result of their dissolution.

In summary, the following key mineral assemblages in the different zones of alteration will be:

1. veins: kaolinite + iron oxides (deposit after transportation: coatings)
2. saprolite pedoplasma zones; illite + kaolinite
3. saprock alteroplasma zones: beidellite (smectite) + kaolinite
4. altered rock grain contacts: illite + orthoclase
5. initial rock grain contacts: orthoclase + muscovite.

The above table was constructed from the data of Meunier (1980) and Dudoignon (1983) for the profile of La Rayrie. Mineralogical observations are given in the studies of Harriss and Adams (1966), Bisdom (1967), Wolff (1967), Sikora and Stoch (1972), Gilkes et al. (1973), Rice (1973), Eswaran and Bin (1978) and Boulangé (1984) from other profiles. These assemblages do not include biotite or its reaction products which contain magnesium and iron. This phase has been left out in order to simplify the chemical system under consideration. It appears that there is little interaction of the biotite with the other minerals in the altered rock, saprock or saprolite zones. Since the mineralogy of the altered biotite is similar to that of altered muscovite (vermiculite + kaolinite), it will be assumed that the exclusion of this mineral will not greatly change the interpretation which follows.

The problem of plagioclase must be considered here. It has been noted in the La Rayrie profile, as well as in others, that the plagioclase is replaced in the early stages of alteration by kaolinite (or halloysite). This introduces the phase which is typical of the latest stage of alteration into the early parts of the alteration sequence. Essentially, the assemblage is one of kaolinite plus silica in solution, which cannot crystallize a phase which would form at silica saturation. Possibly, one can propose that the sodium and the calcium of plagioclase do not seem to play the role of potassium. Neither mica or smectite replace the plagioclase. The only micaceous mineral to form under surface conditions is a potassic one. Neither paragonite nor margarite have been found to crystallize in natural surface environments (Velde 1985). Hence, Na–Ca plagioclases do not form micas. A second observation critical to an understanding of phase relations is the lack of albite found as surface mineral. The Na and Ca tectosilicates observed at surface conditions are normally zeolites (Velde 1985). Since high pH and silica activity is needed to form these minerals, it is possible that lack of silica in solution and the low pH reduces a kaolinite + plagioclase + zeolite field to an aqueous solution–kaolinite tie-line. Silica activity would not be high enough to establish a smectite mineral in the local plagioclase alteroplasma environment. At the early stages of alteration, the alkali potential (sodium and calcium in this case) is still too high to form a smectite, but the potential of Si released into solution is too low to produce a zeolite. The kaolinite forms as a metastable mineral, relative to the others, forming at the same level of alkali (potassium) potential in the sequence. In the alteroplasma environment of the plagioclase grains, the ratio Na + Ca to K is such that the system is no longer one of K + Na but of Ca + Na, and the phase relations change in that there is no mica phase present (see Helgeson et al. 1969).

3.3.1.2 Construction of Phase Diagrams

Composition Diagram

Initially, one must proceed to place the mineral compositions of the phases observed into a chemical framework. Given the phases which occur in weathering as a result of the basic granite composition, it is possible to restrict the chemical parameters necessary to describe the reaction observed. Three components are used: silica, alumina and alkalis. This simplification has been used for quite some time (Garrels and Christ 1965). The assimilation of calcium with alkalis is possible because no calcic silicate (zeolite or margarite) forms during granite weathering. Magnesium has a minor role in the system. Iron is initially present in the granite primary minerals, but it quickly becomes an unreactive element as far as the silicates are concerned because it forms an insoluble oxide Fe_2O_3. In using the three components of Si, Al and alkalis (Na + K), we then automatically ignore the existence and evolution of biotite, if it occurs in the granite. As it turns out, this phase does not appear to react with the aluminous minerals to a great extent in the weathering process, and its exclusion should not greatly affect the phase relations described (Meunier 1980). In fact, the biotite forms the silicate phases, as do the muscovites in the granite. The presence and chemical activity of quartz poses a problem to the interpretation of phase diagrams at low temperature. As has been pointed out by Velde (1985), the very low reactivity of quartz at surface conditions excludes it from the active components in most silicate systems. Silica saturation of a solution normally gives a solid phase in the form of a precipitate of amorphous material. This occurrence has not been noted in the weathering of granites, at least in temperate and tropical climates. Thus, in the system studied, the solutions behave as if they were undersaturated with respect to

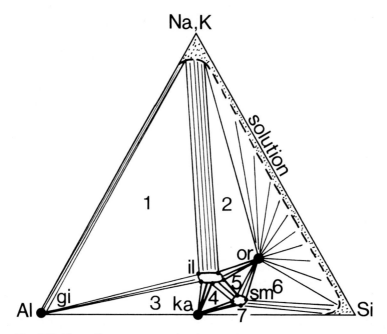

Fig. 3.38. Phase diagram representing the mineral relations in surface conditions in the Si-Al-K,Na system (see text for detailed explanation). *il* Illite; *or* orthoclase; *sm* smectite (expandable minerals including vermiculite); *ka* kaolinite; *gi* gibbsite. Triphased assemblages are indexed from *1* to *7*. *Heavy* tie-lines represent the mineral assemblages observed in the La Rayrie profile

Table 3.2. Mineral assemblages in weathered granite classified according to the microsystem under which they formed

	Mineral. Assemblages	Microsystems
1	Illite + K-feldspar	Contact microsystems
2	Illite + Expandable minerals	Contact microsystems
3	Kaolinite + Expandable minerals	Alteroplasma microsystems
4	Illite + Kaolinite	Pedoplasma microsystems
5	Kaolinite (+ Fe-oxides)	Fissural microsystems
6	Kaolinite + Gibbsite or Gibbsite (+ Fe-oxides)	Residual microsystems (Tropical climate)

amorphous silica in their SiO_2-rich portion. The low reactivity of the stable form of silica (quartz) will have a consequence on the phase diagram derived from the observation of the natural assemblages: since silica is not found as a mineral phase, the assemblages will be buffered by the next most stable siliceous mineral. Gibbsite, which is commonly found in weathered rocks under tropical climates, was not seen in the studied profile. Thus, it represents a boundary phase only.

Figure 3.38 shows the compositions of the mineral phases found in Table 3.2 as projected into the Si–Al–Na,K system. The petrographic observations made above allow one to place the tie-lines between the minerals to form the basic mineral assemblages. The mineral pairs illite + kaolinite, kaolinite + smectite and orthoclase + illite are thus established. This excludes the destabilization of plagioclase (see explanations above).

The relations between illite, orthoclase and quartz in the alkali–silica–alumina system are not established. Since, as has been mentioned previously, the reactivity of quartz is very low and the solutions are never saturated with SiO_2 it is probable that the mineral equilibria in solution will be controlled by another mineral, such as the smectite instead of quartz. In this way, one establishes the active or effective tie-line orthoclase–smectite instead of the illite–quartz tie-line which might represent a more stable mineral pair. The illite–smectite tie-line is based upon reports in literature (cited above) of the coexistence of these minerals. Its justification by our petrographic observations will follow in the section concerning the chemical potential–composition diagrams.

It is useful at this stage to establish the lines of equipotential for the elements which are likely to become mobile or intensive variables in the system of chemical weathering (Korzhinskii 1959; Spear et al. 1982; Meunier and Velde 1986). These lines should be based upon the observations of relative mineral instability at a given level of alteration intensity. In other words, when one sees illite forming with smectite, do other phases form at the same time, or after this mineral assemblage? Such an analysis is a cross-check on the use of a given representation of mineral assemblages. In the composition diagram of Fig. 3.38, the key equipotential is number 7 which includes the illite–kaolinite tie-line and the illite–smectite tie-line assemblages. This is justified by the observed destabilization reaction, illite → kaolinite + smectite (vermiculite), found at level 8. This reaction implies that the two-phase field or tie-line occurs below the illite stability or the equipotential line number 7. This renders an orthoclase–kaolinite tie-line impossible. Such a situation can be found in Helgeson et al. (1969) whereas the illite–smectite tie-line which intersects the orthoclase–kaolinite tie-line can be found in Garrels (1984). The problems of the geometric configuration as a function of the tie-lines in the illite–orthoclase–kaolinite–smectite quadrilateral is discussed at length in Drever (1982). The observation that the alteroplasma of muscovite is made up of smectite and kaolinite, substantiates the illite–smectite tie-line. It should be noted that there is no evidence of a continuous solid solution of the smectite–illite series, such is found in mixed layering of the two components, in the weathering minerals. The I/S mixed layered minerals are typical of diagenesis.

Composition–Chemical Potential Diagrams

In our weathering system, it is evident that the rock is affected by a loss of material. In general, two elements have been considered in the past as mobile species: Si and alkalis (Na, K). This is the classic approach given by Garrels and Howard (1957). If one considers both elements to be intensive variables of the system, i.e. their chemical potential to be variable, the phase diagram which results presents large mono-mineral fields with two or three phase assemblages found at very restricted conditions of the variable chosen. One can say that the system is buffered by the solutions present. In our granite, it was found that there are usually two-phase clay mineral assemblages present. This suggests that there are not two but only one chemical element acting as a fully intensive variable. Such an eventuality leads to a chemical potential–composition diagram such as is represented in Fig. 3.39.

In the weathering of the surfaces of minerals, it has been noted that hydrogen is introduced into the solid phases. As a result, the chemical potential of H^+ (pH) should be considered in the weathering process, as was proposed many years ago. If the compositional representation of Fig. 3.39 is considered as a combination of Si–Al–(Na + K)/H^+, the use of aqueous equilibria is much easier. In this way, one can conceive of an aqueous solution which dissolves orthoclase by adding (OH)$^-$ to the silica and alumina components, so that all species are dissolved. One can then approach natural conditions of alkali and hydrogen ion concentration as independent variables; one must assume that all reaction boundaries are orthogonal to the projection of the chemical system, i.e. there are no field boundaries which can exist over a range of K^+ to H^+ ratios. Also, the solids, such as SiO_2 and Al_2O_3 must not be destabilized by a change in pH alone, such as is known at extreme values of pH. In our representation of the chemical phase relations we will use the representation of μ_x–X, as proposed by Korzhinskii (1959). In fact, the weathering sequence produced in granites is strongly buffered by the minerals present and the initial pH of the rain water which is introduced into the rock–clay system. The pH could vary by two or three whereas the concentration of alkalis reduces to an infinitely small quantity in the upper zones of weathering. Therefore, the variable of greatest interest is the activity of the alkalis, which will have a greater effective range than that of hydrogen ions.

Here, we have chosen to represent the silica and alumina as extensive variables and the alkalis (less hydrogen-ion activity) as the intensive parameter, where their chemical potential is the variable to be considered. The equipotential lines traced in Fig. 3.39 use the alkali chemical potentials which occur between the values where major mineral transformations take place. The important feature of the diagrams is the compression of the three-phase fields of the triangular diagram into a single equipotential (Korzhinskii 1959; Spear et al. 1982). We cannot distinguish, at present, between the relative positions of certain phase reactions as a function of chemical potential on either side of a single phase in Al–Si dimensions; i.e. the position of the illite–gibbsite–solution line with respect to the illite–smectite–orthoclase line because our data do not sufficiently cover the aluminous assemblages. Gibbsite has been reported in early stages of granite weathering (Velde 1985) but was not found in this example. Because the phase associations have not been determined, the mineral occurrence will not be discussed further. Several of

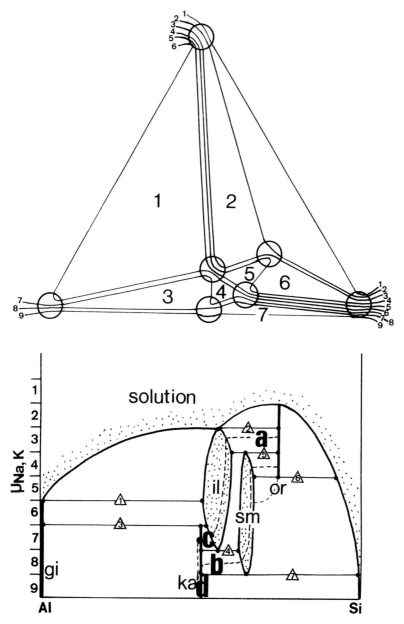

Fig. 3.39. Derivation of the $\mu_{Na,K}$–Si–Al phase diagram using the 9 equipotentials (see text for detailed explanation). *Abbreviations* are identical to those used in Fig. 3.38. Triphased assemblages are indexed by *numbers in triangles*; compositions of phases in triphased assemblages are indicated by *dots*. Solutions and solid solutions are represented by *dotted areas*. The chemical domains of the different microsystems are indicated by letters: *a* contact; *b* alteroplasma; *c* pedoplasma; *d* clay deposits on fracture walls. Alteration processes are indicated by downward trajectories, *stippled lines*, which converge toward a monophased kaolinite assemblage

the reactions do not necessarily occur at the same equipotential but they will be near one another. Thus the general relations of the diagram are the probably correct, but some of the details are not established.

In the potential–composition representation we can fix the position of illite in the presence of orthoclase as a function of Si–Al content from the chemical data (electron microprobe analyses). Also the existence of illite in contact with kaolinite is known. Therefore, the illite–smectite boundary is fixed by the smectite composition. The composition of the smectite at the illite–smectite–orthoclase boundary is fixed by the maximum Si content of illite found in this assemblage. The extension of the smectite composition to the silica-rich side of the diagram is not established, nor is the alumina content of illite in the presence of gibbsite. However, it is clear that the compositions of both illite and smectite become more aluminous as the chemical potential of alkalis decreases.

If we assume the phase diagram to be correct, it is possible to make a certain number of general deductions concerning the behavior of minerals and solutions in which they occur. In the instance of the first stages of alteration, i.e. where the solution just penetrates the magmatic minerals and is therefore almost in equilibrium with the previously formed clay minerals present, the composition of the solution is alumina-rich with respect to the mineral. This is seen in the series of solution–solid curves in the upper part of the diagram. The muscovite–orthoclase contacts are active because the muscovite buffers the K^+ to H^+ ratio of the solution at a level where orthoclase is no longer stable. Illite, the siliceous low-temperature equivalent of muscovite, crystallizes. The quartz–orthoclase contacts are not active because the buffering effect of muscovite does not exist. The solution reaches equilibrium with orthoclase.

Once the solution is buffered by the active clay minerals, the change in the Si/Al content is less important, but there is still a tendency to lose silica. The active part of the system, excluding quartz, becomes one in which the clay minerals illite, smectite (vermiculite) and kaolinite control the Al/Si content of the solution. In this zone, the chemical potential of alkalis is at an intermediate value. If we assume that the system is in the late stages of weathering, the initial solutions are dilute in the content of all dissolved solids. If quartz remains inactive, the tendency for the system will be to produce the next more siliceous mineral from the assemblage illite–smectite–kaolinite which is present. In the zone of lowest chemical potential of alkalis, this phase is kaolinite. If one recalls the mineralogy of the fracture zones in the rock, one finds that kaolinite is the only clay mineral present. Thus, in the situation of dilute solutions, one finds a monomineral assemblage. However, this might simply be due to the fact that the quartz will not react rapidly enough to buffer the assemblage.

The fact that the saprolite pedoplasma contains the assemblage illite + kaolinite, indicates that there is a general loss of silica in the weathering sequence, since the bulk composition of the granite should initially lie somewhere in the area between orthoclase and muscovite, excluding the inactive quartz component. Thus it is evident that the loss of silica is gradual but persistent throughout the process of granite alteration in a temperate climate.

As a counter proof of the correct choice of the alkali chemical potential as the chemical component which is an intensive variable, we can attempt to construct the phase diagram using the Si component as the intensive variable, keeping the alkalis as extensive variables. This analysis is presented in Fig. 3.40.

It can be seen immediately that the first phase to be found in such a situation, at maximum silica chemical potential in solution, will be the smectite mineral. As one will remember, the first mineral to form was seen to be the illite mica. Thus the representation of silica as the intensive variable is not correct.

Chemical Potential–Chemical Potential Representation

The most conventional method of representing the clay mineral assemblages found to be modified by aqueous solutions, is one of chemical potential against chemical potential (or the ratios of these values in some cases). The arguments for the use of potential–composition diagrams in the case of the granite weathering in temperate climate conditions have been given. However, other conditions prevail for surface alteration processes in the tropics where soils and altered rocks tend to form mono-mineral zones as horizons instead of veins, as in temperate climates. The most intense weathering, forms concentrations of gibbsite at the surface which grades into kaolinitic zones (Nahon 1991). Boulangé (1984) has further identified a smectite-rich zone at the base of the kaolinite zone. This horizon is much smaller in extent, centimeters, compared to the normally plurimetric sequences of kaolinite. Here, there is a good evidence that the system is guided by a multi-chemical potential variability where probably only the alumina is an inert, extensive variable. If we transpose the diagram of alkali potential–composition into coordinates of silica potential and alkali potential, we obtain the result as given in Fig. 3.41.

Again, it is assumed that all of the phases are stable over the common ranges of pH found in granite weathering. In the figure, the reaction boundaries are shown as curved lines. It is certain that phase reactions between minerals of variable composition will produce curved phase boundaries (Garrels 1984; Meunier and Velde 1986). However, those presented here could not be calculated due to a lack of chemical data, since not all of the mineral compositions are known for the different reaction boundaries in the granite system under study. It seems reasonable to omit the details for the time being. The topology is correct, respecting the interdiction of a phase-boundary intersection of an invariant point at a sector angle of more than 180°. The applicability of the diagram to weathering in tropical zones has been mentioned. In the case of the weathering in a temperate climate, only the open fissure zones found throughout the profile will be adequately described by such a diagram.

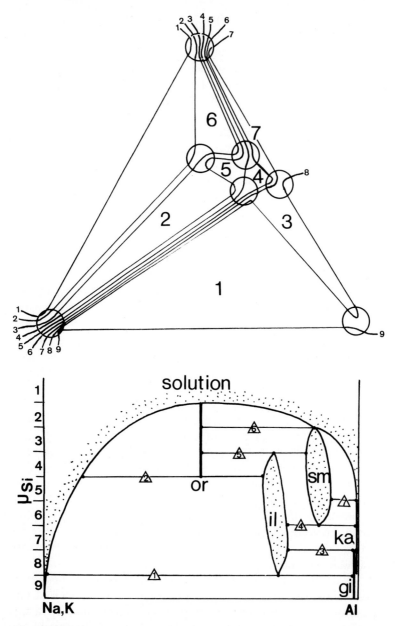

Fig. 3.40. Derivation of the μSi–Al–Na,K phase diagram using the 9 equipotentials (see text for detailed explantion). Abbreviations are identical to those used in Fig. 3.39. Triphase assemblages are indexed by *numbers in triangles*; composition of phases in triphase assemblages are indicated by *dots*

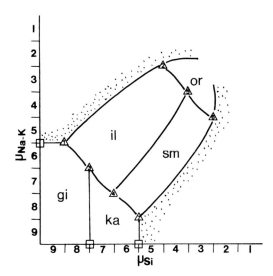

Fig. 3.41. Construction of the μ_{Si}–$\mu_{Na,K}$–Al phase diagram by combination of the $\mu_{Na,K}$–Si–Al and μ_{Si}–Al–Na,K phase diagrams represented in Figs. 3.39 and 3.40 respectively. Triphase assemblages are indexed by *numbers in triangles*. The shape of the curved tie-lines is hypothetical

Reaction paths for tropical weathering will be as follows: the unsaturated water which enters the upper profile will form either kaolinite or gibbsite, depending on which minerals are present in the soil, illite or smectite. If silica activity is very low, illite could transform into gibbsite. However, the normal succession is from kaolinite to gibbsite. Thus the initial minerals must be in equilibrium with a solution which is richer in silica. In the initial stages of weathering, in the rocks at points of microfracturing, the first reaction upon dilution of a solution in equilibrium with orthoclase will be to form either mica (illite) or smectite. Exact reaction paths can only be determined with more information concerning the precise mineral assemblages. The final product of granite weathering in tropical conditions is a soil composed of Al and Fe oxides and hydroxides; Al and F^{3+} being the remaining inert components in the system. This new material was shown to be sensitive to water activity by Trolard and Tardy (1989). The variation of compositions of oxides and hydroxides was calculated and presented in a μ_{H_2O}–Al–Fe^{3+} diagram.

3.3.1.3 Summary

It is apparent that a complete sequence of phase diagrams can be constructed on the basis of the mineral compositions found in the active part of an alteration sequence. The relative order, in the altering sequence, in which each reaction takes place will determine the position of the phases in a hierarchy of equipotential lines. Observed mineral assemblages will relate the phases in compositional space. The equipotential lines must follow the tie-lines between the minerals. Such tie-lines are established by petrographic observation aided by X-ray micro-diffraction (Rassineux et al. 1988).

It has been seen that the weathering reactions of granite under temperate climate are best represented in the alkali chemical potential versus Si–Al composition coordinates. The potential–potential type of diagram is best employed in cases of weathering in tropical zones or highly drained areas where mono-mineral horizons (or nearly so) are encountered. In the case of weathering under temperate conditions, the lateral inhomogeneity gives various assemblages which indicate local control of the solutions by the silicates present.

3.3.2 Soil Clays Developed on Granite Saprolite in the Temperate Zone

Because they are young soils in which the transformation of initial minerals into clays is far from complete, clay transformation in soils of the temperate zone is certainly a difficult case to study. Moreover, granite (but also sedimentary rocks or superficial deposits) contains phyllosilicates such as mica and chlorite, the structures of which are very similar to those of clay minerals. Thus, it is often very difficult to distinguish inherited fine-grained phyllosilicates from clay minerals that formed in place. However, it is a frequent case encountered in Europe and North America. In these areas soils have been intensively cultivated for centuries and are subjected to pollution from farming or industrial activities. Studies of the nature and properties of clays from these soils should be done to answer properly the environmental problems facing the industrialized countries, even if the clay mineralogist would have preferred more academic examples.

The methodology presented above for studying weathering must be modified for the study of soils. Indeed, if the aim is the same (how mineral reactions proceed?), the particular structure of the soils and the very small size of most of the minerals (less than 10μm) do not allow us to apply exactly the same sequence of analytical procedures. Direct observation of reaction sites under microscope is nearly impossible. Thus, the basic technique to employ is the fractionation of the material in different size classes. Each class is composed of polyphase mineral assemblages which can be accurately determined using mathematical decomposition of the complex bands of X-ray diffraction diagrams.

3.3.2.1 Methodology

Clay formation in the temperate zone will be exemplified by a detailed study of clays from a soil developed from granite saprolite in the western part of the Massif Central, France (Righi and Meunier 1991). The soil is a typical Cambisol i.e. a soil with slight morphological differentiation, it is a good example of those encountered in granitic areas and, can be described as follow:

Soil Description

0–8 cm: *A12*, very dark greyish-brown sandy loam, strong, very fine-granular structure, friable, many medium and fine roots, bleached sand grains, few granite gravels, diffuse smooth boundary.

8–15 cm: *A2*, dark yellowish-brown sandy loam, weak, fine-granular structure, friable, common fine roots, few granite gravels, clear smooth boundary.

15–25 cm: *Bw*, dark yellowish-brown sandy loam, weak, subangular blocky structure, friable, few medium roots, granite gravels, diffuse wavy boundary.

> 25 cm: *C*, yellowish-brown sandy loam, weak, subangular blocky structure, friable, gravely, few roots.

The soil pH is about 5, cation exchange capacity (CEC) is low, ranging from 8 to $18\,cmol_c kg^{-1}$ ($1\,cmol_c kg^{-1} = 1\,meq/100\,g$), organic carbon content is rather high, up to 6% in the A1 horizon. The clay content (2 μm) increases upward from 5 to 13% from the C to the A1 horizon. This clearly shows that clay is formed in the soil, other sources of clays, such as wind deposition or colluviation being negligible. Moreover, Fig. 3.42 shows that the clays that formed are predominantly fine sized clays (<0.2 μm).

Another interesting point is the increase in citrate-bicarbonate-dithionite (CBD) extractable iron, which represents iron as oxy-hydroxides. In the soil and parent rock (granite) iron is mainly present in silicates. The weathering of these iron-bearing primary minerals produces amorphous (short-range ordered) Fe-oxy-hydroxides which generally occur adsorbed onto the clay particle surfaces, and which are extractable by the CBD treatment. Therefore, an increase in the amount of CBD extractable Fe measures the increase in weathering of Fe-bearing silicates.

Extraction and Preparation of Soil-Clay Samples

The clay fraction (<2 μm) is separated from the bulk soil sample by sedimentation following destruction of organic matter with hydrogen peroxide and removal of the Fe-oxyhydroxides with CBD. The whole clay fraction is then

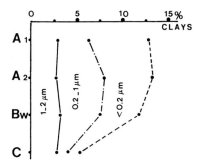

Fig. 3.42. Proportion of total clay in the 1–2 μm, 0.2–1.0 μm and <0.2 μm fractions in the different horizons of a soil

subjected to size fractionation according to the following sizes: <0.1 μm, 0.1–0.2 μm, 0.2–1.0 μm, and 1–2 μm. Size fractionation allows one to separate newly formed soil clays, generally very fine, from their precursors which have larger sizes. The 2–5 μm and 10–20 μm silt fractions are also investigated as characteristic of the inherited minerals.

Moreover, within the coarse fractions, magnetic separation allows the separation of Fe-bearing phyllosilicates, such as biotite and chlorite, from quartz, feldspars and muscovite. The combination of particle-size fractionation with magnetic separation, leads to fractions that contain only phyllosilicates; this is an important step for their further study. For example, a soil sample that contains quartz, feldspars, muscovite, chlorite and clay minerals can be separated in a fine fraction (<0.2 μm) that contains only the clay minerals. Quartz, feldspars, muscovite and chlorite (inherited from the rock) are collected in a coarse fraction. From this coarse fraction, chlorite can be isolated by magnetic separation and subjected to specific investigations.

X-Ray Diffraction: Identification of Minerals

In the silt and coarse-clay (1–2 μm) sub-fractions, the phyllosilicates inherited from the granite saprolite are identified by X-ray diffraction. These minerals are mica, chlorite, vermiculite and interstratified mica/vermiculite and chlorite/vermiculite. Vermiculite and interstratified minerals originate from early stages of the alteration of biotite (a granite-forming mineral).

A series of X-ray diffraction diagrams obtained from subfractions of decreasing size is given in Fig. 3.43. For the finest subfractions, X-ray diffraction produces broad and overlapping bands instead of sharp and separated peaks found in the coarse fractions. Thus, a special treatment of the diffraction signal must be used to improve the reading of the diagrams. The broad bands are decomposed into elementary curves, using a microcomputer program (DecompXR, Lanson and Besson 1992). When the decomposition program is used, it is evident that the soil clays studied here are a complex mixture of three to five individual phases. However, a clear trend of evolution can be seen.

3.3.2.2 Observations

A complete description of decomposed XRD diagrams from the A1 clay subfractions is now given in order to illustrate how one can work with this complex material.

Coarse Clays. in the range of 4–11° 2θ, only two major reflections are obtained from the 1–2 μm subfractions. They are attributed to mica (1 nm) and vermiculite (1.4 nm) respectively, these two minerals being inherited from the saprolite. The peak of mica is decomposed into two curves, one very sharp, indicative of a well-crystallized mineral, the second, broader and less

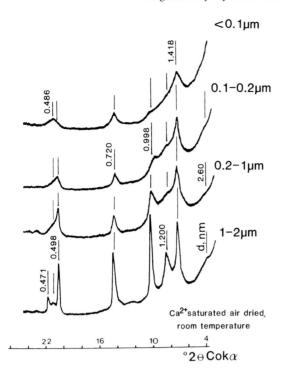

Fig. 3.43. X-ray diffraction diagrams for clay fractions of decreasing particle size for the top horizon of the soil studied. Smectites are found at 1.42 nm, mica or illite at 0.99 nm, mica/vermiculite mixed layers at 1.20 nm, and kaolinite at 0.72 nm

intense, indicating less material and some interstratification of vermiculite with mica.

Fine Clays. compared to the coarse-clay fraction, the pattern of the XRD diagrams from the fine fractions (<1 µm) are different. These diagrams give four or five basic curves instead of two. From the 0.2–1 to the <0.1 µm subfractions the following changes are observed (Fig. 3.44):

1. The sharp peak at about 1 nm, still present in the 0.2–1 µm subfractions, disappears from the finer fractions. Only the second broad mica band, present as a smaller band in the coarse fraction, persists, but its position is shifted toward 1.02 to 1.07 nm and its width is increased;
2. A peak at about 1.20 nm shows increased intensity and decreased width from the 0.2–1 to the <0.1 µm subfractions;
3. A peak at about 1.40 nm becomes the most intense peak in the finest fractions. In the 0.2–1 and 0.1–0.2 µm subfractions, the 1.40 nm reflection is made of two basic curves with either a large or a small width. This indicates the presence of two types of vermiculite particles, one with few layers in the stacking sequence (large width) the other with a better ordered stacking (small width);
4. Smectite layers, or interstratified minerals with a smectitic component (basic curve with a maximum of intensity at about 1.60 nm), are observed in the <0.1 µm subfraction only;

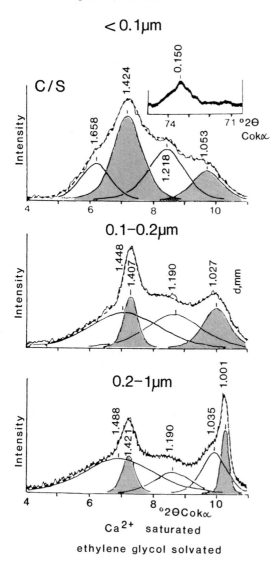

Fig. 3.44. X-ray diffraction patterns of the fine clay fractions. Diagrams show decomposition peaks for the multiphase assemblage. Smectites are found at 1.658 nm, vermiculite at 1.424 nm, illite/vermiculite mixed layers at 1.218 nm and illite at 1.027 nm, < 0.1 µm subfraction: 060 band at 0.150 nm for dioctahedral minerals

5. A rather strong peak at 2.38 nm is present in the <0.1 µm subfraction indicating the presence of an ordered interstratified mica/vermiculite (first-order reflection).

The same general trend is observed for the sub-fractions from the A2 and Bw horizons, except for the finest subfractions from the Bw horizon in which the relative intensity of the peak at 1.40 nm is greater than for the equivalent A1 horizon sample (less smectite and more vermiculite layers in the Bw sample than in the A1 sample). All the <0.1 µm subfractions contain mainly dioctahedral minerals as indicated by a broad reflection near 0.150 nm.

3.3.2.3 Clay Genesis in Temperate Acid Soils

The silt and clay fractions from the C horizon contain chlorite as individual particles, or clustered or separated layers, within interstratified minerals. Compared to the C horizon, the proportion of chlorite layers in the A1 horizon is much reduced in the fine-silt fraction (2-15 μm). Moreover, chlorite has totally disappeared from the coarse clay of that horizon (1–2 μm), but it is found in the finer-clay subfractions (<1 μm). The same remark can be made for the mica/vermiculite interstratified and vermiculite layers. These are present in the fine-silt fractions but have disappeared from the coarse clays of the A1 horizon. The distribution pattern of the mica layers in the different subfractions is quite the opposite: as they are the major component of the A1, 1–2 μm subfractions, their proportion progressively decreases with the subfraction size. Thus, it is deduced that the large particles of chlorite, mica/vermiculite and vermiculite layers are preferentially affected by mechanical fragmentation and dissolution. This results in the concentration of more resistant mica layers in the coarse clay subfractions. Weathered residues of chlorite, mica/vermiculite and vermiculite are accumulated in the fine clays.

This can be seen at the scale of a single clay particle using high resolution microscopy. Figure 3.45 shows a complex clay particle made of mica layers associated with vermiculite and chlorite mixed layers. The mica core is apparently unweathered, whereas the chloritic and vermiculitic mixed layers are partly dissolved, leaving smaller residual particles which were analyzed in the fine clay fraction by X-ray diffraction. Mica cores are recovered mainly in the coarse clay fraction.

Within the fine-clay fraction, if one makes the assumption that the finest subfractions are also the most weathered, the changes from the 0.2–1 μm to the <0.1 μm subfractions indicate that weathering proceeds via the transformation of mica layers through mica/vermiculite and vermiculite layers. Smectite layers are finally formed in the finest subfraction as end products of the mineralogical evolution.

3.3.2.4 Composition and Properties of the Subfractions

Cation Exchange Capacity

CEC varies from $8.5\,cmol_c kg^{-1}$ to $34.4\,cmol_c kg^{-1}$ and increases from the large to the fine fractions (Fig. 3.46). A good agreement is found between the mineralogy of the subfractions, their K_2O content and CEC values. XRD shows a decrease in the proportion of mica layers, from the large to the fine subfractions, whereas the proportion of interstratified, vermiculite or smectite layers increases. This is consistent with the decrease in the K_2O content and the increase in the CEC. Thus, soil evolution induces the formation of fine clays with an increased CEC.

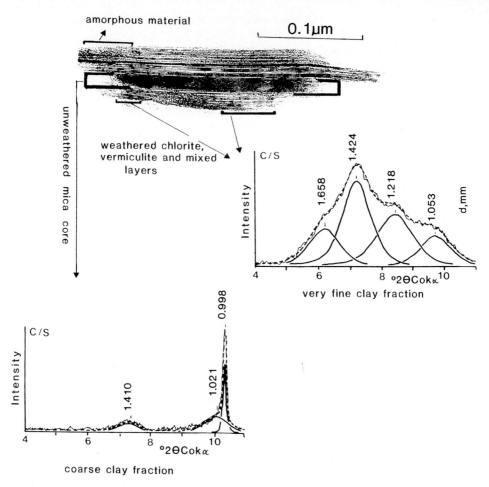

Fig. 3.45. Preferential weathering and fragmentation of chlorite, vermiculite and mica/vermiculite or mica/chlorite mixed layer minerals in a complex particle. Smectites found at 1.658 nm, vermiculite and/or chlorite at 1.424 nm, illite/vermiculite(or chlorite) mixed layers at 1.218 nm, illite at 1.053 nm (broad band) and mica at 0.998 nm (sharp band)

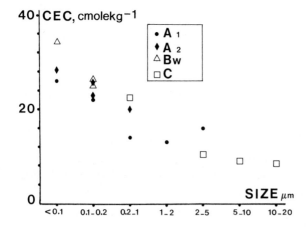

Fig. 3.46. Cation exchange capacity (CEC) for clay and silt fractions in different horizons of a soil profile

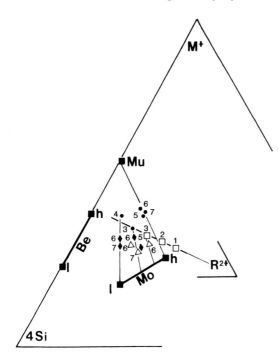

Fig. 3.47. Chemical composition of the separated fractions of different horizons in a soil profile, as plotted in a ternary system M^+–4Si–R^{2+}. *Squares* C horizon; *triangles* Bw horizon; *diamonds* A_2 horizon; *dots* A_1 horizon. *Numbers* represent size fractions: *1* 10–20 M, *2* 5–10 µm, *3* 2–5 µm, *4* 1–2 µm, *5* 0.2–1 µm, *6* 0.1–0.2 µm, *7* <0.1 µm

Chemical Composition

The chemical analyses of the silt and clay fractions are plotted in the M^+–4Si–R^{2+} ternary system (Fig. 3.47) but, because XRD shows that both the coarse and the fine fractions constit of more than one phyllosilicate species, the chemical analysis of such a fraction represents the bulk composition of the mixture. In the diagram, the chemical analyses of the silt and coarse-clay fractions are distributed along a line joining the R^{2+} pole at one side and the composition of a high–charge beidellite at the other side. However, the high-charge beidellitic pole is not reached. Interpretation of Fig. 3.47 confirms the mechanism of preferential dissolution of the vermiculite and chlorite layers, leaving a residue consisting of K-depleted, charge-reduced mica layers with essentiall Al in the octahedral sheet, seen as the sharp 0.99 nm peak in Fig. 3.45.

The chemical compositions of the fine clay subfractions from the Bw and A2 horizons are distributed along several lines joining the composition of montmorillonite in the M^+–4Si–R^{2+} diagram, indicating a change in the weathering pathway. The decrease in the layer charge and increase in the Si atom content are the new directions involved. Each transformed coarse mineral seems to produce a montmorillonite with a layer charge dependent upon its composition. The composition of the fine clays from the A1 horizon are different from the others. They are on a line joining muscovite and high-charge montmorillonite. This could be due to a greater contribution of weathered dioctahedral micas in these fractions.

Evidence for Hydroxy-Interlayered Minerals

In acid soils, the interlayers of vermiculite or smectite are often occupied by non-exchangeable contaminants, these being hydroxy-Al, Fe or Mg polycations. Such clays are known as intergrade or hydroxy-interlayered minerals (Jackson 1963; Barnhisel and Bertsch 1989). Evidence for interlayered contaminants is generally obtained by heating the K-saturated clay sample to increasing temperatures: 110, 300, 550 °C. In response to the heat treatment, interlayered contaminants induce a progressive and incomplete collapse of the interlayers. This results in an asymmetric XRD peak enlarged toward the small 2θ angles. This is observed with the clays in this study; the collapse of the interlayers is progressive and, even after heating to 550 °C, incomplete for all the samples. An improved identification can be done using decomposition of the diagrams, as shown in Fig. 3.48.

For the coarse clay (1–2 μm) subfractions, decomposition of the diagrams gives a sharp curve, with its maximum intensity at about 1 nm, attributed to completely collapsed interlayers. A broad curve with its maximum at about 1.04 nm is associated with the 1.00 nm sharp curve. The intensity of this broad curve increases from the coarse to the fine fractions where it is prominent. Moreover, a third curve, centered at slightly higher *d*-spacings (1.08 to 1.11 nm), is an important component of the XRD diagrams of the finest subfractions. These curves are attributed to interstratification of minerals having either collapsed or uncollapsed interlayers with hydroxy-contaminants. These appears to be more abundant in the finest subfractions.

A chemical treatment (Na-citrate) can be performed to extract interlayered hydroxy-Al (or Fe) contaminants. XRD diagrams of the treated samples are compared to those of the corresponding untreated sample. If interlayered materials are extracted, the collapse of the mineral is greatly improved. The composition of the extract can be analyzed, giving information about the nature of the interlayered materials. When this done on the Bw, <0.1 μm subfraction (Fig. 3.48) only a small decrease in the relative intensity of the 1.09 nm peak is observed and the diagram for the sample which was K-saturated and heated to 550 °C is still strongly asymmetrical: the interlayered material is poorly extracted. For the A1, <0.1 μm subfraction an improvement of the collapse is observed, but the diagram is still asymmetrical (Fig. 3.49). However, compared to the Bw, <0.1 μm subfraction, the effect of the treatment is greater for the A1, < 0.1 μm subfraction. Moreover, the XRD behavior of the glycolated samples from either the A1 or Bw, <0.1 μm subfractions is different. Compared to the untreated sample, the proportion of swelling layers (1.65 nm) is strongly increased in the A1, <0.1 μm subfraction. This is an indication that most of the interlayered material is extracted by the treatment.

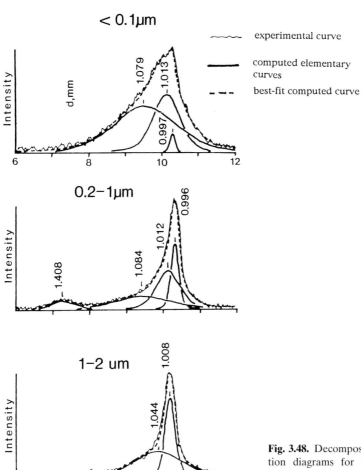

Fig. 3.48. Decomposed X-ray diffraction diagrams for K-saturated and heated samples. Intergrade minerals exhibit incomplete collapse to 1 nm interlayer spacing

The composition of the extract is also different according to the sample from which it is obtained. From the Bw, <0.1 μm subfractions large amounts of MgO (30% of total MgO) are extracted together with Al_2O_3 and Fe_2O_3. Conversely, only small amounts of MgO are extracted from the A1 subfraction. As the interlayered material is quite well extracted from the A1 sample and, as the extract contains almost exclusively Al and Fe, it is presumed that hydroxyl-Al and Fe are the interlayered materials in this sample. Despite the fact that Fe and Al are extracted from the Bw horizon at the same levels as from the A1 horizon fraction, only slight changes are observed in the XRD behavior of the treated Bw sample. A large part of the interlayered

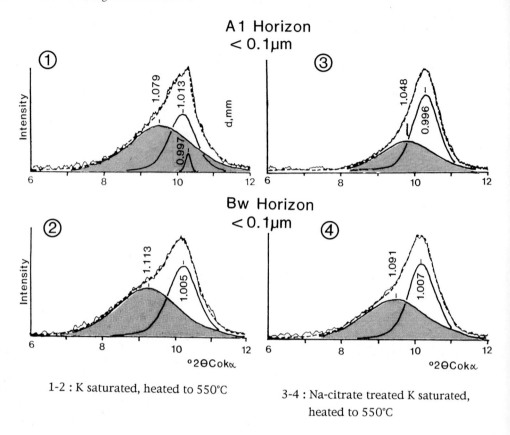

Fig. 3.49. Decomposed X-ray diffraction diagrams for K-saturated and heated samples before and after Na-citrate treatment which extracts the interlayer hydroxy-Al contaminants

compounds is apparently not extracted. Compared to the A1 fraction, far larger amounts of Mg are extracted from the Bw fraction by the Na-citrate treatment. This is the result of partial dissolution of chlorite layers. The poorly crystalline parts of the minerals are dissolved by the treatment which, however, does not result in its conversion to vermiculite. Therefore, it is likely that a chloritic mineral with a more or less stable brucite interlayer is a component of the Bw fine-clay fraction. As the Na-citrate treatment does not selectively dissolve the interlayered hydroxide sheet, the remaining undissolved minerals may contribute to the XRD intergrade behavior still observed for the treated sample.

In conclusion, the origin of the intergrade minerals in the finest subfractions, either from the A1 or the Bw horizons, appears to be quite

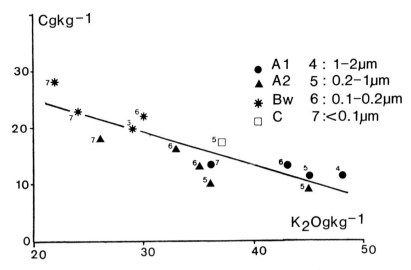

Fig. 3.50. Organic carbon resistant to peroxidation vs non-exchangeable K_2O for separated size fractions. $Cgkg^{-1}$ organic carbon per kg soil

different. In the A1 horizon this fraction is identified mainly as a hydroxy-Al (Fe) intergrade smectite that may have been derived from the weathering of mica layers. In the Bw horizon, the major component is an intergrade vermiculite originating, at least in part, from chlorite layers. These differences are likely to be linked to the physico-chemical conditions which prevail in the horizons; the A1 horizon being more acid and richer in organic matter than the Bw horizon. In strongly acid and organic soils, chlorite may be completely dissolved, whereas when soil is weakly acid and poor in organic matter it transforms to vermiculite.

Evidence for Clay-Organic Complexes

Even after extended treatment with hydrogen peroxide, soil clay samples may still contain rather large amounts of organic carbon (Theng et al. 1986; Schnitzer et al. 1988). For example, organic carbon was found in the clay sample described above. There is a clear negative correlation between the K_2O and organic carbon content (Fig. 3.50). As shown by the XRD study, these clays are a mixture of mica (illite) and vermiculite or smectite, i.e. minerals with no K in the interlayers but hydrated exchangeable cations. So, the decrease in K_2O content means a decrease of the proportion of mica (illite) layers and an increase in vermiculite or smectite layers. The fact that the organic carbon content increases as the K_2O content decreases is a strong indication that organic carbon is associated with vermiculite and smectite and is probably located in the interlayers of these minerals.

3.3.2.5 Conclusion

Clays in a soil which has developed from a granite saprolite under temperate climate originated largely from the mechanical fragmentation of pre-existing phyllosilicates, the clays produced from weathering of feldspars being far less abundant in soils than in the weathered granite. Phyllosilicates are granite forming minerals (e.g. muscovite, biotite) or are produced by alteration. Vermiculite was produced by weathering of biotite in the saprock. Chlorite was formed at an early stage by hydrothermal alteration of biotite. Kaolinite produced by granite weathering is found in the soil but only as a minor component of the clay mineral assemblage.

In the soil, the mechanical fragmentation of inherited phyllosilicates preferentially affects chlorite and vermiculite (Fe-bearing minerals). Consequently, chlorite residues, with a typical XRD pattern of intergrade minerals, are the major component of the fine clays in the deeper soil horizon (Bw). In the surface horizon (Al), where weathering is more aggressive, a greater contribution of pre-existing dioctahedral micas is observed in the formation of fine-clay soils. These micas are mica/vermiculite, mica/smectite or vermiculite/smectite interstratified minerals. These clays also contain interlayered materials that would not be residual brucite layers, such as is found in the Bw horizon but hydroxy Al (or Fe) polycations. Figure 3.51 illustrates schematically the evolution of muscovite, vermiculite and chlorite in an acid soil.

3.3.3 Summary of Weathering Effects

Mineral reactions in weathered profiles and soils occur within a few types of microsystems: grain contacts, alteroplasma, pedoplasma, fissures and soil aggregates. In these microsystems the local water/rock ratio and the chemical composition of primary mineral both control the clay phases that are formed by weathering. Contact and plasmic microsystems are rather closed systems with low water/rock ratios. Thus, the formation of the new clays is essentially controlled by the local chemical conditions which are dictated by the altering mineral chemistry: for instance, illite grows at the mica-orthoclase grain contacts, a high-potassium medium. Given the importance of rock chemistry, we will look in detail at the clays formed in weathering profiles of several rock types (Sect. 3.4). The profiles to be studied are all in the same temperate climates. Soil microsystems are totally open. They are directly exposed to climatic wetness fluctuations and the clay minerals formed are largely controlled by local water/rock ratios. Consequently it will be more convenient to describe clays of pedogenetic formation in various climatic conditions later (Sect. 3.5).

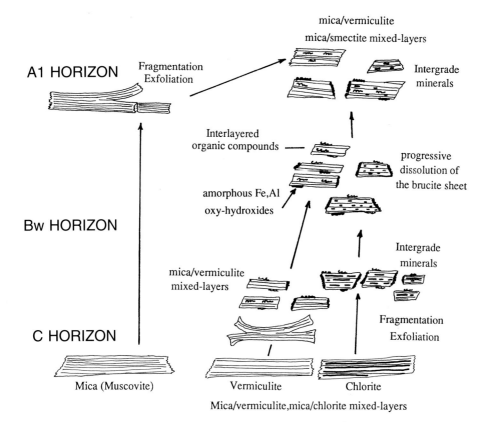

Fig. 3.51. Schematic evolution of primary phyllosilicates in an acid soil

3.4 Clays Formed During Rock Weathering

The granite example has provided a pattern of the weathering mechanism. The question is to know if this pattern is applicable to any rock whatever its particularities. Several profiles of magmatic, metamorphic and sedimentary rocks will be presented here, in order to show the effects of the chemical and mineralogical properties of the parent rock on the formation of the weathered profiles. Some examples were chosen because of their high contrast in chemical or mineralogical compositions with granites: e.g. Fe, Mg-rich crystalline rocks and clay-bearing sedimentary rocks.

3.4.1 Weathering of Basic and Ultrabasic Rocks

3.4.1.1 Weathering Profiles

In temperate climates, the weathering profiles which have developed on basic (amphibolite, gabbro) and ultrabasic rocks (serpentinite, peridotite) normally

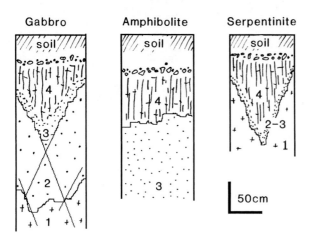

Fig. 3.52. Schematic representation of weathering profiles developed on basic to ultrabasic rocks. *1* Unaltered rock; *2* saprock; *3* saprolite; *4* clay-rich horizon with prismatic structure. (Ducloux et al. 1976; Ildefonse 1980; Proust 1983)

show four distinct zones: *1* altered coherent rock; *2* saprock; *3* saprolite; *4* prismatic level (which is a neo-structured clay-rich level under the soil) (Ducloux et al. 1976; Proust and Velde 1978; Ildefonse 1980; Fontanaud 1982; Proust 1983). These profiles are usually 1 to 3 m thick (Fig. 3.52).

In tropical countries, such profiles frequently reach several tens of meters in thickness and produce large oxide and hydroxide deposits (Trescases 1973; Nahon and Colin 1982; Nahon et al. 1982; Delvigne 1983; Colin 1984; Boukili et al. 1983). Excepting amphiboles (hornblende), the major primary minerals, in such rocks are anhydrous; e.g. calcic plagioclase, pyroxenes (CPx and OPx), olivine and magnetite. However, most of these minerals are partialy or totally transformed during hydrothermal alteration or low-grade metamorphism events into hydrated phases; e.g. chlorite, antigorite, talc. Both hydrous and anhydrous phases react under weathering conditions to produce clay minerals associated with oxides or hydroxides.

The weathering processes appear to be strongly dependent on the initial structure of basic to ultrabasic rocks. Altered profiles of macrocrystalline rocks (gabbros, amphibolites, lherzolites) have a similar organization to the granite profile, but those developed on serpentinite are obviously different. We shall distinguish these two kinds of rocks in spite of the fact that their chemical compositions are quite similar from the alteration view point.

3.4.1.2 Weathered Macrocrystalline Basic Rocks

As is the case for granitic rocks, the weathering process of basic rocks operates in microsystems. Contact microsystems dominate in the altered coherent rock level, producing saponite (hornblende–chlorite contact) or Fe-beidellite (plagioclase–actinolite contact). In the saprock, the great contrast in composition between the coarse-grained primary phases leads to the juxtaposition of different microsystems, producing the alteroplasma inside the primary minerals. The secondary phases which crystallize are controlled by the chemical composition of the mineral host: aluminous clays are found in plagioclase (dioctahedral vermiculite, beidellite, halloysite); Al–Fe–Mg clays in amphiboles (trioctahedral vermiculite, saponite); Mg–Fe clays in pyroxenes and olivines (talc, saponite, Mg-rich gels). In the saprolite, the collapse of the original rock structure induces the reorganization of mineral assemblages. As a consequence, the local contrasts of chemical composition tend to be reduced. New secondary clay phases appear (di- and trioctahedral vermiculites) while the pre-existing ones such as saponite or chlorite/vermiculite mixed-layer change in chemical composition. The clay content becomes high enough to provoke the formation of prismatic structure. New clay minerals crystallize: Al and Fe^{3+}-rich smectites + kaolinite in the saprolite of weathered aluminous amphibolite profiles; Mg-trioctahedral vermiculite or Mg gels + nontronite in saprolites developed on magnesian rocks (gabbro, peridotite).

In open fractures which are found everywhere in the profiles (fissural microsystems), no magnesian clay phases are observed. Whatever the chemical composition of the parent rock, the secondary phases produced are nontronite or Fe-beidellite associated with large amounts of Fe-oxide. Weathering parageneses are summarized in Table 3.3 (as the parental chlorite weathers to chlorite/vermiculite mixed layer, the secondary product is considered to be vermiculite for simplification of phase analysis).

3.4.1.3 Weathered Macrocrystalline Ultrabasic Rocks

The weathered profiles developed on ultrabasic rocks are organized differently according to the size and crystalline structure of the parent minerals. In the case of a macrocrystalline rock which is partly or not serpentinized, such as the lherzolite studied by Fontanaud and Meunier (1983), the saprock represents about half of the profile thickness (decimeters). The early alteration stages are controlled by chemical microsystems. The initial serpentinized lherzolite is composed of clinopyroxene, calcite, chrysotile, spinel and some orthopyroxene relicts. Two alteration stages can be distinguished in the saprock. The first weathering reaction is: pyroxene → talc + Fe-oxide + Si and Ca in aqueous species. Talc is a highly siliceous and magnesian low-temperature phase. At this stage, chrysotile and calcite remain unweathered. The second weathering reaction is the formation of saponite at the expense of pyroxene, talc and chrysotile. Calcite is dissolved and spinel is oxidized, but the initial structure of the rock is still conserved. Saponite is highly magnesian and Fe-rich. The new assemblage shows a bulk composition poorer in R^{2+} ions than the previous one. There is a tendency to form a single-silicate phase assemblage in the saprock, because most of the Fe ions remain in the ferrous state and behave similarly to magnesium; Fe^{3+} is fixed in oxides.

Table 3.3. Weathering reactions in different macrocrystalline basic rocks. These examples all contain hydrous, high-temperature minerals, as is commonly the case. The minerals which react and their reaction products are indicated

	Amphibolite	Amphibolitized gabbro	Serp. lherzolite	Serpentinite
Altered coherent rock (contact microsystems)	Horn + Chl → Sap And + Chl → Kaol	Labr + Act → Fe-Beid		Chl1 + Antig + Talc → Nont + Chl2
Saprock (alteroplasma microsystems)	Horn → Sap And → Kaol (Halloy) Chl → Chl/Verm3	Labr → Verm2 + Zeol Act → Ta + Nont + Ox	OPX + CPX → Ta + Ox Ta + Chry → Fe-Sap Spinel → Ox	
Saprolite (pedoplasma microsystems)	Horn → Verm2 + Verm3 And → Kaol (Halloy) Chl → Chl/Verm3	Act → Verm3 Labr → Verm2 + Fe-Beid Plasma → Verm3	OPX + CPX → Fe-Sap	Chl1 + Antig + Talc → Nont + Chl2
Prismatic clay-rich Horizon (solifluxed pedoplasma microsystems)	Horn → Verm2 + Fe-Beid And → Kaol (Halloy) Chl → Verm3 Plasma → Verm3	Plasma + Verm3 → Fe-Beid + Ox	Plasma → Al-Sap + Nont	Chl2 + Nont → Corr
Fissure coatings (fissural microsystems)	Nont + Fe-Beid	Fe-Beid + Ox	Nont (Fe-Gel) + Ox	

In the upper neo-structured clay-rich zone, a new assemblage appears. It is composed of aluminous saponite and nontronite which form big flakes. As oxidation of Fe progresses, a bi-phased clay mineral assemblage crystallizes in which each phase fixes preferentially one or other of the two inert components (Al and Fe^{3+}). Fissures in the saprock and the restructured zones are filled by a $Mg-Fe^{3+}-Si$ gel and Fe-oxides; saponite becomes unstable. At this stage, the oxidation of Fe ions is total and the Mg loss is at its maximum. It is highly probable that aluminous saponite recrystallizes into nontronite via a concentrated solution phase in which the gel precipitates. Nontronite can then be formed as Harder (1976) showed. For simplification, because the gel is a metastable phase, nontronite will be considered here as the stable one (Table 3.3).

In summary, the important parameters which control the mineral reactions occurring during the weathering of ultrabasic rocks are μ_{Mg} and the oxidation of Fe^{2+} ions. Their individual effects are difficult to distinguish since the decreasing of μ_{Mg} is concomitant with the oxidation increase during the weathering process. Nevertheless, these parameters are independent: μ_{Mg} is an intensive variable wheras the oxidation modifies the Fe^{3+} composition of the microsystems (inert components).

3.4.1.4 Phase Diagrams

The bulk chemical composition of basic to ultrabasic rocks depends on the relative proportions of five major elements; Si, Al, Mg, Ca and on Fe (whose valence state changes with oxidation). Chemical mass balance calculated for each weathering profile shows that the variation in silica content is quite low from unweathered rock to the prismatic level and that the less soluble elements (Al and Fe, when in the Fe^{3+} state) are concentrated in the solid phases. Calcium and magnesium are the most exported elements in surface waters (Barnes et al. 1967, 1978; Tardy 1969; Barnes and O'Neil 1971; Nesbitt 1974; Pfeiffer 1977). However, there is an important difference between calcium and magnesium. Indeed, they can both fill the exchangeable sites in the interlayer of expandable secondary phases but only Mg is fixed in the crystalline lattice of most of the secondary phases (talc, saponite, vermiculite). As a consequence, the Mg activity in solution and the oxidation of Fe^{2+} ions are the determinant variables for the crystallization of the dioctachedral or trioctahedral clay phases. Therefore, it is possible to reduce the chemical composition of the solution-rock system to a three component one: Mg–Al–Fe.

When plotted in the Mg–Al–Fe system, the chemical composition domains of the weathering minerals do not overlap in spite of extended solid solutions. For simplification, it should be noted that, kaolinite and iron oxides, which have very low Fe for Al substitutions, are represented by the Al-pole and the Fe-pole of the diagram respectively. Taking the amphibolitized gabbro

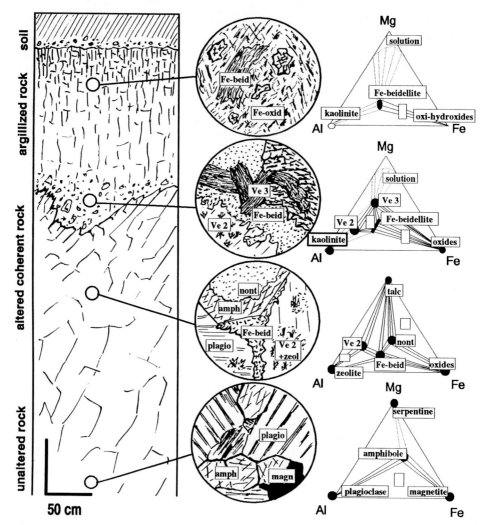

Fig. 3.53. Weathering profile on a gabbroic rock (Ildefonse 1980). Detailed petrographic views were drawn from microphotographs; the diameter represents 400 µm. Primary and secondary mineral assemblages are represented by *open rectangles* in phase diagrams (the choice of the Mg–Al–Fe system is explained in the text). No purely magnesian solid phase was identified in the argillized horizon; the Mg-pole is represented by solutions

from Le Pallet (Ildefonse 1980) as an example, the sequence of alteration minerals is represented in Fig. 3.53.

A general phase diagram (no mobile components) can be drawn considering the stability or instability relationships between phases as observed in thin sections from different weathering profiles developed on macrocrystalline rock (Table 3.3; Fig. 3.54). Applying the phase rule for a three-inert-component system, the maximum number of coexisting phases will be three; conse-

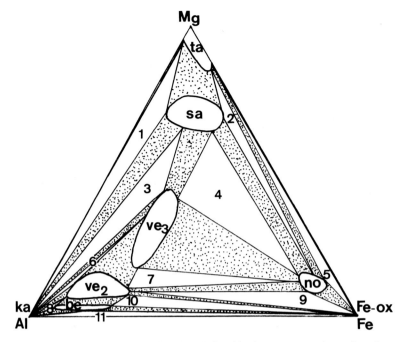

Fig. 3.54. Phase diagram depicting the relationships between secondary minerals crystallizing in weathered basic to ultrabasic rocks in the Mg–Al–Fe chemical system. Triphase assemblages are indexed from *1* to *11*. Biphase assemblages are represented by *dotted areas*. *ta* talc; *sa* saponite; *ve₃* trioctahedral vermiculite; *ve₂*: dioctahedral vermiculite; *be* beidellite; *no* nontronite; *ka* kaolinite; *Fe-ox* iron oxides or hydroxides

quently subdivisions of the Mg–Al–Fe diagram must be sub-triangles. These sub-triangles are limited by tie-lines joining the composition domains of two stable coexisting phases (stippled area in Fig. 3.54).

The degree of opening of the dominating microsystems increases from altered coherent rock (contact microsystems), through saprock and saprolite (plasmic microsystems) to open fractures (fissural microsytems). Magnesium, which could be considered as an inert component (acting by its concentration) in the most closed microsystems, becomes an intensive variable in the open ones (acting by its chemical potential). In the first case, Mg concentration determines the weight of Mg phases; in the second case, Mg chemical potential controls the stability fields of secondary phases. Consequently, the Mg–Al–Fe triangular system is no longer adapted and must be replaced by the μ_M–Al–Fe phase diagram, which can be derived from the former using μ_{Mg} equipotentials.

The chemical potential of Mg ions in solutions (μ_{Mg}) varies from high to low values as alteration intensity increases. This is represented by the sequence of equipotentials from the Mg-pole (higher possible value) to the Al-Fe line (lower possible value). Portions of some equipotentials are determined by the observed parageneses: for example saponite–talc, saponite–trioctahedral vermiculite, nontronite–Fe-oxide. When all the observed parageneses are represented, complete equipotentials can be established (Fig. 3.55a). Then, the set of equipotentials is completed in order to respect the phase rule corollaries, i.e. the number of equipotentials (N_{eq}) is given by the number of tri-phased assemblages (N_{Tr}) by the relation: $N_{eq} = N_{Tr} + 2$ (Meunier and Velde 1979). Finally, a chemical potential–composition phase diagram can be derived by positioning the weathering phases according to their relative inert element contents (Al and Fe) and the μ_{Mg} range in which they are stable (Fig. 3.55b). We must keep in mind that the stability fields of the different phases observed in the profile are established in relative coordinates in such a diagram. Indeed, it can be surprising at first glance to see talc represented as a complete solid solution between the Al and Fe poles. Do not forget that Al and Fe ions are highly diluted by Mg ions (talc is represented here as the phase stable in the highest μ_{Mg} conditions): Al and Fe are contained in talc as traces.

The different mono or bi-phased secondary mineral assemblages are described by the μ_{Mg} trajectories for each primary mineral and for the secondary plasmic systems in the neo-structured areas (Fig. 3.55b). The trajectories are clearly distinct for Al-rich amphibolites, amphibolitized gabbro and lherzolite. In Al-rich amphibolites (23% Al_2O_3), kaolinite (or its hydrated equivalent: halloysite) crystallizes in the very early alteration stages. Its presence in the alteration system buffers the amounts of Al in amphibole secondary products. The two trajectories converge toward the kaolinite + Fe-beidellite assemblage which composes the reacting portion of the neo-structured areas (pedoplasmic microsystems). This buffering effect disappears in fissural microsystems where the unique stable phyllosilicate is the Fe-beidellite associated with Fe-oxides. Kaolinite becomes unstable in the highly diluted and highly oxidizing solutions which flow in open fractures. Indeed, Barnes and O'Neil (1971) shown that spring waters in ultrabasic massifs are supersaturated with respect to magnesian phases.

Alteration reactions in the amphibolitized gabbro differ from those described above because the parent rock is less aluminous (16% Al_2O_3). The more aluminous secondary phase produced in the weathering profile is a Fe-beidellite which appears in the altered plagioclase; kaolinite or halloysite were not observed. The destabilization of actinolite first produces a talc + nontronite assemblage. Nontronite is stable until large amounts of aluminum are dissolved by the dissolution of plagioclase. Then, the next stable phase is a trioctahedral vermiculite. At this alteration stage, the mechanical resistance of the rock collapses, producing a new material. The pre-existing clay phases (primary as well as secondary) are involved in another set of reactions which produce the final Fe-beidellite + Fe-oxide assemblage. This assemblage is also stable in the fissural microsystems.

The third rock (lherzolite) is very Al-poor (3% Al_2O_3). The first phyllosilicate to form during the earlier alteration stages is talc associated with Fe-

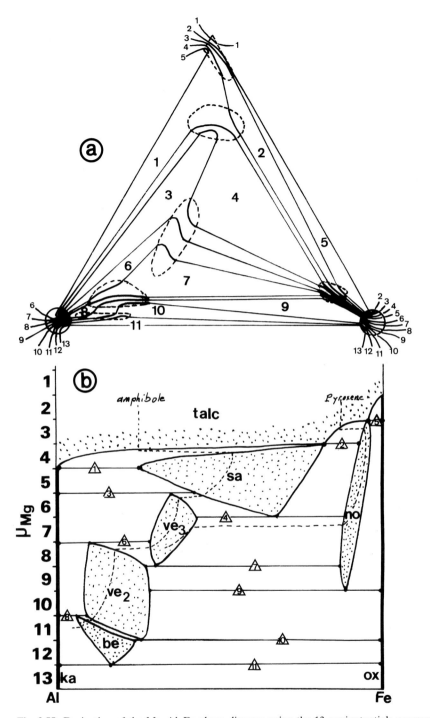

Fig. 3.55. Derivation of the Mg–Al–Fe phase diagram using the 13 equipotentials necessary to depict all the assemblages represented in Fig. 3.53. Triphase assemblages are indexed by *numbers in triangles*. Composition of minerals in the triphase assemblages are represented by *dots*. The sequence of mineral reactions due to amphibole and pyroxene are represented by trajectories in the μ_{Mg}–Al–Fe space (*stippled lines*)

oxides, because only the ortho-pyroxenes are reacting. Then, clino-pyroxenes are dissolved and saponite precipitates as the most aluminous secondary phase. The intense weathering in the upper horizons produces an abundant clay mineral fraction. The initial rock structure disappears. In these clay-rich levels, the iron oxidation increases with the dissolution of primary silicates and a new mineral assemblage is stable: saponite + nontronite. The Fe^{3+} bearing phases, nontronite (gel) + Fe-oxides, form the ultimate assemblage observed under temperate weathering conditions in the fissural microsystems.

3.4.1.5 Weathered Serpentinite

The serpentinite taken as an example here has a chemical composition close to that of the lherzolite: it is very poor in aluminum (less than 2% Al_2O_3). Nevertheless, in spite of such similarity, the weathering processes are completely different from that described above for two major reasons. First, the primary mineral size is frequently less to 1 mm. The minerals are intimately mixed and at a small volume scale (mm^3) the rock can be considered as homogeneous. Secondly, serpentinite is a phyllosilicate bearing rock which is composed of talc, serpentine, chlorite and chrysotile. Because of its homogeneity at small scale and the high reactivity of magnesian phyllosilicates in surface conditions, the rock appears to weather as a whole with no apparent microsystem effect at the thin-section scale (Fig. 3.56).

The rock is converted into a clay-rich horizon through a very narrow saprock of about 1–2 cm thick (Ducloux et al. 1976). A magnesian smectite (saponite) and a silica-rich secondary chlorite are produced in the thin zone

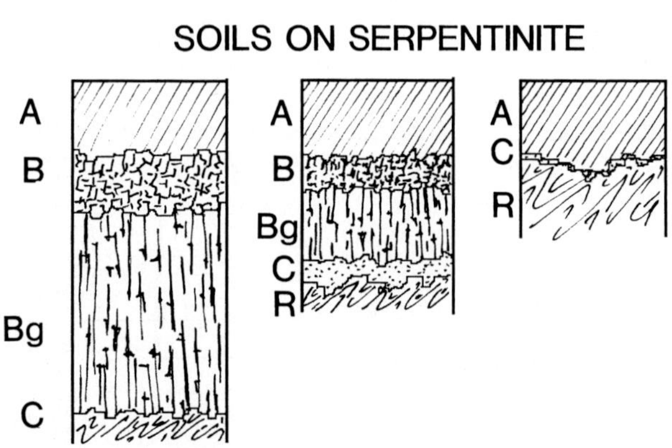

Fig. 3.56. Soil profiles observed on a weathered serpentinite (Ducloux et al. 1976). The illuvial horizon is composed of two parts showing a prismatic structure at the bottom (*Bg*) and a polyhedric one at the top (*B*). The thickest profile shown is formed on a fractured area; the two other profiles are formed on unfractured serpentinite at different slopes

where the internal destabilization of primary minerals does not change the parent structure (saprock). The saprolite is observed as a prismatic clay-rich horizon (about 50% day by weight). The magnesian saponite disappears and a ferric smectite (nontronite) becomes predominant in the <0.1 µm fraction. A secondary chlorite is still observed in the 0.1–2 µm fraction. In the upper part of the profile, the structure changes from prismatic to polyhedric (Fig. 3.56) and the clay fraction proportion decreases to about 30–40% in weight. The soil chlorite disappears and is replaced by a new Mg-rich and Al-poor phase, corrensite (ordered chlorite/vermiculite mixed-layer mineral). This suggests a reaction such as chlorite + smectite → corrensite + smectite, in which an overall balance in the Si–$3R^{2+}$–$2R^{3+}$ coordinates in maintained between the two pairs of alternate assemblages. Tie-lines between the two mineral pairs in Fig. 3.57 suggest this reciprocal relation.

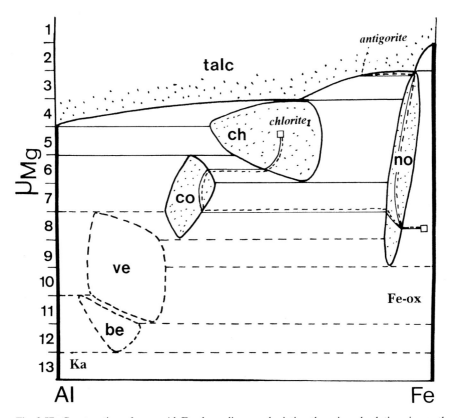

Fig. 3.57. Construction of a μ_{Mg}–Al–Fe phase diagram depicting the mineral relations in weathered profiles developed on serpentinite. *ch* Secondary chlorite; *co* corrensite (ordered chlorite/trioctahedral vermiculite mixed layer); *ve* dioctahedral vermiculite; *be* beidellite; *no* nontronite; *ka* kaolinite; *Fe-ox* iron oxides or hydroxides. The alteration process of the primary chlorite *chlorite I* and antigorite is controlled by two sequences of mineral reactions by which trajectories *continuous* and *stippled lines* in the phase diagram converge toward the nontronite + Fe-oxyhydroxide assemblage

The change in mineral assemblage in the polyhedric horizon is probably due to the decrease of Fe^{3+} content which stabilizes the corrensite at the expense of the pre-existing chlorite phase. The loss of iron is due to the presence of organic matter which reduces the Fe^{3+} ions to soluble Fe^{2+}. Weathering of serpentinite leads to an intense leaching of the dominant chemical components (Si and Mg) and to the oxidation of Fe^{2+} ions. The remarkable fact is the formation of Al–Fe^{3+} bearing secondary phyllosilicates in a rock which is very aluminum-poor (less than 2% Al_2O_3). Indeed, nontronite, as well as soil chlorite or corrensite, fix Al atoms in their crystalline lattice even when they are in such low concentrations in the system. Expandable magnesian clay phases (saponite and vermiculite) do not crystallize in the weathering profile because the chloritic phases (secondary-soil chlorite and corrensite) are stabilized by the high concentration of Mg ions in solutions. In spite of the fact that Si and Mg components are leached out from the profile, the reciprocal chemical balance between the iron-rich smectite and the chloritic phases suggests a relation which is found in chemical systems closed to chemical migration. Phase relations depicted in Fig. 3.55 must be modified as shown in Fig. 3.57.

In spite of the fact that it was impossible to observe directly the mineral reactions at the scale of a thin section, it is necessary to consider two different microsystems in order to depict the sequence of secondary parageneses as weathering increases: these are Al-depleted minerals (talc, serpentine, chrysotile) and Al-richer ones (chlorite). The reactions for both microsystems converge toward a final assemblage composed of nontronite + iron oxyhydroxides.

3.4.2 Weathering of Basaltic Rocks

Basaltic plateaus, tuff formations and ash deposits around volcanoes belong to the more fertile areas in the world because volcanic rocks are particularly unstable in surface conditions; glass is extremely reactive with water. Its dissolution is as intense as the reactivity surface is important: highly permeable ash deposits are much more rapidly transformed into soils compared to basalt flows. Soils on ash deposits are rich in organic matter; they belong to the Andosol class (Sect. 3.5). In spite of the high rate of their formation, andosols are never thicker than 2 m wheras soils on basalts are formed more slowly and become thicker (5–15 m). Because of the high reactivity of glass to ground solutions, there is no transition horizon between the fresh ash deposit and the soil. In contrast, weathering profiles on basaltic rock are frequently several meters thick and show an organization similar to those described on other crystalline rocks: saprock, saprolite and clay-rich zones (Fig. 3.58). There are however some differences due to the presence of glass and to the prismatic structure of basaltic flows (Ildefonse 1987).

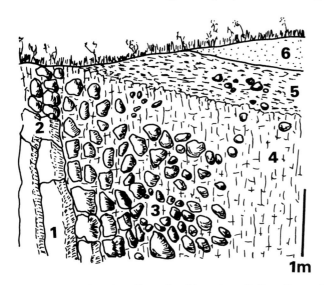

Fig. 3.58. Weathered profile on basaltic rocks at Belbex, France (Ildefonse 1987). *1* Unaltered prismatic basalt; *2* fractured basalt; *3* weathered basalt (ball-shaped coherent rock); *4* saprolite (polyhedral structure); *5* clay-rich horizon (lamellar structure); *6* allochthonous material

Usually, glass and olivine crystals in basaltic flows have been altered by hydrothermal alteration processes during the cooling period at the Earth's surface. They are partially or totally replaced by a microcrystalline clay mineral + Fe-oxide assemblage, classically identified as iddingsite (Eggleton 1984) Consequently, weathering reactions concern not only the primary minerals (unaltered olivine, pyroxenes and plagioclases) but also hydrated glass and hydrothermal alteration products. These components do not react at the same rate; some of them are transformed during the earliest stages of meteoric alteration, such as olivine, hydrated glass and hydrothermal products; others do not react until the rock becomes a saprolite (plagioclases and pyroxenes). Each primary component behaves as an individual microsystem, as it does in the case of macrocrystalline rocks. Chemical transfers operate between the different intracrystalline microsystems as early as the first alteration stages: for example, the weathering secondary products of olivine (saponite Fe-beidellite) contain Al ions which, of course, cannot be found in place but necessarily come from the neighbouring altered glass. Secondary parageneses are summarized in Table 3.4.

As happens in basic rocks, alkalis, silica and Mg ions are leached out during the weathering of basalts while Al and Fe^{3+} are concentrated in the upper intensely altered horizons. Celadonite, originating from hydrothermal alteration of glass, is replaced by a Fe-beidellite + halloysite assemblage through an intermediate series of celadonite/smectite mixed-layered minerals. The saponite–nontronite assemblage, originating from olivine hydrothermal alteration (iddingsite), is replaced by the Fe-beidellite + Fe-oxide assemblage.

Table 3.4. Weathering products in the alteration of basalt according to the alteration horizon. The host mineral and the associated weathering minerals are given

	Olivine	Hydr. Miner.	Glass	Labrador	Pyroxene	Fissures
Unweathered basalt	Iddingsite	Celadonite Fe-beidellite Halloysite	Hydrated glass	–	–	Calcite
Fractured basalt	Nontronite Fe-saponite Iddingsite	Fe-beidellite Halloysite Calcite	Hydrated glass Cryptocryst. minerals	–	–	Calcite
Saprock	Iddingsite Fe-beidellite Halloysite	Celadonite-smectite Fe-beidellite Halloysite	Hydrated glass Cryptocryst. minerals	Halloysite	Fe-beidellite Fe-oxides	Halloysite Fe-beidellite Fe-oxides
Saprolite	Fe-beidellite Halloysite	Halloysite	Halloysite	Halloysite	Fe-beidellite Fe-oxides	Fe-beidellite Halloysite Fe-Mn oxides
Clay-rich zones	Halloysite	+	Fe-beidellite	+	Fe-Mn oxides	

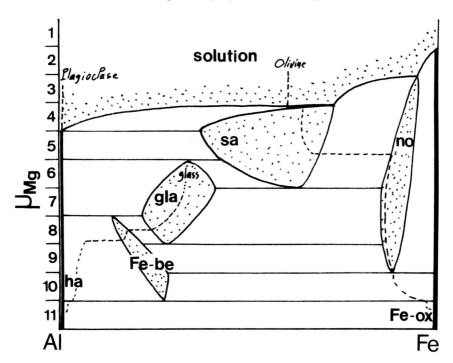

Fig. 3.59. μ_{Mg}–Al–Fe phase diagram depicting secondary mineral relationships in the weathered profile observed at Belbex, France (Ildefonse 1987). *sa* Saponite; *gla* glauconitic mixed layer; *Fe-be* iron-rich beidellite; *ha* halloysite; *Fe-ox* iron oxides or hydroxides. The alteration process of glass and olivine is controlled by a sequence of mineral reactions which end in a biphased halloysite + Fe-oxiyhydroxide (*stippled lines*)

The large homogenization of the rock operated in the neostructured clay-rich horizon leads to the formation of a triphased assemblage: Fe-beidellite + halloysite + Fe-oxides. The sequence of mineral assemblages from unweathered basalt to clay-rich horizons can be represented in a μ_{Mg}–Al–Fe phase diagram (Fig. 3.59). Trajectories are indicated for each individual microsystem; they converge downward to a triphase assemblage. Solutions flowing through fissural microsystems are diluted where μ_{Mg} reaches the lowest values in the profile. As a consequence of this, the stable assemblage is composed of only one phyllosilicate species which concentrates Al ions (halloysite) and an oxide phase which concentrates the Fe^{3+} ions.

3.4.3 Weathering of Clay-Bearing Rocks

3.4.3.1 Weathering of Glauconitic Sandstones

The weathering of glauconitic rocks was investigated by Loveland (1981) and Courbe et al. (1981) using petrographic techniques and, particularly,

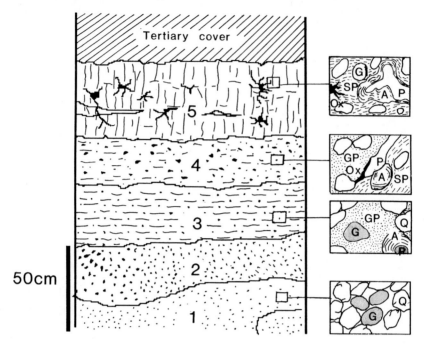

Fig. 3.60. Weathered profile developed on a glauconitic sandstone at Chacé, France (Courbe et al. 1981). *1* Unweathered green sandstone (30% glauconite); *2* reddish level (the initial structure of the sandstone is conserved); *3* transition zone where the sandstone structure disappears (about 10% clay minerals); *4* oxidized level (Fe-oxide concretions, 25% clay minerals); *5* clay-rich horizon (prismatic structure; 35% clay minerals)

microprobe analysis. In spite of the fact that profiles could be differently developed on limestones or sandstones because of large differences in porosity, the glauconite transformation under weathering conditions seems to be identical in all the cases. The weathering profile of Chacé, near Saumur (Maine-et-Loire, France) is taken as a representative example (Fig. 3.60).

The parent rock is a green fine-grained sandstone with 77.3% of the grains between 100 and 200 µm in diameter; glauconite, in rounded grains, represents about 30% in volume of the sand. In the earliest alteration stage, the rock appearance does not change but some of the glauconitic grains are replaced by a green plasma. On moving up the profile, the amounts of quartz and glauconite grains gradually diminish whereas the clay fraction increases and the color becomes reddish. Several active microsystems can be identified: altero-plasmic (internal destabilization of glauconite grains), pedoplasmic (formation of clay-rich areas in place of altered glauconites) and fissural (cutanes). Mineralogical investigations (XRD and electron microprobe microanalyses) show that glauconites are progressively transformed into a smectite + Fe-oxyhydroxide assemblage through illite/smectite mixed-layered minerals. The K_2O amount regularly decreases from unaltered glauconite

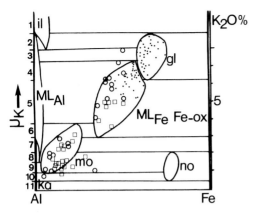

Fig. 3.61. Construction of a μ_K–Al–Fe phase diagram depicting the relations between secondary minerals in the weathered profile of Chacé. The correspondence between relative values of μ_K and K_2O in % weight is given by scales on the left and the right sides respectively. The position of analytical points (microprobe analyses) is indicated by *dots* (green glauconite grains in zone 1); *squares* (glauconitic plasma and cutans, zone 4); *circles* (glauconitic plasma and cutans, zone 5). *il* illite; *gl* glauconite; ML_{Fe} iron-rich mixed-layered phase; ML_{Al} Al-rich mixed-layered phase; *mo* montmorillonite; *no* nontronite; *ka* kaolinite; *Fe-ox* iron oxides or hydroxides

grains to coatings which are composed of kaolinite + Fe-oxide. This is shown in the μ_K–Al–Fe diagram (Fig. 3.61).

The clay fraction formed in the pedoplasmic microsystems is composed of a montmorillonite and nontronite, representing the aluminous and ferric poles respectively. In the highly permeable zones where cutans are observed, the smectites are no longer stable and are replaced by a kaolinite + Fe-oxyhydroxide assemblage. The mineral sequence observed in this profile is the reverse of glauconitization which operates at the seawater–sediment interface (Hower 1961; Velde 1976). This suggests that glauconitization is a reversible process under low-temperature conditions and thus should be considered as a stable reaction

The result of glauconite weathering is to release K and Fe ions into the altering solution. These solutions react to produce new mixed-layer minerals in a step-wise process which gives an iron-rich or aluminous trend depending upon the position in the rock fabric. Materials in close contact with glauconitic grains or plasma change into a montmorillonite–nontronite assemblage (some of the Fe ions are incorporated into clay minerals) while those near pores, in closer proximity to flowing diluted solutions, produce iron-poor phases (kaolinite) associated with Fe-oxyhydroxides.

3.4.3.2 Weathering of Marls

Marls are sedimentary rocks commonly composed of clay minerals intimately mixed with carbonates and sometimes pyrite. Their permeability is very low and the alteration effects are only detectable by a color change from black to ochre in the unaltered and altered levels respectively. The Roumazières profile (Charente, France) used here as an example is developed on a Toarcian age marl composed of detrital mica, quartz, pyrite, dolomite, illite and chlorite (Laffon and Meunier 1982). No modification of the rock structure is observed in the ochre weathered areas. Nevertheless, some mineralogical change occurs:

1. the large white mica flakes are almost unaltered;
2. pyrite crystals are oxidized;
3. dolomite, as well as calcite, grains are dissolved;
4. the initial illite + chlorite assemblage is partly replaced by a kaolinite + illite/smectite mixed layer + Fe-oxide assemblage.

Because of the oxidation of pyrite crystals, the solutions become more acidic and dissolve the Mg-rich trioctahedral phyllosilicates chlorites as well as the carbonates. Consequently, the solutions are locally strongly enriched in Ca, Mg, Si and Al. These new chemical conditions favour the formation of Si–Mg-rich dioctahedral clay minerals at the expense of illite which is replaced by illite-montmorillonite (I/S) mixed layer minerals. The excess Al ions are fixed into kaolinite. The change of the initial illite + chlorite clay assemblage to the I/S + kaolinite assemblage in the weathered zones is represented in Fig. 3.62. As the total loss of dissolved component is quite small and can be neglected, the major effect of weathering is the oxidation of Fe^{2+} ions which explains why the bulk rock composition (large dot) moves in the MR^3–$2R^3$–$3R^2$ coordinates from the right to the left side of the triangle. In summary, one can consider that weathering is the reverse process of the diagenesis which had transformed the initial sediment into the present marl (Velde 1985).

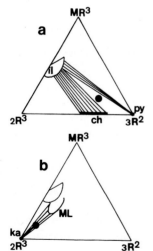

Fig. 3.62. Representation in the MR^3–$2R^3$–$3R^2$ phase diagram of the mineral assemblages of **a** the unweathered marl and **b** the oxidized zones. *il* illite; *py* pyrite; *ch* chorite; *ML* mixed-layered minerals; *ka* kaolinite. (Laffon and Meunier 1982)

It is remarkable that the weathering of sedimentary rocks, i.e. glauconitic sandstone and marl, follows the reverse step-wise pathway of glauconitization and diagenesis respectively. This could indicate that reactions involving clay minerals in low-temperature conditions are stable, or at least reversible metastable reactions.

3.4.4 Summary and Conclusions

Under weathering conditions, rocks, whatever their origin, are progressively transformed into a final assemblage of Fe and Al oxyhydroxides. This endpoint is a attained under tropical or equatorial climate where the reactions between minerals and meteoric water are intense and proceed for long periods of time. This confirms that Al and Fe are true inert components. In contrast, silica, alkalis, calcium and magnesium are dissolved. Their chemical potential can act as active intensive variables in permeable zones of the profile where they control the formation or the dissolution of mineral phases. Under temperate climates, where the weathering processes are less aggressive and where they have usually been active only since the last glaciation, one can observe a series of mineral reactions which progressively transforms the parent rock into a soil.

Weathering of rocks under temperate climate forms the soil saprolite–saprock–rock sequence which is found as one descends into the profile. Thickness and geometrical characteristics of each of these zones depend on chemical composition and structure of the parent rock, topography, permeability, and the age of the alteration. Whatever the profile, there are essentially two parts: (1) that where high temperature phase assemblages react with meteoric solutions and (2) that where clay minerals react with the solutions. Most of these reactions proceed inside microsystems in which local equilibria are reached. The new phase assemblages produced in these microsystems are in their turn destabilized, when local conditions change because of modifications of the structure of the weathered rock or the soil. In summary, one could say that, under temperate climates, weathering processes lead the primary phases to a kaolinite + Fe-oxyhydroxide assemblage via a sequence of reactions which depends on the local structural properties. The chemical transfers are controlled by this sequence, the steps of which are represented in chemical potential–composition phase diagrams.

In summary, the weathering processes under temperate climatic conditions depend on the chemical and mineralogical properties of the parent rock. The microsystem effects control the mineral reactions. Clay properties are highly variable from point to point in the profiles. This variability disappears when the weathering conditions are more aggressive, i.e. in tropical or equatorial climates. This introduces the major difference between soils and weathering profiles: soils are more sensitive to climate than to the chemical and mineralogical properties of the parent rocks. In contrast to weathered profiles, it is easier to present the great soil groups according to climatic conditions that to rock type. Such a presentations is proposed in Section 3.5 which devoted to clays in soil environments.

3.5 Clays Found in Soil Environments

There the chemical composition of the rock is of great importance in determining the clays which form in the early and intermediate stages of weathering. In soils, the importance of this compositional control is balanced by the effects of other parameters, climate, age, and topography, which are more determining controls. The discussion that follows gives an insight into the clay types formed in soil under various climates. The rate of clay formation, the influence of organic compounds, and the effect of mechanical transportation are also discussed for soils developed in cold or temperate climates. A short chapter is devoted to specific clays (short-range ordered minerals) formed in soils from volcanic ashes.

3.5.1 Clays in Soils from Cold and Temperate Climates

Because of the short time of evolution (<10 000 years) and the low temperatures that reduce the rates of weathering reactions, the formation of clay minerals in soils in the cold areas of the world is limited to very small amounts of change. The processes of weathering and soil formation are strongly influenced by complexing organic acids which are produced by slowly decaying plant residues. This process of weathering, in which organic compounds act not only as proton donors but also as complexing anions, is known as acidocomplexolysis. It makes weathering in northern areas quite distinct from that which occurs in other parts of the world (Ugolini and Sletten 1991).

Acido-complexolysis is characterized by organic acidity, which generates a pH in the range of 3–5, and organic complexation of metals such as Al or Fe, which leads to the complete solubilization of all the minerals, except quartz, without any formation of clay. However, soil solutions contain mixtures of organic acids which have strong, weak, or even no tendency towards complexation. Therefore, depending on the type of acid which predominates, the weathering trends actually observed are intermediate between complete destruction by acido-complexolysis, and a less drastic attack on initial minerals leading to the formation of interstratified clays (illite/vermiculite, illite/smectite). Moreover, smectites (or illite/smectites) are the major clay minerals observed in the eluvial horizon of Podzols, which are the typical soils in northern areas.

3.5.1.1 Nature and Rate of Clay Mineral Formation

A very demonstrative study has been made of a sequence of soils of increasing age in the Hudson Bay area, Canada (Protz et al. 1984, 1988). Along the coast of Hudson Bay, periodic storms have built sequential beach ridges made of sand and fine gravel materials including up to about 50% carbonates. From these materials, a sequence of soils, with ages ranging from less than 100 years to about 5500 years, from the coast to the inner country, has developed. The morphology of the soils clearly indicates the development of podzolization: a typical Podzol profile is formed within 2000 years of soil development. The soil formation starts with the dissolution and leaching of calcium carbonate. The depth of complete leaching of $CaCO_3$ increases with time and reaches 30 cm after 5000 years as shown in Fig. 3.63. A parallel decrease in the soil pH from 8 to 4 is observed.

In the C-horizons of all soils, the clay mineralogy is practically the same, but changes in the clay mineralogy with time are observed in the surface (A) horizons. X-ray diffraction analysis indicates that chlorite and mica contents decrease with increasing age, whereas vermiculite content increases and after 4500 years becomes the only clay mineral present together with a small amount of smectite. The amount of vermiculite that has formed is well correlated with time of evolution as shown in Fig. 3.64. A diffraction peak at 0.150 nm indicates that vermiculite, as well as the associated smectite, are dioctahedral. Thus, the mineralogical evolution over 4500 years involves the destruction of chlorite and the transformation of mica to dioctahedral vermiculite. A few smectite layers are also formed.

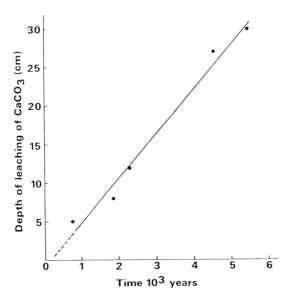

Fig. 3.63. Depth of leaching of $CaCO_3$ as related to time in the Hudson Bay area. (Data from Protz et al. 1984)

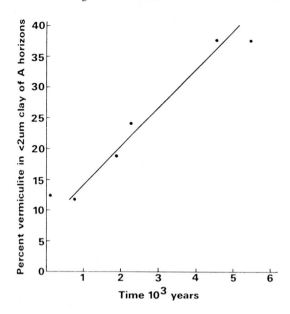

Fig. 3.64. Vermiculite formation related to time of soil formation in the Hudson Bay coastal area. (Data from Protz et al. 1984)

Another chronosequence developed on similar parent material was studied in the southern James Bay area where the mean annual air temperature is 5.5 °C higher than at Hudson Bay. In the C-horizons, the <2 µm fraction is dominated by mica and chlorite, and only minor or trace amounts of vermiculite are present. In contrast, the <2 µm fraction of the E-horizons of the older (>1200 yr) soil is dominated by vermiculite, whereas chlorite is absent. Again, the XRD patterns of the E-horizon clays show the destruction of chlorite and the transformation of mica to vermiculite and eventually smectite. The amount of vermiculite which has formed appears to be correlated with time of evolution, as given by the distance of the soil profile from the coast. Compared with the Hudson Bay coast, the rate of vermiculited formation at Bay James is twice as great (Fig. 3.65). That can be explained by wetter and warmer climatic conditions at this place.

The mean annual drainage is 180 mm of water at Hudson Bay, but 370 mm in the southern James Bay area. When coupled with the 5.5 °C higher annual temperature, an approximately two-fold increase in the rate of biological and chemical reactions in southern James Bay over the Hudson Bay coast may be expected. The mineralogical evolution in these young sandy sediments illustrates the early stages of the formation of soil clays. Chlorite is an easily weatherable mineral which is quickly destroyed or transformed. Transformation of mica to vermiculite is also an early process of soil-clay formation. The rate of these processes is strongly related to climatic factors. The wetter and warmer the climate, the greater is the rate of clay formation.

Another detailed study was done by Gjems (1967) on the clay minerals from more than 200 Podzols in Norway. The relative amounts of the different

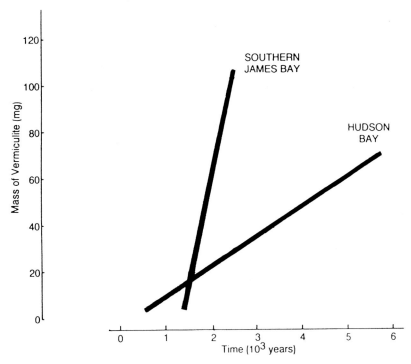

Fig. 3.65. Comparison of vermiculite accumulation, between Southern James Bay and the Hudson Bay coast. (Data from Protz et al. 1988)

mineral phases was estimated on the basis of XRD analyses. From the data obtained it was deduced that:

1. transformation of illite and chlorite to vermiculite occurs in the soil profiles;
2. smectites form in the E horizon of Podzols at the expense of chlorite and vermiculite;
3. in podzolized soils, mixed-layer minerals (mica/chlorite, chlorite/vermiculite) become less chloritic upwards in the profile;
4. the mixed-layer minerals are generally of a more regular interstratified type in the E compared with the B and C horizons.

Therefore, the profiles of the Norwegian Podzols are characterized by enrichment of smectites in the E horizon. Smectite, vermiculite and mica/vermiculite increases upwards in the profile inversely with trioctahedral mica, chlorite, mica/chlorite and mica/vermiculite interstratified minerals. The intensity of weathering in the E horizon of Podzol profiles, expressed as the content of expanding minerals, is related to the environmental factors as follows:

1. recognisable amounts of expanding minerals are formed after a few hundred years of soil formation;

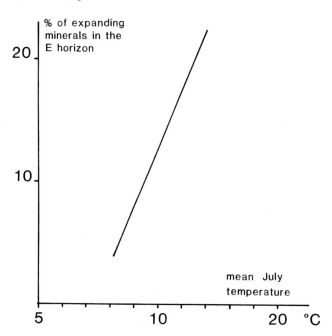

Fig. 3.66. Relationships between climate and rate of weathering in Norway. (Data from Gjems 1967)

2. weathering increases with a warmer climate; this is clearly shown by the relationships between the amount of smectite in soil and the mean July temperature at the place the soil come from (Fig. 3.66);
3. weathering increases with decreasing acidity of the parent material.

A far shorter period of time is needed here in Norway than in northern Canada to form smectite; once again the role of climate in the rate of clay formation is evidenced.

3.5.1.2 Podzolization and Clay Mineral Evolution in the Temperate Zone: Influence of Organic Matter

In the temperate zone, podzolization never affects the whole soil mantle, but only some specific sites characterized by acid, sandy parent materials. Usually Podzols are associated with acid Cambisols. The clay-mineral associations appear to be different in the two types of soil. Mineralogical transformations and their relation to soil-forming processes are studied in a short toposequence in the Landes du Medoc (France) (Righi et al. 1988). In this sequence (Fig. 3.67). Both Podzols and Cambisols are occur within a small area. Except for the dynamics of organic matter, all other factors are the same for the two types of soil. The parent material (a coarse quartzitic sand) contains few clay minerals: a mica phase, a vermiculite and a randomly interstratified chlorite–vermiculite mineral. In the Cambisol, the mica phase is transformed into a vermiculite mineral. After potassium saturation and

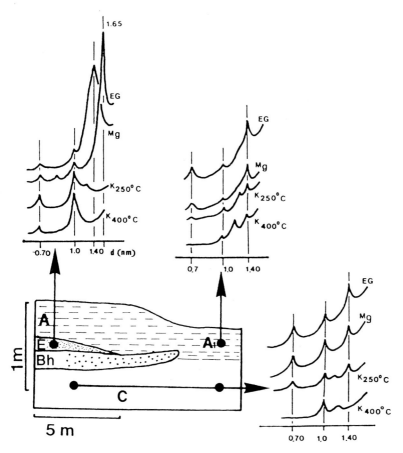

Fig. 3.67. Mineralogical changes of clay minerals in a podzol and in an acid brown soil from the same short toposequence (Médoc, France). *EG* Ethylene glycol saturated, *Mg* Mg-saturated, air dried, *K* heat treatment at different temperatures, as indicated. (Data from Righi et al. 1988)

heating this mineral exhibited a progressive and incomplete collapse. Such a behavior is generally that of intergrade minerals with an incomplete hydroxide interlayered sheet.

In the E horizon of the Podzol, the mica phase is transformed to a 1.4 nm mineral. The 1.4 nm reflection moved to 1.65 nm after ethylene-glycol solvation, indicating the presence of smectitic layers. This example shows that in the temperate zone, an acidic and complexing medium favors the formation of smectitic minerals, with only small amounts of interlayered material being formed, whereas intergrade minerals, with a more developed interlayered hydroxide sheet, are found in acid soils where complexing organic acids are less aboundant. Exportation out of the weathering medium of Al and Fe as organic complexes is certainly the reason for a greater alteration of initial minerals in podzolized soils.

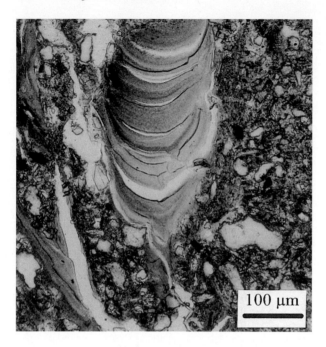

Fig. 3.68. Microphotograph of clay accumulations in soils under plain light

3.5.1.3 Clay Illuviation in Soils Developed from Glacial Loess Deposits: Movement by Transport of Solids

Loess is an aeolian deposit dominated by silt-sized particles (2-50 µm), which covers about 10% of the land surface. Because only glacial grinding can produce appreciable quantities of silt, most of the loess deposits are close to glaciated areas. Loess covers were built-up from silty material which was spread across the landscape by wind moving over barren areas. In northern Europe and North America thick loess deposits were spread over large areas during the Quaternary period. Soils developed from loess material exhibit very clear features of clay illuviation, such as clay coatings filling small pores (Fig. 3.68). Clay illuviation is the movement of clay particles by transport in suspension in the soil water. It is a major process that affects the clay fraction in soils: large amounts of the clay produced by weathering reactions may be lost from the surface horizon of soils through illuviation. On the other hand, redeposition of translocated clay causes the formation of clay-rich horizons at a certain depth in the soil profile. This process has been extensively studied, and so it is possible to form a general description in terms of a chronosequence of soil development, as illustrated in Fig. 3.69 (Jamagne and Begon 1984).

Because the loessic material is generally calcareous, the first weathering stage involves the dissolution of calcium carbonates which are leached out of the soil profile. A change to a more brownish colour, the development of a polyhedric soil structure and limited mechanical fragmentation of silt-sized

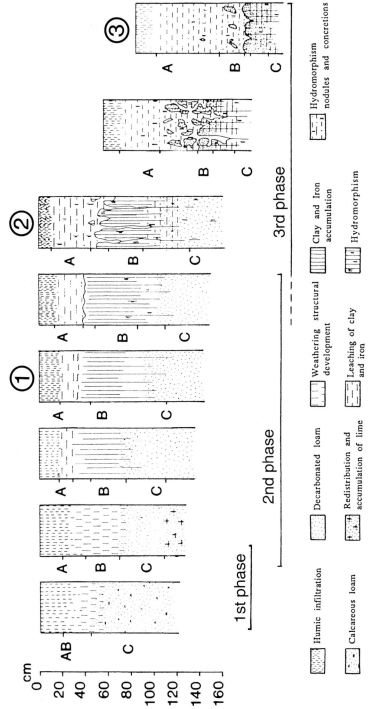

Fig. 3.69. Schematic evolution of soil-morphology sequence on loamy materials. (After Jamagne and Begon 1984)

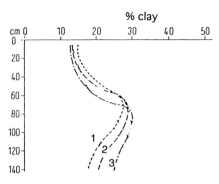

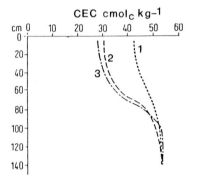

Fig. 3.70. Distribution of clay and CEC of clay in three soil profiles with increasing clay illuviation. *1, 2, 3* refer to profiles in Fig. 3.69. (After Begon and Jamagne 1972)

phyllosilicates is associated with this stage. Decarbonation and a subsequent decrease of Ca concentration in the soil solution, favors the dispersion of clays and hence their suspension in solution. The movement of clays begins and their redeposition occurs deeper in the soil profile, producing the typical distribution of clay with depth shown in Fig. 3.70. The finest clay fraction (0.1 μm) is most affected by this dispersion and translocation wheras coarse clays are less mobile. Generally, for soils from the temperate zone, fine-clay fractions contain a higher proportion of expandable clay minerals (vermiculite, smectite) than coarse fractions, which are dominated by mica, kaolinite and quartz. This explains why the clay fractions in the B horizon have a higher CEC (Fig. 3.70).

Considering the spectacular evolution of soil morphology, surprisingly, only slight mineralogical changes are observed (Jamagne et al. 1984). Illite, chlorite, kaolinite and smectite are the phyllosilicates that generally constitute the clay fraction in the loess material. During soil formation, illite is transformed into illite/vermiculite interstratified minerals, while chlorite is quickly weathered to give smectite layers as weathering products. Kaolinite would be the most stable phyllosilicate. The most important change is certainly found in the interlayer spaces of vermiculite and smectite, which progressively desaturate, i.e. exchangeable basic cations (Ca, Mg) are removed and replaced by Al polycations; hydroxy-Al interlayered minerals are formed. This was demonstrated by the isotopic analysis of the clay minerals in two Luvisols from western Canada (Spiers et al. 1985). The parent material, a glacial till, contains smectite, kaolinite, dioctahedral mica and chlorite as initial phyllosilicates. The following changes were induced by pedogenesis in the mineral suite: kaolinite remains in the E horizons whereas mica and smectite are enriched in the B-horizon as the result of illuviation. In addition, smectites, which are a mixture of montmorillonite and beidellite in the parent material, are almost exclusively beidellite in the E horizon. A gradual increase of the beidellite proportion in the smectite mixture is observed from the parent material to the E-horizon. Thus, either neogenesis of beidellite in the E horizon or preferential illuviation of montmorillonite over beidellite are possible processes. As clays inherited from the parent material should be isotopically distinct from any neo-formed one, the hypothesis of beidellite neo-formation was tested using oxygen isotope analysis (^{18}O). Isotope data demonstrated that beidellite was not neo-formed in these soils. Therefore, only differences with regard to susceptibility upon translocation can explain the changes observed in clay mineral distribution between the different soil horizons.

A further stage of evolution is induced by the reduction in permeability as the result of clay illuviation. The accumulation of illuvial clays in a B horizon (medium part of the soil profile) causes a progressive filling of the pores in this horizon, so that it becomes more compact and less permeable. As a consequence, the amount of water percolating through the B-horizon is strongly reduced and a perched water table forms during the rainy season. The compact, clay-rich B horizon develops a typical coarse prismatic structure. Prismatic aggregates are separated by large and deep cracks which are progressively enlarged and deepened, forming a typical morphology called a glossic (tonguing) feature. These are the main pathways for water circulation. The clays in suspension are not pure clay minerals but, more exactly, particles comprising clay minerals and amorphous iron-oxyhydroxides, the later forming coatings on the clay mineral surfaces. The origin of these amorphous iron coatings is mainly in the weathering of the initial Fe-bearing phyllosilicates by means of oxidation of ferrous iron, exsudation out of the phyllosilicate structure, and subsequent precipitation on the surface of particles. The seasonal hydromorphic conditions in the upper part of the soil profile induce the reduction of ferric oxy-hydroxides. As iron is far more soluble in the ferrous state it can be removed, causing the dissociation of the iron-oxy-hydroxide–clay complex. Subject to large local changes in redox conditions, dissolved ferrous iron reoxidizes and precipitates forming iron nodules.

Clay minerals released by the dissociation of the iron-oxyhydroxide–clay complex are very dispersible, and so a secondary stage of clay illuviation is observed, leading to the accumulation of clay in the bottom of the tongues (DeConinck et al. 1976). The result is a strongly contrasted soil profile with an upper horizon, greatly impoverished in iron and clays, overlaying a lower compact clay-rich B-horizon. Water flow is strongly reduced in the vertical direction and water moves laterally over the impervious floor made by the B-horizon.

3.5.1.4 Clays in Soils from Heavy Clay Rocks: Selective Transport of Clays

In addition to loessic materials, other sedimentary clay rocks also have a large extension in northern Europe and America. Soils developed from these materials are initially clay-rich soils, and their clay minerals are mostly directly inherited from the parent material. Only slight mineralogical evolution occurs. Soil formation on these rocks generally induces a strong structure of the material within the first meter below the surface. This structure determines the pattern of water circulation in the soil. Although only slight mineralogical changes occur, soils developed from clay-rich sedimentary rocks often exhibit large differences in clay content between the upper and lower horizons of the soil profile, upper horizons having lower amounts of clays. Illuviation of clay fails to explain the contrasting depth function of clay content in these soils, as no absolute accumulation of clays in the deeper horizons can be proved.

Mass-balance studies show that the decrease of the fine-clay content in the upper horizons accounts for most of the matter which is lost. Losses as high as 400 kg/hectare/year of fine clays have been measured in heavy clay soils in France (Nguyen Kha et al. 1976). Moreover, very stable suspensions of fine clays can be collected in drainage waters from these soils. This process is described as superficial and selective erosion. Loss of fine clays occurs both through runoff and seepage waters, according to rainfall intensity and the soil structure which can change with the season. During winter and early spring, water content of soils is close to field capacity and water moves slowly through micropores. Heavy rains cannot penetrate totally into the soil and runoff occurs. During summer and autumn, soil desiccation leads to the opening of large cracks. Rainwater enters these large cracks and mechanically disperses fine clays which finally join seepage waters and eventually streams and rivers. No redeposition of the dispersed clay occurs within the soil profile. As the different clay mineral species do not have the same ability to be dispersed, the change in their proportions in the soil profile occurs. Smectite and expandable minerals (vermiculite, interstratified minerals) are the most affected by selective transport. Kaolinite, mica, and also fine quartz accumulate relatively in the upper horizons.

3.5.1.5 Summary

Most of the soils in the temperate zone are developed from rocks which contain phyllosilicates. This is the case for soils formed on granite, sedimentary rocks and a great majority of those formed on metamorphic rocks. Fine-grained sediments like loess or glacial tills, which cover large areas in Europe and north America, also contain phyllosilicates similar to clay minerals such as muscovite, biotite and chlorite. In these soils the clay fraction is constituted of a mixture of relicts of the initial phyllosilicates, with the new clay minerals formed by weathering. However, not all of the initial phyllosilicates have the same stability: some are more easily weathered than others. For instance chlorite is certainly one of the most unstable phyllosilicates in the soil environment. Weathering of chlorite in soil conditions proceeds through intense physical fragmentation leaving clay-sized particles. Mineralogcial transformations of the chlorite remnants leads to the formation of complex chlorite/vermiculite and chlorite/smectite mixed layers. The brucite sheet of the chlorite is partially dissolved, and intergrade minerals between vermiculite and chlorite are formed. The end result is the formation of a fine clay fraction (<0.2 μm) with smectitic characteristics associated with coarser fractions (>0.5 μm) which are relicts of the initial biotite and muscovite.

Physical transport of clays from the surface to the B horizon is the other process that affects clay minerals in soils. This induces large changes in soil permeability at the depth at which clays are redeposited, and water flow from the soil surface to groundwater is strongly disturbed.

3.5.2 Clays in Soils on Volcanic Rocks: The Short-Range-Ordered Minerals, Allophane and Imogolite

Tephra deposits of rhyolitic, dacitic and andesitic composition form superficial formations in many parts of the world. Highly weatherable, they are of major importance as soil parent material. The presence of allophane and imogolite (short-range-ordered clay minerals) constitutes one of the specific features of soil derived from volcanic ash and pumice, and called Andisols. Thus, the geographical distribution of allophane and imogolite is associated with areas of recent volcanic activity: e.g. Pacific ring of volcanism, the West Indies, Africa, Italy and Australia. However, allophane and imogolite are also found in soils derived from rocks other than volcanic ash, specifically soils derived from basalts. Imogolite is a paracrystalline unique nesosilicate consisting of a tube unit with a diameter of about 2 nm (Fig. 3.71). Its structural formula is $(OH)SiO_3Al_2(OH)_2$. Allophane is X-ray amorphous and consists of hollow, irregular spherical particles with diameters of 3.5 to 5.0 nm and with compositions $(SiO_2)_{1-2}Al_2O_3(H_2O)^+_{2.5-3.0}$ (Wada 1989). The formation of imogolite and allophane occur during weathering of volcanic ash under humid, temperate or tropical climatic conditions. The large proportion of glass in the volcanic materials favors the formation of short-range-ordered minerals which are metastable phases and are characteristic of recent soils. Because of the high activity of glass, nearly all varieties of volcanic ash (basaltic, andesitic or rhyolitic) produce allophane–imogolite as weathering products. With time

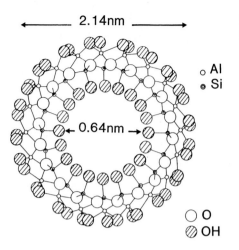

Fig. 3.71. Cross section of an imogolite tube. (After Cradwick et al. 1972)

allophane–imogolite transforms to gibbsite or halloysite, depending on whether the environment favors desilication or not. However, allophane–imogolite may persist in soil for long period of time if conditions are favorable. Their greatest stability occurs under a climate with constant humidity. A dry season induces soil desiccation and increases silica concentrations in the soil solution. This results in the formation of halloysite and cristobalite at the expense of allophane–imogolite minerals (Lowe 1986).

There are factors other than age and periodical dessication which contribute either to the persistence of allophane–imogolite or to its transformation into halloysite. Because rhyolitic material has a lower Al/Si ratio, it favors the transformation into halloysite, whereas andesitic material with a higher Al/Si ratio promotes allophane and imogolite stability. The accumulation of large amounts of humus in andisols is generally ascribed to its reaction with allophane, forming humus-allophane complexes. However, the absence of allophane–imogolite in horizons in which the accumulation of humus is very active, suggests that the formation of Al-humus complexes inhibits the formation of allophane–imogolite through competition for Al released by weathering of volcanic ash.

In contrast with 2/1 phyllosilicates (smectite, vermiculite) allophane and imogolite do not have a permanent CEC. They develop a variable electric charge on their surface. Sign and amount of charge are governed by pH and concentration of ions (anions and cations) in the soil solution. Cation exchange capacity or anion exchange capacity (AEC) can develop. Figure 3.72 shows that allophanic soil material has an increasing CEC and a decreasing AEC as pH increases from 4 to 8. At pH 4, CEC is nearly zero and an AEC of $20\,cmol_c kg^{-1}$ has developed. At pH 8, AEC is strongly reduced and CEC is at its maximum ($30\,cmol_c kg^{-1}$). Probable reactions for charge development on allophane–imogolite surfaces are given below:

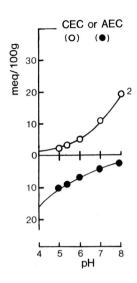

Fig. 3.72. Charge characteristic vs. pH of an allophanic material showing the strong variation of surface characteristics dependent on pH. (After Wada 1989)

high pH; $Si(OH) + OH^- = SiO^- + H_2O$

low pH; $Al^{VI}(OH)(H_2O) + H^+ = [Al^{VI}(H_2O)_2]^+$.

This has important implications on the status of plant nutrients in soils containing allophane and imogolite. Specific adsorption of anions such as phosphate occurs extensively on these materials. Organic anions such as citrate or acetate are also strongly adsorbed and thus protected against biological degradation.

3.5.3 Clays in Soils Formed Under Tropical Climate Conditions

Because of the higher mean annual temperature and the longer time of evolution, the formation of clay minerals in soils is generally more advanced in tropical areas than in temperate and cold regions. In tropical areas the tendency is to produce simple clay assemblages made of only one or two clay phases. The phases produced are controlled for a large part by the chemistry of the soil solution, which itself depends on the prevailing climatic conditions, especially the amount of rainfall and its distribution throughout the year. High rainfall, regularly distributed throughout the year induces soil solutions with low concentrations of silica and basic cations: this favors the formation kaolinite clay. Low rainfall, followed by dry periods with intense evaporation, induces soil solutions concentrated in silica and basic cations. In this latter environment smectite clays are formed. From the equator to the tropics a sequence of increasing dryness is observed. Clay minerals that have formed in

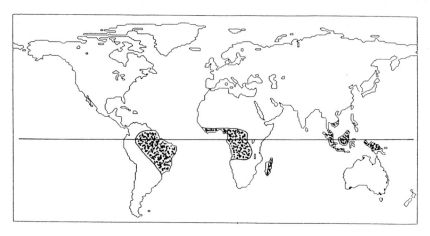

Fig. 3.73. Distribution of Ferralsols in the world. These are soils with kaolinite as the dominant clay mineral. (After FitzPatrick 1980)

soils reflect this change: kaolinite and Al,Fe-oxyhydroxides are found in Ferralsols of the wet equatorial zone, whereas smectites are found in Vertisols of the tropical dry zone. In arid areas with extreme evaporation, a fibrous caly mineral, palygorskite, can form.

3.5.3.1 Equatorial Wet Zone: Kaolinite and Al,Fe-Oxyhydroxides in Ferralsols

Humid tropical climatic zones in Africa, South America and Indonesia are covered by the so-called lateritic formations or Ferralsols (Fig. 3.73). When they have not been strongly eroded, lateritic soil profiles show an ordered succession of three main horizons from the bottom to the top (Fig. 3.74) (Bocquier et al. 1984; Muller and Bocquier 1986):

1. A lower weathering horizon (saprolite) is present, in which the weathering products maintain the structure of the original rock. The clays are mainly macrocrystalline kaolinite, hematite and goethite, associated with stable primary grains such as quartz or muscovite. This horizon is friable and very porous. In the upper part of the horizon, blocks of saprolite are isolated in a ferruginous clay material which is characterized by the disappearance of the original rock structure and is composed of poorly ordered microcrystalline kaolinite associated with iron oxides. A gradual transition may be observed between a discontinuous red and compact matrix which contains hematite, and a continuous and loose yellow matrix which contains goethite.
2. An intermediate nodular horizon occurs in which two types (coarse and fine) of indurated nodules are found. The coarser (20–80 mm in diameter)

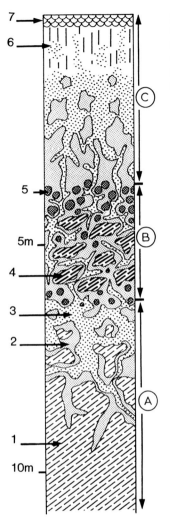

Fig. 3.74. Schematic representation of a Ferralsol profile (Muller and Bocquier 1986). *1* Saprolite, *2* red, compact matrix, *3* yellow friable matrix, *4* ferruginous lithorelict, *5* clay-rich nodules, *6* yellow, compact matrix, *7* zone of accumulation of organic matter. *Circled A, B, C* indicate major soil horizons

are irregular ferruginous relicts of the parent material and are characterized by a more or less preserved rock structure. They are mainly composed of macrocrystalline kaolinite associated with hematite. These nodules are more abundant in the central part of the horizon and display a gradual transition with their surrounding matrix. The finer nodules are sub rounded clayey nodules characterized by a total disappearance of the rock structure. They are mainly composed of microcrystalline kaolinite associated with hematite. These nodules are more abundant, and even occur exclusively, in the upper part of the horizon. Here the internodular matrix has the same characteristics as the ferruginous matrix in the underlying horizon, but the red hematitic matrix becomes the most abundant and is very compact.

3. An upper loose ferruginous clay horizon occurs. From the bottom to the top of the horizon the yellow matrix is again more abundant: the red matrix becomes progressively discontinuous and then disappears at about one meter depth. The upper part of this horizon is affected by organic matter accumulation.

From the brief description given above it appears that the weathering products are essentially kaolinite with Fe-oxides (hematite) and oxyhydroxides (goethite), plus some gibbsite found in the intermediate nodular horizon. Iron compounds concentrate in small soil volumes; the nodules. These are separated by internodular material which is constituted by kaolinite and smaller amounts of iron oxyhydroxides. Physical properties, especially permeability, in the upper loose horizon are attributed to the specific organization of kaolinite clays in that horizon. Indeed, in spite of a large clay content, the upper horizon remains highly permeable because the clays have built up microaggregates which are small rounded volumes of soil, 10 to 200 μm in diameter. The microaggregates (Fig. 3.75) are made up of kaolinite particles associated through edge–face contacts which are made very stable by thin coatings of iron oxides. They induce a large interaggregate porosity at the micrometer μm scale and also a large intra-aggregate microporosity at a submicron scale (0.1 μm).

These microaggregates are stable enough to behave like undissociable particles and they are often called pseudo-sand or pseudo-silt according to their size. For such microaggregated clays swelling–shrinking processes due to wetting–drying cycles are restricted: this material has quite a rigid behaviour. As the kaolinite crystal surfaces are covered to a large extent by iron-oxyhydroxide coatings, the surface properties of the kaolinite clay are partly hidden and inactive. For instance, CEC is far lower than for a reference or uncoated kaolinite. The specific association between kaolinite and iron-oxyhydroxides produces a material with a rigid structure and a low physico-chemical activity.

The critical role attributed to iron-oxyhydroxides in the inactivation of the kaolinite clay can be demonstrated by using a treatment that removes iron compounds from the soil sample (NH_4-oxalate in the dark). After this treatment the usual properties of the kaolinite clays are restored: the soil material swells and shrinks following wetting and drying, and the CEC is greatly improved, compared to the untreated initial material. Moreover, only a small fraction of the iron-oxyhydroxides (the fraction which is amorphous or poorly crystallized) is removed by the treatment; however, that is the crucial fraction which is actually responsible for the microaggregate stability (Pédro et al. 1976).

The dissociation of these microaggregates may occur in nature by any process that causes the removal of iron-oxyhydroxide coatings, as can be done experimentally. Any process that provokes the removal of amorphous iron-oxyhydroxides will cause the destruction of the microaggregates and thus

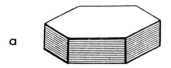
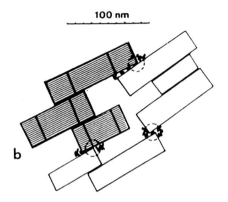
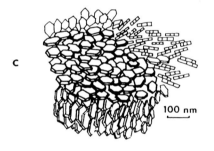

Fig. 3.75. Organization of kaolinitic material in a ferralsol. *a* Kaolinite crystal, *b* particle aggregate composed of kaolinite crystals connected at plate ends, *c* micro aggregate structure made of edge–face crystal associations. (After Robert and Herbillon 1990)

will completely change the soil structure and soil physical properties. Possible processes include, reduction and complexing of iron by organic acids; reduction of iron through water stagnation; and strong dessication. The latter process needs some explanation. In areas where a marked dry period occurs, Ferralsols with typical microaggregated structure show a progressive transformation of that structure to form soils with a compact and impervious horizon. This is attributed to the deep and strong desiccation of the soil that occurs during the dry season. The extreme hydric conditions generate constraints responsible for evacuation of water contained in the microsites and for breaking of interparticle bonds. Moreover, it has been shown that when the water films become increasingly thin, water molecules dissociat more than in free water generating a very low local pH (pH < 2). This strong surface acidity is responsible for the dissolution of iron-oxyhydroxides and consequently the disruption of the microaggregates. That irreversible change in the kaolinite–

iron relationship is induced when climate shifts from humid towards dry or arid as has been demonstrated in West Africa (Chauvel 1976).

The disruption and collapse of the microaggregates leads to a more compact and impervious new structure which restricts the vertical flow of water. Runoff and lateral water flow are favoured with possible mechanical translocation of the clay particles which can now be dispersed. The upper soil layers become impoverished in regards to clay and so become more sandy.

3.5.3.2 Tropical Dry Zone: Smectites in Vertisols

So called Vertisols are formed in areas with tropical wet–dry climates (climates with a marked dry season of 4 to 8 months). In this climate, drainage and consequently leaching of basic cations (Ca, Mg) are at a minimum. They accumulate in the soil, providing favorable conditions (pH, Ca, Mg concentrations) for the formation of smectites. However, the parent material from which the soil develops, must contain sufficient amounts of basic cations, therefore development of Vertisols is restricted to basalt, shale, limestone or volcanic-ash rocks. Cation deficiencies in the parent material can be made good by seepage and subsequent accumulation in lower parts of the landscape where soil water becomes concentrated by evaporation. The largest areas covered by Vertisols occur in India, Australia and the Sudan (Fig. 3.76).

Smectites from Vertisols show something of a continuum between montmorillonite–beidellite–nontronite with respect to octahedral composition and charge distribution, but these compositions clearly tend to cluster around that of the iron-rich beidellite (Wilson 1987; Bradaoui and Bloom 1990). The reason is that Vertisol smectites are derived directly or indirectly

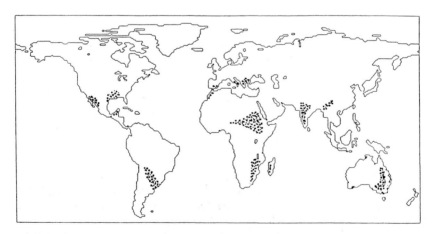

Fig. 3.76. World distribution of smectite-dominated soils (Vertisols). Compar with distribution of ferralsols, Fig. 3.73. (After FitzPatrick 1980)

from parent materials containing ferro-magnesian minerals. A positive correlation exists between the amount of structural iron in the smectite and the amount of iron in the parent rock, especially for Vertisols directly derived from weathering of basic parent rocks. Smectites in Vertisols are poorly ordered structures. They are rarely pure smectites. XRD diagrams most often exhibit features typical for interstratification and/or interlayering. In many Vertisols the smectites show higher basal spacings than might be expected from Ca-saturated montmorillonite. A spacing of 1.6 nm for material in the air-dry state and 1.8–2.0 nm after ethylene-glycol or glycerol treatment is common. Such values have been shown to be characteristic of Vertisols in many countries, such as Kenya, South Africa, Turkey and Uruguay (Rossignol 1983). This high spacing is thought to be the result of organic matter in the interlamellar space. Although evidence for smectite–organic complexes has been produced, this is only for clay minerals exhibiting normal spacings and normal swelling behavior. Most of the published XRD diagrams clearly show features that have patterns similar to randomly interstratified I/S. This indicates that at least a part of the smectitic clay in many Vertisols is inherited from the rock.

The coexistence of interstratified I/S minerals and clay-sized micas with high charge beidellite in Vertisols developed from sedimentary or alluvial materials, is often taken as an indication that bedeillite is a transformation product of mica. However, SEM (scanning electron microscope) pictures have shown very clearly that the growth of smectite layers occurs on the dissolved edges of mica (Kounestron et al. 1977). This indicates that the process of smectite formation from mica is more than a simple transformation through K-depletion, oxidation of octahedral Fe and subsequent lowering of the layer charge, but a process of dissolution–recrystallization. If chlorite is present in the initial material, it may also be a potential source mineral for smectite. In this case smectite seems to be montmorillonite rather than beidellite. Also, repeated wetting and drying cycles by K-rich solutions on high charge smectite may lead to interstratification of expanded and collapsed 2:1 layers. Moreover, interstratified I/S minerals can be obtained by wetting and drying montmorillonite samples placed initially in K-bicarbonate solutions. The presence of large amounts of smectitic clays induces the features and properties of Vertisols. Their main characteristic is a specific soil structure that develops as the result of swelling and shrinking of the soil material, which causes pressures and stresses. If the pore space is not sufficient to accommodate the increase of volume, internal movements of the soil material are induced which are demonstrated by shear planes. The combination of wide and deep cracks from the surface downward to a depth of 50 to 150 cm, and a loose, granular surface mulch that sloughs down into the cracks during the dry season, seems to start a process of mechanical pedoturbation. Surface soil reaches lower parts of the soil profile in the cracks. When the soil swells and the infilled cracks can no longer provide the volume required for swelling of adjoining soil bodies, subsurface soil is pushed upwards.

The specific soil structure of Vertisol, which changes according to the season, also induces a large degree of heterogeneity of water content in the soil. Rainwater coming on a dry, deeply cracked Vertisol percolates first through the wide open cracks leading to rewetting of deep soil layers, while the bulk volume of the surface soil layers remains dry. In spite of large gradients of water tension, homogenization of water content throughout the soil takes a long time, because water conductivity is low in the clayey material.

3.5.4 Arid and Semi-arid Zones: Palygorskite in Saline Soils and Calcareous Crust

Pedogenetic formation of palygorskite, a fibrous clay mineral, is reported from arid areas of the world (Fig. 3.77) in which saline or alkaline ground water frequently affect soil formation, such as the case of the flat, saline plains called sebkhas. Stability of palygorskite requires alkaline pH conditions and high activities of Si and Mg in solution. Palygorskite would not be stable under a pH of 8 at high Mg concentration, or under a pH of 9 at low Mg concentration. In soils, the required concentrations of Mg and Si are not commonly reached, except those affected by rising ground water and submitted to strong and continuous evaporation. This is the case in the saline sebkha soils (Singer 1989). In the absence of rising ground water supplying concentrated solutions, calcareous crusts are the other pedogenic environment which allows palygorskite formation. Calcareous crusts (horizons of calcite accumulation) are commonly associated with soil formation in arid and semi-arid zones. In the crust, local increase of pH levels to greater than 9 are observed. In such conditions, Si dissolution from quartz can be high. The destabilization of Mg-bearing primary minerals provides adequate Mg activity. The required condi-

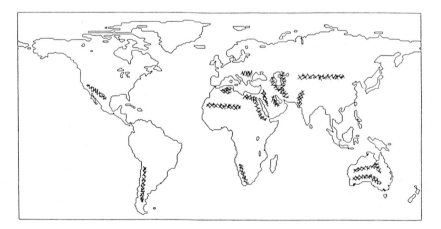

Fig. 3.77. World distribution of soils with palygorskite clays. (After FitzPatrick 1980)

tions of soil-solution chemistry are reached for palygorskite formation (Paquet 1983).

Smectite is the other clay mineral associated with palygorskite in these soils. The mutual stability relationships of these two clay minerals show that palygorskite formation is favored over smectite, by an increase in either pH, Mg or H_4SiO_4 activities. Accordingly, clay minerals which develope in calcareous crusts from Mg deficient parent materials, such as granites, are smectites and not palygorskite. In conclusion palygorskite has a wide distribution in soils of semi-arid and arid areas, where it appears in small to moderate amounts. Because this mineral rarely dominates the clay fraction, no specific soil properties are associated with its presence.

3.6 General Conclusions

Weathering of rocks and pedogenesis are major processes for clay mineral formation at the Earth's surface. They are essentially water–rock interaction processes. In this specific surface environment, only a limited number of variables govern the amount and the type of clays which form. These are *rock composition, water/rock ratio, temperature and time*. Water/rock ratio and rock composition are certainly the most important in determining the type of clay mineral which forms. Time and temperature are kinetic factors which determine the rate at which the chemical process proceeds. Basically, the water/rock ratio is determined by the amount of rainfall. Large amounts of rainfall produce high water/rock ratios and, in a freely drained system, solutions are diluted as they are rapidly renewed. This stands only as a general statement. Indeed, the effective water/rock ratio can be strongly modified by permeability and topographic effects. In the early stages of weathering, water penetrates the rock only through small fissures, permeability is low, water flow is slow and the residence time of solution in contact with rock is long: a low effective water/rock ratio is reached, even in regions with large rainfall. With increased weathering, porosity increases, leading to higher water/rock ratios. A maximum porosity is generally reached by which the soil is a totally open system.

In the soil system topographic effects are important. In addition to permeability, slope also controls the effective water/rock ratio. The displacement of the aqueous solution over the soil profile is more rapid if the slope is greater. This shortens the solution residence-time and prevents extensive water–rock interaction, thus the soil solution chemistry appears to be one of a high water/rock ratio. In flatter areas, the residence-time of solutions increases and low water/rock ratios are reached. Moreover, in lowland positions, under dry climate, evaporation may concentrate solutions and make them similar to those which circulate in the slightly weathered rock. Therefore, the importance of rock composition is greater in sites where the effective water/rock ratio is low, i.e. sites with either, low permeability, a flat topographic position or low rainfall. With reference to these variables (porosity, slope, climate) it is possible to explain schematically the occurrence of some major clay species. Although in variable quantities, *kaolinite* is found in most of the weathering and soil profiles. This mineral forms where solutions have low silica and alkali cation activities. This can be reached in the early stages of weathering by the destabilization of an individual aluminous primary mineral, such as plagioclase in granite or garnet in amphibolite. Increasing chemical weathering (progressive leaching of silica and alkali cations) and higher water/rock ratios also favor the formation of kaolinite: this clay mineral is found in the large fissures where the water flow is high, and also in soils on steep slopes in which the residence time of solutions is low. The thick mantle of kaolinite which covers the tropical wet zones is the result of very humid climatic conditions acting over a long period of time.

Except in tropical dry zones, *smectites*, either di- or trioctahedral, are found only in the lower parts of the weathering profiles, i.e. sites where the rock composition has a greater influence on the chemistry of the whole system. For example, saponite, a trioctahedral, highly magnesian smectite, is formed by destabilization of serpentine, pyroxene and talc in ultrabasic-rock saprocks. In the soil where magnesium is leached to a great extent, saponite is replaced by nontronite, indicating the greater stability of dioctahedral smectites in sites with higher water/rock ratios. In acid rocks, *beidellite–montmorillonite* (aluminous smectites) commonly forms by destabilization of orthoclase in the slightly weathered-rock zone. In tropical dry and arid climates, low water/rock ratios and high silica and alkali cation activities are the caused by low rainfall and the concentration of solutions through evaporation. Beidellite is the dominant clay mineral in soils in downslope and lowland areas.

Because most of soils in the temperate zone are developed from phyllosilicate-bearing rocks, they contain complex mixtures of inherited and partially transformed primary phyllosilicates. *Complex mixed layers and intergrade clay minerals* dominate the clay assemblage in temperate soils. Sediments like loess or glacial tills, from which many temperate soils have developed, contain large proportions of muscovite, biotite and chlorite. These phyllosilicates are unstable under the surface conditions but tend to remain in the soil-clay fraction as metastable relicts. The result is a mixture of primary minerals with newly formed clays which have a rather similar structure and chemistry. The primary phyllosilicates are destabilized in a stepwise process, the interlayer spaces being first affected. Biotite is progressively transformed into vermiculite through extraction of interlayered potassium. The brucite sheet of chlorite progressively dissolves and minerals with intergrade properties between vermiculite and chlorite are found as relicts of the initial chlorite. Because, in a single phyllosilicate particle, not all the layers are weathered, mixed layer minerals are formed; e.g. mica/vermiculite, illite/smectite, chlorite/vermiculite(smectite). These minerals are difficult to identify because several stages of change are found in the same soil clay sample. However, they are the dominant clays in temperate soils, which are the most concerned with agricultural and/or industrial pollution. Their chemical (exchange capacity, adsorption) and physical (swelling) properties must be established in order to assess their control on the environment.

References

Barnes I, O'Neil JR (1971) The relationship between fluids in some fresh alpine-type ultramafics and possible modern serpentinization, western United States. Geol Soc Am Bull 80: 1947–1960
Barnes I, Lamarche VC, Himmelberg G (1967) Geochemical evidence of present-day serpentinization. Science 156: 830–832
Barnes I, O'Neil JR, Trescases JJ (1978) Present-day serpentinization in New-Caledonia, Oman and Yugoslavia. Geochim Cosmochim Acta 42: 144–145
Barnhisel RI, Bertsch PM (1989) Chlorites and hydroxy interlayered vermiculite and smectite. In: Dixon JB, Weed SB (eds) Minerals in soil environments, 2nd edn. Soil Science Society of America, Madison, pp 729–788
Baudracco J, Bel M, Perami R (1982) Effets de l'altération sur quelques propiétés mécaniques du granite du Sidobre (France). Bull Int Assoc Engin Geol 25: 33–38
Begon JC, Jamagne M (1972) Sur la genèse des sols limoneux hydromorphes en France. In: Schlichting E,. Schwertmann U (eds) Pseudogley and gley, genesis and use of hydromorphic soils. Verlag Chemie, Weinheim, pp 307–318
Berner RA (1980) Early diagenesis: a theoretical approach. Princeton University Press, Princeton 241 pp
Bisdom EBA (1967) Micromorphology of weathered granite near the ria de Arosa (NW Spain). Leiden Geol Med 37: 33–67
Bocquier G, Muller JP, Boulangé B (1984) Les latérites. Connaissances et perspectives actuelles sur les mécanismes de leur différenciation. In: Livre Jubilaire du cinquantenaire, AFES, Plaisir, pp 123–140
Boukili H, Novikoff A, Besnus Y, Soubies F, Queiroz C (1983) Pétrologie des produits de l'altération des roches ultrabasiques à chromite de Campo Formoso, état de Bahia, Brésil. Sci Géol Mém 72: 19–28
Boulangé B (1984) Les formations bauxitiques latéritiques de Côte-d'Ivoire. Les faciès, leur transformation, leur distribution et l'évolution du modelé. Mem ORSTOM 175 pp
Bradaoui M, Bloom PR (1990) Iron-rich high-charge beidellite in Vertisols and Mollisols of the High Chaouia region of Morocco. Soil Sci Am J 54: 267–274
Chamayou H, Legros JP (1989) Les bases physiques, chimiques et minéralogiques de la science du sol. Agence de coopération culturelle et technique. PUF, Paris, 594 pp
Chauvel A (1976) Recherches sur la transformation des sols ferrallitiques dans la zone tropicale à saisons contrastées. Evolution et réorganisation des sols de Moyenne Casamance (Sénégal). ORSTOM, Paris, 495 pp
Colin F (1984) Etude pétrologique des altérations de pyroxènite du gisement nickèlifère de Niquelandia (Brésil). Thèse de spécialité, Univ Paris VII, 137 pp
Courbe C, Velde B, Meunier A (1981) Weathering of glauconites: reversal of the glauconitization process in a soil profile in western France. Clay Min 16: 231–243
Cradwick PDG, Farmer VC, Russell JD, Masson CR, Wada K, Yoshinaga N (1972) Imogolite, a hydrated aluminium silicate of tubular structure. Nature (Lond) Phys Sci 240: 187–189
DeConinck F, Favrot JC, Tavernier R, Jamagne M (1976) Dégradation dans les sols léssivés hydromorphes sur matériaux argilo-sableux. Exemple des sols de la nappe détritique bourbonnaise (France). Pédologie XXVI: 105–151

Delvigne J (1983) Micromorphology of the alteration and weathering of pyroxenes in the Coua Boca ultramafic intrusion, Ivory Coast, Western Africa. Sci Géol Mém 72: 57–68
Drever JI (1982) The geochemistry of natural waters. Prentice Hall, Englewood Cliffs, 388 pp
Duchaufour P (1991) Pédologie. Sol, végétation, environnement, 3eme ed. Abrégés, Masson, Paris, 289 pp
Ducloux J, Meunier A, Velde B (1976) Smectite, chlorite and a regular interlayered chlorite-vermiculite in soils developed on a small serpentinite body, Massif Central, France. Clay Min 11: 121–135
Dudoignon P (1983) Altérations hydrothermales et supergène des granites. Etude des gisements de Montebras (Creuse), de Sourches (Deux-Sèvres) et des arènes granitiques de Parthenay. Thèse de spécialité, Univ Poitiers, Poitiers, 117 pp
Ebelmen M (1847) Recherches sur la décomposition des roches. Ann Mines 12: 627–654
Eggleton RA (1984) Formation of iddingsite rims on olivine: a transmission electron microscopy study. Clays Clay Min 32: 1–11
Eggleton RA, Boland JN (1982) Weathering of enstatite to talc through a sequence of transitional phases. Clays Clay Min 30: 11–20
Eswaran H, Bin WC (1978) A study of a deep weathering profile on granite in peninsular malaysia. III. Alteration of feldspars. Soil Soc Am J 42: 154–158
Feth TH, Roberston CE Polzer WL (1964) Sources of mineral constituents in waters from granitic rocks, Sierra Nevada, California and Nevada. US Geol Surv Water Supply Pap 1535, 70 pp
FitzPatrick EA (1980) Soils. Their formation, classification and distribution. Longman, London, 352 pp
Fontanaud A (1982) Les faciès d'altération supergène des roches ultrabasiques. Etude de deux massifs de lherzolite (Pyrénées, France). Thèse de spécialité, Univ Poitiers, Poitiers, 103 pp
Fontanaud A, Meunier A (1983) Mineralogical facies of a serpentinized lherzolite from Pyrénées, France. Clay Min 18: 77–88
Garrels RM (1984) Montmorillonite/illite stability diagram. Clays Clay Min 32: 161–166
Garrels RM, Christ LL (1965) Solutions, minerals and equilibria. Freeman, Cooper, San Francisco, 450 pp
Garrels RM, Howard DF (1957) Reactions of feldspar and mica with water at low temperature and pressure. Prol 6th Natl Conf on Clays and clay minerals, Pergaluon, New York, pp 68–88
Gilkes RJ, Scholz G, Dimmock GM (1973) Lateritic deep weathering of granite. J Soil Sci 24: 523–536
Gjems O (1967) Studies on clay minerals and clay mineral formation in soil profiles in Scandinavia. Med Nor Skogsgorsoeksues 21: 303–345
Harder H (1976) Nontronite synthesis at low temperatures. Chem Geol 18: 169–180
Harriss RC, Adams JAS (1966) Geochemical and mineralogical studies on the weathering of granitic rocks. Am J Sci 264: 146–173
Helgeson HC, Garrels RM, Mackenzie FT (1969) Evolution of irreversible reactions in geochemical processes involving minerals and aqueous solutions. Geochim Cosmochim Acta 33: 455–481
Hillel D (1980) Føndamental of soil physics. Academic Press, New York, 413 pp
Hochella MF, White AF (1990) Mineral-water interface geochemistry. Reviews in Mineralogy 23. Mineralogical Society of America, Washington
Hower J (1961) Some factors concerning the nature and origin of glauconite. Am Min 47: 886–896
Ildefonse P (1980) Mineral facies developed by weathering of a metagabbro, Loire Atlantique (France). Geoderma 24: 257–273
Ildefonse P (1987) Analyse pétrologique des altérations prémétéoriques et météoriques de deux roches basaltiques (basalte alcalin de Belbex, Cantal et Hawaiite de M'Bouda, Cameroun). Thèse doctorat, Université de Paris 7, 323 pp
Jackson ML (1963) Interlayering of expansible layer silicates in soils by chemical weathering. Clays Clay Min 11: 29–46

Jamagne M, Begon JC (1984) Les sols léssivés de la zone tempérée. In: Livre Jubilaire du cinquantenaire, AFES, Plaisir, pp 55–76

Jamagne M, DeConinck F, Robert M, Maucorps J (1984) Mineralogy of clay fractions of some loess in northern France. Geoderma 33: 319–342

Korzhinskii DS (1959) Physicochemical basis of the analysis of the paragenesis of minerals. (Trans) New York Consultant Bureau, New York, 143 pp

Kounestron O, Robert M, Berrier J (1977) Nouvel aspect de la formation des smectites dans les Vertisols. CR, Acad Sci Paris 284: 733–736

Laffon B, Meunier A (1982) Les réactions minérales des micas hérités et de la matrice argileuse au cours de l'altération supergène d'une marne (Roumazières, Charentes) Sci Géol 35: 225–236

Lanson B, Besson G (1992) Characterization of the end of smectite-to illite transformation: decomposition of X-ray patterns. Clays Clay Min 40: 40–52

Loveland PJ (1981) Weathering of a soil glauconite in southern England. Geoderma 25: 35–54

Lowe DJ (1986) Controls and rates of weathering and clay mineral genesis in airfall tephras: a review and New Zealand case study. In: Colman SM, Dutie DP (eds) Rates of chemical weathering of rocks and minerals. Academic Press, New York, pp 265–330

Macias F, Chesworth W (1992) Weathering in humid regions with emphasis on igneous rocks and their metamorphic equivalent. In: Martini IP, Chesworth W (eds) Weathering, soils and paleosols. Elsevier, Amsterdam, pp 283–306

Meunier A (1980) Les mécanismes de l'altération des granites et le rôle des microsystèmes. Etude des arènes du massif granitique de Parthenay (Deux-Sèvres). Mèm Soc Géol Fr 140: 80

Meunier A, Velde B (1979) Weathering mineral facies in altered granite: the importance of local small-scale equilibria. Min Mag 43: 261–268

Meunier A, Velde B (1986) A method of constructing potential-composition and potential-potential phase diagrams for solid-solution type phases: graphical considerations. Bull Min 109: 657–666

Mogk DW, Locke WW (1988) Application of Auger electron spectroscopy (AES) to naturally weathered hornblende. Geochim Cosmochim Acta 52: 2537–2542

Muller JP, Bocquier G (1986) Dissolution of kaolinites and accumulation of iron oxides in lateritic-ferruginous nodules: mineralogical and microstructural transformations. Geoderma 37: 113–136

Nahon D (1991) Introduction to the petrology of soils and chemical weathering. John Wiley, New York, 313 pp

Nahon D, Colin F (1982) Chemical weathering of orthopyroxenes under lateritic conditions. Am J Sci 282: 1232–1243

Nahon D, Colin F, Tardy Y (1982) Formation and distribution of Mg, Fe, Mn-smectites in the first stages of the lateritic weathering of forsterite and tephroïte. Clay Min 17: 1–9

Nesbitt HW (1974) The study of some mineral-aqueous solution interactions. PhD Thesis, John Hopkins University Baltimore, 180 pp

Nguyen Kha, Rouiller J, Souchier B (1976) Premiers résultats concernant une étude expérimentale du phénomène d'appauvrissement dans les Pélosols. Sci Sol 4: 259–268

Niggli P (1938) La loi des phases en minèralogie et pétrographie. Herman, Paris, 2 vols

Paquet H (1983) Stability, instability and significance of attapulgite in the calcretes of mediterranean and tropical areas with marked dry season. Sci Géol 72: 131–140

Parneix JC, Meunier A (1983) Les transformations de la microfissuration du granite de Mayet-de-Montagne (Allier, France) sous l'influence des réactions minérales d'altération hydrothermale. Ann Géophys 38: 203–310

Pédro G (1968) Distribution des principaux types d'altération chimique à la surface du globe. Présentation d'une esquisse géographique. Rev Géog Phys Géol Dyn X 5: 457–470

Pédro G (1985) Les grandes tendances des sols mondiaux. Cultivar 184, numero spécial: Sols et Sous-sols

Pédro G, Chauvel A, Melfi J (1976) Recherches sur la constitution et la genèse des terra roxa estructurada du Brésil. Ann Agron 27: 265–294

Petit JC, Della Mea G, Dran JC, Schott J, Berner RA (1987) Mechanism of diopside dissolution from hydrogen depth profiling. Nature 325, 6106: 705–707
Pfeiffer HR (1977) A model for fluid in metamorphosed ultramafic rocks. Observations at surface and sub-surface conditions (high pH spring waters). Schweiz Mineral Petrogr Mitt 57: 361–396
Prigogine I, Defay R (1954) Chemical thermodynamics. Longmans, New York, 960 pp (English translation by Everett)
Protz R, Ross GJ, Martini IP, Terasmae J (1984) Rate of podzolic soil formation near Hudson Bay, Ontario. Can J Soil Sci 64: 31–49
Protz R, Ross GJ, Shipilato MJ, Terasmae J (1988) Podzolic soil development in the southern James Bay lowlands, Ontario. Can J Soil Sci 68: 287–305
Proust D (1983) Mécanismes de l'altération supergéne des roches basiques. Etude des arènes d'orthoamphibolite du Limousin et de glaucophanite de l'île de Groix. Thèse doctorat, Université de Poitiers, Poitiers, 197 pp
Proust D, Velde B (1978) Beidellite crystallization from plagioclase and amphibole precursors: local and long-range equilibrium during weathering. Clay Min 13: 199–209
Rassineux F, Beaufort D, Meunier A, Bouchet A (1987) A method of coloration by fluorescein aqueous solution for thin section microscopic observation. J Sediment Petrol 57: 782–783
Rassineux F, Beaufort D, Bouchet A, Merceron T, Meunier A (1988) Use of a linear localization detector for X-ray diffraction of very small quantities of clay minerals. Clays Clay Min 36: 187–189
Rice RM (1973) Chemical weathering on the Carnmenellis granite. Min Mag 39: 429–447
Righi D, Lorphelin L (1987) Structure des microagregats des sols podzolises sur micaschistes d'un versant type de l'Himalaya (Népal). In: Bresson LM, Courty MA (eds) Soil micromorphology, N. Fedoroff. AFES, Paris, pp 295–302
Righi D, Meunier A (1991) Characterization and genetic interpretation of clays in an acid brown soil (Dystrochrept) developed in a granite saprolite. Clays Clay Min 39: 519–530
Righi D, Ranger J, Robert M (1988) Clay minerals as indicators of some soil forming processes in the temperate zone. Bull Miner 111: 625–632
Robert M, Herbillon A (1990) Genèse, nature et role des constituants argileux dans les principaux types de sols des environnements volcaniques insulaires. In: Decarreau A (ed) Matériaux Argileux. SFMC and GFA, Paris, pp 539–576
Rossignol JP (1983) Les Vertisols du nord de l'Uruguay. Cah ORSTOM Ser Pédol 20: 271–291
Schnitzer M, Ripmeester JA, Kodama H (1988) Characterization of the organic matter associated with a soil clay. Soil Sci 145: 448–454
Sikora W, Stoch L (1972) Mineral forming processes in weathering crusts of acid magmatic and metamorphic rocks of lower Silesia. Miner Pol 3: 39–52
Singer A (1989) Palygorskite and Sepiolite group minerals. In: Dixon JB, Weed SB (eds) Minerals in soil environments, 2nd edn. Soil Science Society of America, Madison, pp 829–872
Spear FS, Ferry JM, Rumble III D (1982) Analytical formulation of phase equilibria: the Gibbs' method. In: Ferry JM (ed) Characterization of metamorphism through mineral equilibria. Reviews in Mineralogy 10, Mineralogical Society of America, Washington, pp 105–122
Spiers GA, Dudas MJ, Muehlenbachs K, Pawluck S (1985) Isotopic evidence for clay mineral weathering and authigenesis in Cryoboralfs. Soil Sci Soc Am J 49: 467–474
Sposito G (1981) The surface chemistry of soils. Oxford University Press, New York
Stevenson FJ (1982) Humus chemistry. Genesis, composition, reactions. John Wiley, New York, 443 pp
Tardy Y (1969) Géochimie des altérations. Etudes des arènes et des eaux de quelques massifs cristallins d'Europe et d'Afrique. Mem Ser Carte Géol Als Lorr, 31, 187 pp
Tardy Y, Roquin C (1992) geochemistry and evolution of lateritic landscapes In: Martini IP, Chesworth W (eds) Weathering, soils and paleosols. Elsevier, Amsterdam, pp 407–443
Theng BKG, Churman GJ, Newnam RH (1986) The occurrence of interlayered clay-organic complexes in two New Zealand soils. Soil Sci 142: 262–266

Thompson JB (1955) The thermodynamic basis for the mineral facies concept. Am J Sci 253: 65–103

Trescases JJ (1973) L'évolution géochimique supergène des roches ultrabasiques en zone tropicale. Formation des gisements nickèlifères de Nouvelle-Calédonie. Mèm ORSTOM 78: 259

Trolard F, Tardy Y (1989) A model of Fe^{3+}-kaolinite, Al^{3+}-goethite, Al^{3+}-hematite equilibria in granites. Clay Min 24: 1–21

Ugolini F, Sletten RS (1991) The role of proton donors in pedogenesis as revealed by soil solution studies. Soil Sci 151: 59–75

USDA (1975) Soil Taxonomy: a basic system of soil classification for making and interpreting soil surveys. Agric Handbook 436. US Government Printing Office, Washington DC

Veblen DR, Busek PR (1980) Microstructures and reaction mechanisms in biopyriboles. Am Mineral 65: 599–623

Velde B (1976) The chemical evolution of glauconite pellets as seen by microprobe determinations. Min Mag 30: 753–760

Velde B (1985) Clay minerals. A physicochemical explanation of their occurence. Developments in Sedimentology 40 Elsevier, Amsterdam, 427 pp

Wada K (1989) Allophane and Imogolite. In: Dixon JB, Weed SB (eds) Minerals in soil environments, 2nd edn Soil Science Society of America, Madison, pp 1051–1087

Wilson MJ (1987) Soil smectites and related interstratified minerals: recent developments. In: Schultz G, van Olphen H, Mumpton FA (eds) Proc Clay Conf, Denver, 1985. The Clay Minerals Society, Bloomington, pp 167–173

Wolff RG (1967) Weathering of Woodstock granite near Baltimore, Maryland. Am J Sci 265: 106–117

4 Erosion, Sedimentation and Sedimentary Origin of Clays

S. Hillier

4.1 Introduction

Of all the various types of sediment, fine-grained sediments with a high proportion of clay minerals are by far the most abundant. In the marine environment, terrigenous (land-derived) muds cover about 60% of the continental shelves, and almost 40% of the deep ocean basins, amounting in total to about a third of the Earth's surface (Fig. 4.1). Similarly, although proportionally much smaller in area, continental aquatic environments are often dominated by fine-grained sediments, and these environments include many of the rivers, lakes, deltas and estuaries used extensively by man.

However, their wide distribution is not the only feature that makes fine-grained sediments an important component of the environment, it is also the properties that their fine grain size and chemistry impart to them. Large surface areas (up to $800 m^2 g^{-1}$) and surface properties such as adsorbtion, make them important transporters and sinks for pollutants. Surface properties also contribute to many engineering problems, such as the silting up of harbours, and landslides.

In addition to their modern-day environmental importance, the sedimentology of clay minerals can be used to learn about past environments on both historic and geological time scales. Frequently, clay minerals are used to infer sediment dispersal patterns, palaeoclimatic variations, and stratigraphic correlations, all with varying degrees of success. Their importance is further emphasised by the fact that mudstones and shales account for about 70% of the sedimentary geological record (Blatt et al. 1980) and they have an outcrop area of about 33% of the worlds land surface (Meybeck 1987), twice as much as any other rock type, on both counts.

Clay sedimentology is concerned with the interpretation of clays and clay minerals in terms of sedimentary environments, provenance and tracers of sediment dispersal, palaeoclimatology, palaeocirculation, and even geotectonics (Chamley 1989; Weaver 1989). All of these aspects are based on how reliably we can determine the processes that have resulted in the observed assemblages and distribution of clay minerals in sediments. This brief account of the sedimentology of clays is mainly concerned with the processes that affect their sources and fluxes to aquatic environments, mechanisms of

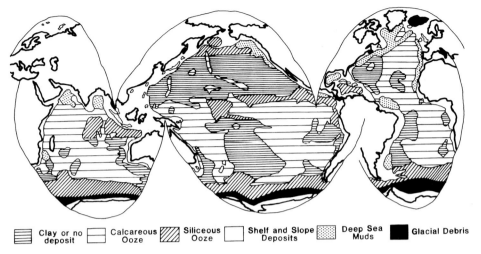

Fig. 4.1. Distribution of recent sediment types in the world ocean. Modified from figures in Berger (1974) and Davies and Gorsline (1976)

erosion, transport and deposition, and the mechanisms and products of transformation and in situ formation of clay minerals in sediments. Some processes are common to many environments while others are often, but not always, environmentally specific. A final part of this account considers the environmental information that can be obtained from clays and clay minerals if we are able to establish their origins, and some of the problems that may confront an investigator attempting to do this.

Much of what concerns the sedimentology of clays as fine-grained minerals (largely phyllosilicates) also concerns the sedimentology of clays as fine-grained (<2µm) materials, as well as that of muds and shales, as fine-grained sediments containing a high proportion of clay-sized material. As a consequence of such close connections and overlap, no attempt is made to stick to a precise terminology, but instead to try to convey what is meant by its context.

4.2 Origins, Sources and Yields, and Global Fluxes of Clay Minerals

4.2.1 Origins of Clay Minerals in Sediments

The clay minerals present in a sediment or sedimentary rock may have one of two main origins; they may be detrital or they may be authigenic (Fig. 4.2). Detrital clays are inherited from another environment to the one in which they are now found. Authigenic clays are formed in situ, and this may occur by one,

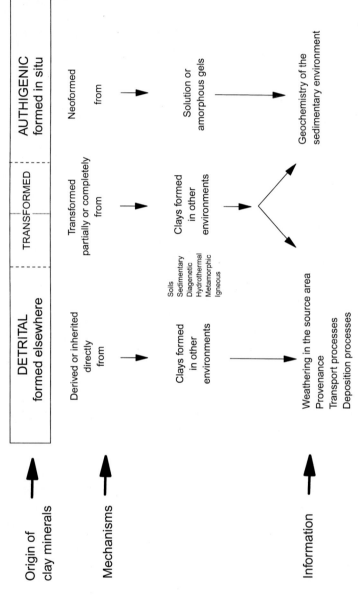

Fig. 4.2. The various origins of clay minerals in the sedimentary environment and the information that may be obtained from them. Most clay minerals, probably more than 90%, are detrital. In areas starved of detrital inputs authigenic clays may make an important contribution to the clay fraction. Clays formed by transformation are intermediate in character. In much of the early literature on clays in sediments, transformation processes were believed to be more important than they are generally considered to be now

or some combination, of three different processes; direct precipitation from solution (a process known as neoformation.), reaction or ageing of amorphous materials, and transformation of some precursor mineral, i.e. where some structural elements are retained. Clay minerals themselves are usually the precursor minerals for transformation processes, so that a third origin known as transformed lies between the poles of detrital and authigenic origins and has characteristics attributable to both (Fig. 4.2).

Transformation processes may be degradative or aggradative (Millot 1970, 1978). According to Millot (1970), the majority of transformation processes in the sedimentary environment involve aggradative transformations of clay minerals formed by previous degradative transformations in soil environments. Although there is evidence for some forms of transformation processes, such as the aggradation of soil vermiculites back to more micaceous clays on exposure to the marine environment (Weaver 1989), many of the mineral trends attributed in the early literature (Millot 1970) to transformations in the sedimentary realm, particularly the more dramatic ones, such as chlorite formation in estuaries and in hypersaline basins, have since been shown to be the result of other processes. Estuarine clay patterns are now largely attributed to mixing of different detrital sources (Chamley 1989), and the chlorites in ancient hypersaline facies are probably formed by authigenic processes during deep-burial diagenesis (Hillier 1993). Nevertheless, transformation processes in the sedimentary realm may be widespread, but difficult to document clearly because they are easily masked by more dramatic changes due to other processes.

One of the major problems involved in studying the sedimentology of clays is that of trying to distinguish between detrital and authigenic clay minerals. The problem arises because of the fine grain size of clay minerals and because clay minerals are often recycled from one environment to another (Eberl 1984). Fine grain size makes direct, origin-diagnostic, observations of clay minerals difficult, while recycling of clay minerals between the weathering, sedimentary, diagenetic and hydrothermal environments, means that a clay mineral observed in the sedimentary realm may well have formed in another. Although authigenic clays do not normally form in great abundance in the sedimentary environment, they tend to become increasingly abundant during burial diagenesis. Consequently, where ancient sediments are concerned, the distinction between authigenic and detrital clay minerals becomes an even more important issue. Often firm evidence for the origin of a clay mineral can only be obtained by combining data from a variety of sources and techniques. Even then the evidence is often circumstantial.

4.2.2 Sources and Yields

Most clay minerals, probably more than 90%, are detrital and they are supplied to the sedimentary system from two main sources; one is rocks, the other is the soils that develop on them by weathering. According to Potter et al. (1980) most mud (clay minerals) comes from the erosion of pre-existing mudstones. On the other hand, Curtis (1990) asserts that most muds are derived from the erosion of soils. Such apparently different statements are no doubt due to the boundary problems of precisely defining what constitutes a rock source and what constitutes a soil source. For instance, clay minerals derived from a rock, frequently pass through a soil with little or no obvious modification by soil processes. Others are only partly modified (transformed) so that they have features of both a rock and a soil source (see Sects. 3.2.2.2, 3.2.2.4, 3.2.3). The difficulties of such definitions are largely avoided if the problem is restated in terms of whether most clay minerals are derived by physical weathering, in which precursor minerals and aggregates are simply reduced in size, or chemical weathering, in which clay minerals may be transformed and new ones neoformed.

In absolute terms, the production of mud is greatest from areas of both high relief and high rainfall, especially where bedrocks are dominated by mudstones and shales. Low-lying arid regions, regardless of bedrock, and carbonate regions, produce the least. Considering that only about one fifth of the Earth's land surface is subjected to intense chemical weathering, that the conditions conducive to the greatest mud production are not conducive to well-developed soil formation, that erosion typically samples whole soil profiles not just the potentially most altered surface horizons, and that mudstones and shales cover 33% of the land surface, then it is probable that most mud (clay minerals) is derived by physical weathering of pre-existing mudstones and shales. This, however, is the present-day situation, but with changes in the distribution of climatic belts and the presence or absence of mountain ranges in the geological past, it may not always have been the case. Similarly, vegetation strongly affects sediment yield, by both sheltering and binding soil constituents together. Clearly, changes in vegetation through time, such as the evolution of grasses during the Cretaceous, or the development of agriculture, must have had substantial effects on soil development and erosion, and therefore the relative importance of physical versus chemical weathering, and hence rocks versus soils as sources of clay minerals.

Sediment yield depends on a complex interplay of many factors, amongst the most important of which are climate, relief, vegetation and lithology. On a local and on a basin scale, sediment yields vary greatly and there is little correlation between drainage basin size and yield (Fig. 4.3). Nor is there any general statistical relationship between single variables such as runoff and yield (Selby 1994). Variation of sediment yield in the Amazon basin (Fig. 4.4) was studied by Gibbs (1967) who showed that 80% of the sediment is derived from only 12% of the basin area comprising the mountainous Andean regions. Land use by man has also greatly affected sediment yields, which in many regions have probably increased two to three times since late Pleistocene times (Gorsline 1984).

Erosion, Sedimentation and Sedimentary Origin of Clays   167

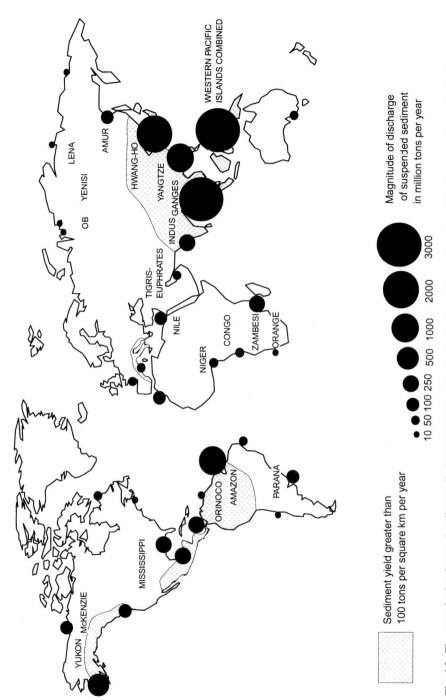

Fig. 4.3. The magnitude of annual sediment discharge of the world's major rivers and the worldwide distribution of regions with the highest sediment yields. Note the huge amounts of sediment erosion in southeast Asia and the western Pacific islands due the combination of high relief and high rainfall in these regions. Based on data and figures from Strakhov (1967), Stoddart (1971), Milliman and Meade (1983) and Jannson (1988)

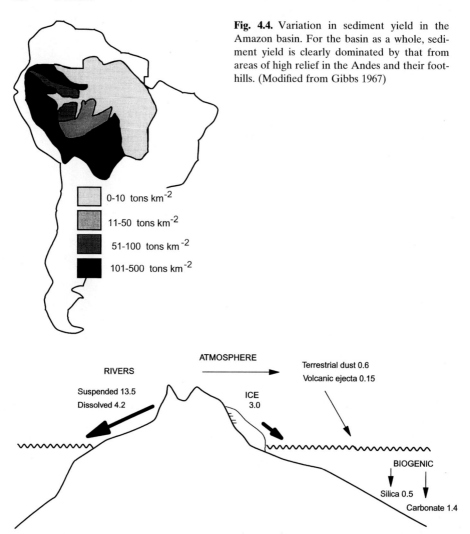

Fig. 4.4. Variation in sediment yield in the Amazon basin. For the basin as a whole, sediment yield is clearly dominated by that from areas of high relief in the Andes and their foothills. (Modified from Gibbs 1967)

Fig. 4.5. Masses of sedimentary materials transported to the oceans annually. Most of this sedimentary material is both clay sized and dominated by clay minerals. Modified from Garrels and Mackenzie (1971) with updated estimates of fluxes from Milliman and Meade (1983) and Gorsline (1984)

4.2.3 Global Fluxes

Whatever the source of clay minerals, through the agents of erosion they are transported away from their sources to be deposited elsewhere, and eventually may find their way to the oceans. The bulk of the mud supplied to the oceans comes from the worlds twenty largest rivers (Milliman and Meade 1983). Estimates of the annual global fluxes of materials to the ocean indicate that rivers supply about 13.5 billion tons yr^{-1} of suspended load. Ice transport is second in importance supplying about 3 billion tons yr^{-1}, while supply from atmospheric transport is estimated at about 0.75 billion tons yr^{-1}, mainly from dusts of terrigenous origin (Fig. 4.5). Assuming, as did Weaver (1989), that about 60% of all this material is phyllosilicates, then about 10 billion tons yr^{-1} of phyllosilicates are delivered to the oceans. Of this amount, Weaver (1989) suggests that about half is in the <2 µm clay-size fraction. For completeness, mention must also be made of the fine-grained biogenic contribution that comes from the remains of planktonic carbonate and siliceous micro-organisms. Together these contribute about 2 billion tons yr^{-1} to oceanic sediments, mainly in areas of high biological productivity, and their accumulated deposits cover large areas of the ocean floor (Fig. 4.1).

4.3 Erosion, Transport and Deposition of Clay Minerals

As with all other sedimentary materials the agents of erosion, transport and deposition of clay minerals are water, wind, and ice. On the continents rivers are the principal agents of transport and deposition of clays. Then, through estuaries and deltas, the suspended clay-mineral loads of rivers reach the ocean where they are further dispersed by currents, and in many instances also by mass flows after a period of temporary deposition. Wind also erodes, transports and deposits large quantities of clay-sized material from many areas. Of the three agents, water, wind and ice, processes involving water are by far the most important (Fig. 4.5). However, there are significant areas of the oceans where the majority of the clay minerals in the sediments have been transported there by wind or by ice.

4.3.1 Transport by Rivers

By far the majority of clay mineral transport in fluvial systems occurs in suspension, and suspended loads, and therefore clay minerals, account for by far the largest proportion of sediment moved by rivers. At the mouths of large rivers such as the Mississippi, the Amazon and the Congo, typically more than 90% of the total sediment transport is suspended load. The suspended load is defined as those grains whose settling velocity is greatly exceeded by the upward component of turbulence, and although this is obviously a loose definition, suspended loads generally consist of particles finer than coarse silt (about 30µm). In deep rivers, because of the greater settling velocity of larger and denser particles, a concentration gradient may develop where the mass of sediment per unit volume decreases with height above the bed. These changes in mass are often also accompanied by changes in the composition of the

170 S. Hillier

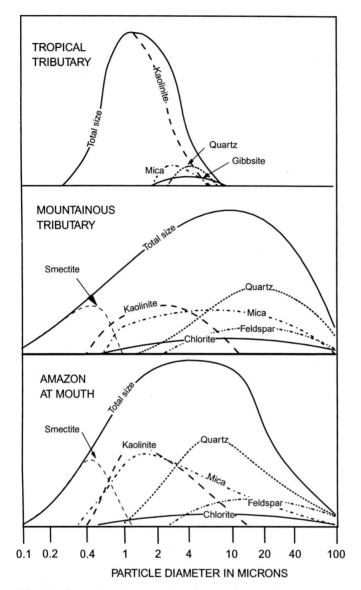

Fig. 4.6. Comparison between mineralogy and grain-size distribution of suspended sediments from a lowland tropical tributary and a mountainous region tributary to the Amazon river, and at the mouth of the Amazon. (Modified from Gibbs 1967)

suspended load. Furthermore, particles which are carried in suspension in one part of the flow may be carried as bed load in another part due to changes in the turbulence of the flow, i.e. more turbulent flows can carry more and larger particles.

The way in which the suspended sediments in the Amazon river vary between tributaries which rise in tropical regions compared to mountainous regions is shown in Fig. 4.6. Tributaries which rise in low-relief tropical areas carry predominantly kaolinite, and the vast majority of the suspended load is <2 µm in size. In contrast, tributaries which rise in mountainous regions carry more varied loads, and a much wider grain-size distribution concentrated in the silt-size fraction. The abundant mica, chlorite, quartz and feldspar reflect the dominant origin by physical weathering. Because of the dominance of the mountainous regions in terms of sediment yield (Fig. 4.4) the composition of the suspended sediment at the mouth of the Amazon (Fig. 4.6) is also dominated by the minerals derived from the rocks and soils of the mountainous regions (Gibbs 1967; Meade 1988).

Much of the suspended load carried by rivers originates outside the river channel and is supplied as wash load from overland flow on hill slopes. However, erosion of the river banks can also be a very important, often the dominant, source of sediment (Reid and Frostick 1994). Concentrations of suspended sediment typically range from a few tens to several thousand mg l^{-1}, although values as high as six hundred thousand mg l^{-1} (60% sediment by weight) have been observed (Reid and Frostick 1994).

Generally, suspended sediment concentration increases with increasing water discharge so that most clay is transported during times of flood. However, the relationship is only a very general one that varies both from one river to another and with time in the same river. This is due to the many factors which affect the availability of sediment, including the length of time since the previous flood. Furthermore, sediment concentrations are normally higher, by one or two orders of magnitude, on the rising limb of a flood hydrograph than at corresponding levels of discharge on the falling limb. This is due to the flushing of most of the available sediment from hill slopes and channel banks during the early stages of a storm. The importance of floods to the transport of clay-sized material is emphasised by the fact that many rivers transport more than half of their annual sediment load in only 5–10 days of the year (Meade and Parker 1985).

4.3.2 Transport in the Sea and Ocean

Most of the fine-grained sediment that reaches the open sea is supplied by rivers, and initially most of this is transported no further than the shallow shelf environments of the continental margins. Locally, this fluvial supply often results in the formation of mud belts on the shelf, their location being controlled by the dynamics of supply and dispersal. Generally, the concentration of suspended sediments in shelf waters shows an exponential decreases from values of 10–100 mg l^{-1} in coastal areas to values of 0.1–1 mg l^{-1} over the outer shelf. McCave (1972, 1985) described the sites of shelf mud belts in terms of five different types, namely muddy coasts, nearshore, mid- and outer-shelf mud belts, and shelf mud blankets (Fig. 4.7). Notionally, their location is explained by the relationship between the general exponential decrease in

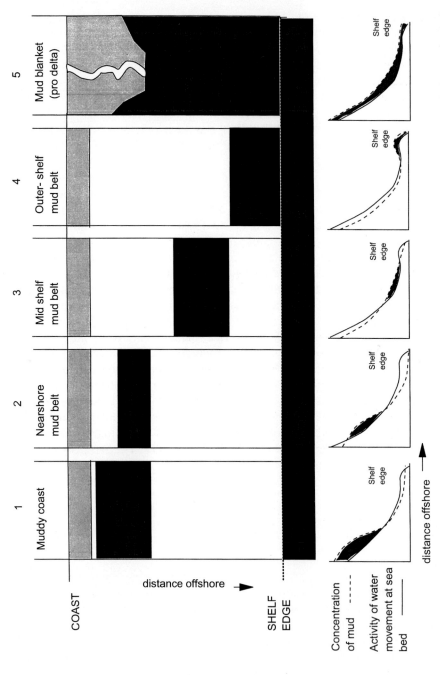

Fig. 4.7. The different locations of mud belts on continental shelves and their schematic explanation in terms of changes in mud concentration and a notional parameter combining wave and tidal activity. (Modified from figures in McCave 1972)

suspended sediment concentration in an offshore direction and the pattern of wave and current activity at the sea bed. Muddy coasts form in areas adjacent to major fluvial inputs, as do nearshore mud belts where supply is less, while mid-shelf mud belts characterise areas of higher wave energy which displaces mud deposition seawards. In areas offshore from major deltas, especially those in the tropics, muds often blanket the whole shelf. Muddy coasts and blanket deposits are essentially due to very high supplies of mud, and demonstrate that muds can be actively deposited in areas of high energy simply if there is enough of it about. An impressive example of this contradiction to long-held assumptions that mud only accumulates slowly in quiet wave-free conditions is provided by the 1600 km of muddy coastline between the Amazon and Orinoco rivers. Here suspended sediment concentrations are typically up to 1000 mg l^{-1}: about 150×10^6 m^3 yr^{-1} of mud is transported to the northwest in suspension and another 100×10^6 m^3 yr^{-1} as migrating mud banks, largely as a result of wave action (Wells and Coleman 1981).

According to McCave (1972) the transport of mud across the shelf is achieved by both diffusion and by advection. Advective plumes associated with fluvial supply points and with areas of convergent tidal currents are the most important, but there is also a gradual seaward diffusive transport of material suspended by wave and current activity. Examples of advective transport are the Rhine mudstream northwards along the Dutch coast, and the Amazon mudstream northwards along the Guinea coast as mentioned above.

In addition to the supply of suspended sediments from rivers to the continental shelves, Meade (1972) emphasised the potential importance of other sources and the assumption that river suspended loads are naturally transported out onto the shelf. For example, on most outer shelves organic material from marine organisms may make up the bulk of suspended matter, and resuspension of relict sediments by wave and current activity may represent an important source. Concerning rivers as sources, on the Atlantic coastal plain of the United States, Meade estimated that about 90% of the fine-grained suspended sediment supplied by rivers is deposited in the estuaries and in wetlands fringing the coast. Based on studies of the clay mineral assemblages it has also been shown that some fine-grained sediments in the estuaries of the eastern United States have been transported into them from the sea (Hathaway 1972; Peaver 1972). The role of coasts in northwest Europe as either sources or sinks of sediments to the marine environment has been discussed by Kirby (1987). In several areas there is evidence that human activity has resulted in estuaries changing from sinks, for both fluvial and marine sediments, to sources of sediments to the shelves (Avoine 1987; Irion et al. 1987). The contribution of clay minerals from the rivers Ems, Weser and Elbe to the mud deposits of the German Bight in the southern North Sea is confirmed by the associated high levels of pollution, and it appears that mean sedimentation rates of between 1–4 m for the last hundred years are also much higher than for older deposits as a result of human activity (Irion et al. 1987).

Further availability of shelf muds to the deep sea depends on sea level and shelf width. High sea levels and wide shelves promote deposition on the shelf, whereas low sea levels and narrow shelves promote the transport into, and deposition of mud in the deep sea environment. According to Stow (1994), the processes that transport and deposit fine-grained sediments in the deep sea are resedimentation processes that involve some form of mass flow, bottom currents, and surface currents in combination with pelagic settling (Fig. 4.8).

Resedimentation processes are event processes driven by gravity, and as far as clays and muds are concerned the most important are turbidity currents. Turbidity currents are mixtures of sediment and water propelled downslope as a result of their density contrast with the surrounding waters. They develop as a result of events that include slumps, debris flows, storms, and river floods, and they may travel up to a few thousand kilometres across the sea floor. Waning turbidity currents that have already deposited the bulk of their sediment may become more buoyant than the deep cold ocean waters they have penetrated. Frequently, this results in the discharge of the remaining fine-grained material into a plume rising up to 1000m above the ocean floor, a process known as flow lofting. By this means fine-grained sediments may be dispersed well beyond the turbidity current deposit itself (Stow 1994).

Bottom, or contour, currents are driven episodically by wind, tides and waves, and more continuously by thermohaline circulation (Faugères and Stow 1993). Unlike turbidity currents they are not driven by gravity and so they may flow along, and even up, slopes as well as down them. The major pattern of deep water circulation in the world ocean is driven by the cooling and sinking of saline surface waters at high latitudes and their slow movement away to lower latitudes. As result of the Coriolis force these water masses are banked up on the western sides of the oceans to form the Western Boundary Undercurrents. Water from these currents spreads slowly eastwards and upwells to the surface layers. Associated with these currents are concentrations of suspended matter forming bottom nepheloid layers, usually extending up to about 1 km above the ocean floor. They are formed by the resuspension of bottom sediment by the current motions of the abyssal circulation. The suspended particles in these layers are on average less than 12 µm in size, and concentrations are only about $0.01-0.3\,\text{mg}\,\text{l}^{-1}$. Net particulate loads of nepheloid layers in the Atlantic were calculated by Biscaye and Eittreim (1977) and were shown to be almost an order of magnitude lower in the eastern Atlantic, at 13×10^6 tons, compared to the western Atlantic, at 111×10^6 tons. This difference is due to the strong western boundary currents. These layers are also well developed in areas where surface kinetic energy may propagate downwards to produce "deep sea storms" (McCave 1986).

Pelagic settling is the slow and continuous vertical settling of fine-grained materials in the open sea. Most material settles at rates of between $1\,\text{cm}\,\text{sec}^{-1}$ to $1\,\text{cm}\,\text{min}^{-1}$ as aggregates that may be flocs, or pellets produced by filter feeders. In the open sea, biological debris, such as the skeletons of calcareous and siliceous microfossils, are often the dominant component of various

Erosion, Sedimentation and Sedimentary Origin of Clays

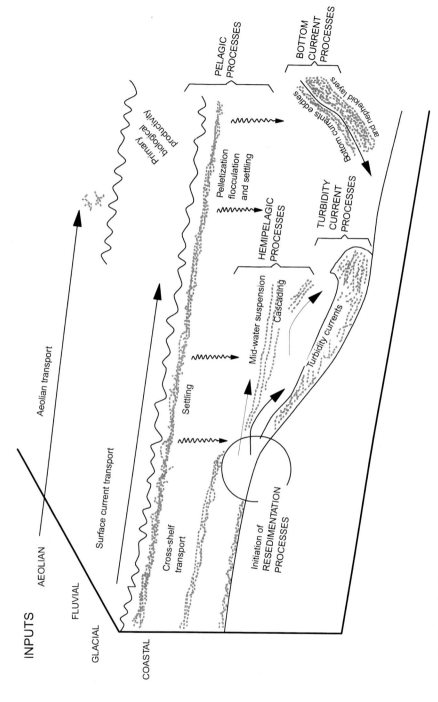

Fig. 4.8. Fine-grained sediment transport and deposition processes in the ocean. (Modified after Stow 1994)

pelagic deposits known as oozes (Fig. 4.1). The accumulation of these deposits is controlled by biological productivity in surface waters and by the carbonate compensation depth (CCD). The CCD is the depth at which the rate of carbonate supply is balanced by the rate of dissolution; thus carbonate does not accumulate below this depth which is typically at about 4 km. Siliceous oozes are found in peri-equatorial zones, the subarctic and Antarctic, and along the margins of some continents. All are areas of important upwelling of waters enriched in nutrients due to the bacterial oxidation of dead planktonic organic matter that has sunk from the surface layers. The deep-sea sediments which accumulate below the CCD, in areas of low biological productivity and away from important terrigenous supply, are known as red clays. The deepest-sea red clays are a combination of mostly aeolian, but also volcanic and cosmic sources (Glasby 1991). Somewhere between pelagic settling and the deposits of low-density turbidity currents there is a range of overlap of processes and deposits that are conveniently termed hemipelagic (Fig. 4.8). A typical distri-

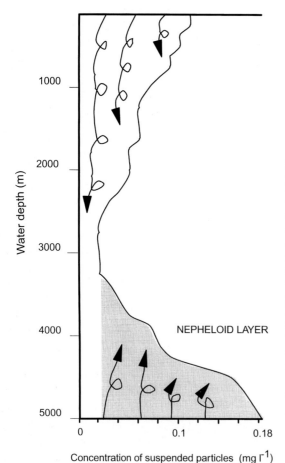

Fig. 4.9. Typical profile of suspended particulate matter in the ocean from an area with a well-developed nepheloid layer. The surface waters contain more biogenic particles than bottom waters and the decreasing concentration depth profile is due to a combination of large-scale equilibrium processes including vertical settling, dissolution and decomposition of biogenic particles, and lateral injection of particles from shelf areas. The increasing concentration in deeper waters towards the ocean floor, known as the nepheloid layer, contains material suspended and mixed vertically or laterally injected by bottom contour currents and turbidity currents. (Modified from Biscaye and Eittreim 1977)

bution of suspended particulate matter in the ocean resulting from the interaction of pelagic settling and all of the various transport processes is shown in Fig. 4.9.

4.3.3 Deposition of Clay Minerals by Settling

The deposition of clay-sized particles from all kinds of water bodies occurs, principally, by settling from suspension. In static water the rate of settling is governed mainly by particle size, and with various assumptions, e.g. that the particles are spheres, it can be estimated from Stokes' law. The effect of different particle shapes has been treated by Lerman (1979). Basically, non-spherical particles settle more slowly than their volume equivalent sphere, especially platy shapes like those of many clay minerals (see Sect. 2.2.1). Under many sedimentary conditions the water in which the particles are suspended is not static but flowing, and this will further determine whether particles remain suspended, or if they are deposited. For clay particles less than 1 µm in size settling times are extremely slow, of the order of about $3\,h\,cm^{-1}$ or longer. In effect, the bulk of sedimentation of fine-grained particles only occurs because of processes of aggregation which increase particle size by one or two orders of magnitude (McCave 1984). Up to concentrations of about several tens of thousands of $mg\,l^{-1}$ settling rates tend to increase because of the greater chance of particle–particle collisions leading to the formation of aggregates. Above these concentrations setting rates begin to decrease due to mutual particle hindrance. The terms, aggregation, agglomeration, flocculation and coagulation are sometimes used interchangeably, and sometimes more specifically to refer to a certain type of process of particle association (van Olphen 1977; Yariv and Cross 1979). From a sedimentological viewpoint, aggregation is probably the best general term.

In order for aggregation to occur, two things must happen: particles must approach each other close enough to interact and they must sometimes stick together as a result. Three physical processes are responsible for bringing the particles together; Brownian motion, velocity gradients due to laminar and turbulent shear, and differential settling, also known as scavenging. The operation of these three different mechanisms is discussed in some detail by Lerman (1979). In estuaries and in coastal waters turbulent shear is dominant, whereas in the deep ocean basins Brownian motion may dominate for particles of < 1µm in size and is succeeded by scavenging and then by shear for increasingly large particle sizes (McCave 1984). The second step of aggregation, that of sticking particles together, depends on the surface properties of the particles. The sticking together part of aggregation may have various causes such as electrochemical flocculation by salts, or biological processes that produce "glues" such as the biopolymers exuded by bacteria (McCave 1984). The obvious aggregation of particles which occurs in estuaries has often been attributed to salt flocculation, but there is a substantial body of evidence which indicates to some workers that biological processes may be much more important than changes in salinity.

4.3.3.1 Salt Flocculation

The process of salt flocculation occurs between fine-grained particles that are charged. The charge may originate in two ways, one is from substitutions within the mineral structure, and the other from surface reactions. Most clay minerals carry a permanent net negative charge due to the substitution of cations of higher valence by cations of lower valence in the crystal structure. In contrast, charge originating at the surface is pH dependent; it is usually positive at low pH and negative at high pH. For a clay particle in suspension its charge is balanced by ions of opposite charge which are loosely held in a "cloud" surrounding the particle. In this cloud the concentration of the so-called counter ions is in excess of that in the electrically neutral bulk solution, and there is a deficiency of ions of similar charge to the particle (Fig. 4.10). In water of low ionic strength this disturbance in electrical neutrality extends

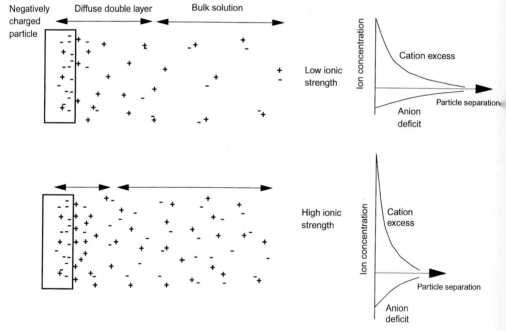

Fig. 4.10. Changes in the diffuse double layer thickness with changes in ionic strength. The net negative charge on a particle in suspension is shown to be balanced by an excess of cations (+) in the solution surrounding the particle. Together, this balanced region of charge is known as the diffuse double layer. In a solution of low ionic strength the diffuse double layer extends out much further from the particle surface than in a solution of higher ionic strength. The distribution of cations (+) and anions (−) in the solution is also shown graphically. (Modified from Arnold 1978)

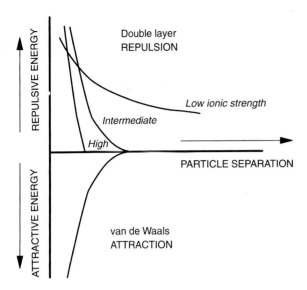

Fig. 4.11. Changes in repulsive energy as a function of particle separation for three different ionic strengths. Attractive energy due to van de Waals attraction remains constant so that at low ionic strength repulsive energy is greater than attractive energy, whereas at high ionic strength attractive energy is greater. (Modified from van Olphen 1977)

much further into the solution than is the case in a solution of high ionic strength (Fig. 4.10). As particles approach each other in suspension their clouds of counter ions interact and cause them to repel each other. In addition to this repulsive force, the particles have an attractive force due the van de Waals attraction. In solutions of high ionic strength the cloud of counter ions becomes so compressed (Fig. 4.11) that the particles may approach each other close enough for the van de Waals attraction between them to stick them together. This process is known as flocculation, or more specifically salt flocculation. The idealised distribution of charge from the clay particle and the cloud of counter ions surrounding it are often referred to as the "double layer", and the dynamic interactions between surfaces, ions in the surrounding "cloud", and the bulk solution are described by various diffuse double layer models (van Olphen 1977; Stumm 1992).

Salt flocculation occurs most obviously when there is a change in salinity, such as at the contact between fresh water and sea water in estuaries, or when fresh water flows into a saline lake. Changes in pH from river water (pH 5–6) to sea water (pH 8) will also affect pH-dependent surface charge and therefore may also influence the flocculation process. The rate at which flocculation occurs depends on the concentration of particles in suspension, the amount of agitation, i.e. the processes that may bring the particles together, and salinity which is the most important variable. Gibbs (1983) found that salt flocculation occurs already at salinities of 0.05 to 0.1%. Different cations also have different flocculating powers related to their hydrated ionic radii and their valence. For monovalent cations flocculating power decreases in the order $Cs > Rb > NH_4 > K > Na > Li$. The effect of valence means that the flocculating power of divalent cations may be 5–200 times, and that of trivalent cations 250–15 000 times, the power of monovalent cations (van Olphen 1977). In the absence of complicating factors, such as specific adsorption of certain ions, the critical concentrations of an ion required for flocculation are in the ratio $1/(\text{valency})^6$.

In river waters, which typically have ionic strengths of <0.01 mol kg^{-1}, the diffuse double layer might extend over a thickness of about 100 Å for a monovalent cation as opposed to 50 Å for a divalent cation. In sea water with an ionic strength of 0.7 mol kg^{-1} the double layer for both monovalent and divalent cations would be less than a few Ångstroms thick.

4.3.3.2 Differential Flocculation and Settling

From theoretical considerations and laboratory experiments with saline solutions, it is well known that some minerals flocculate more readily than others (Hahn and Stumm 1970; Gibbs 1983). This is known as differential flocculation. The flocculation behaviour of a mineral depends on its surface charge density, such that smectites with their typically low charge density (high surface area/low charge) will flocculate less readily than illites and kaolinites which have higher surface charge densities (small surface area and/or high charge). However, the results of various experiments with natural clays are often contradictory. Most probably, this reflects the complex nature of the natural materials, including the presence or absence of surface coatings of organics and hydrous oxides which modify surface properties. Although it can be observed in the laboratory, it is doubtful whether differential flocculation is important in nature. Furhtermore, it is not certain whether salt flocculation is an important process in nature, as it may be overwhelmed by the aggregation effects of biological processes (Meade 1972; Eisma 1986). On the other hand, differential settling, due simply to differences in particle size, does appear to be a mechanism capable of sorting clay minerals in natural environments (Gibbs 1977). In particular, fine-grained smectites often appear to be preserved in suspension longer than other clay minerals (Chamley 1989).

4.3.3.3 Bio- and Organic Flocculation

Eisma (1986) reviewed a large number of studies on flocculation in estuaries and emphasized the paramount importance of biological processes. He concluded that the formation of flocs in estuaries is controlled by the origin of the organic matter present, the organisms producing it, and the (temporary) conditions of deposition and resuspension. According to Eisma (1986) salt flocculation plays only a minor role, if any, affecting only the very fined-grained materials <1 µm in size. Biological processes act to both bring the particles together, often as faeces, and keep them together as a result of the sticky properties of various mucopolysacharides produced by bacteria, algae and higher plants. The size of flocs is controlled by the binding strength of the organic matter, and they may also be broken up by organisms that consume the organic "glues". Furthermore, many natural clay particles are coated with organic, principally humic, substances. Much of the negative charge of such organic-coated clay particles may be due to the organic coatings themselves, which are weak polyelectrolytes. Just as the extent of the electric double layer varies with pH and ionic strength, so too does the extent and the configuration of the organic layer (known as the macromolecular absorbed layer). The stability, i.e. flocculation or dispersion, of much particulate matter may therefore be largely controlled by the interactions between such organic layers as the coated particles collide with one another (O'Melia and Tiller 1993). The nature of the coatings are themselves related to the origins of the clay particles, such as soil versus rock origins. Changes in clay-mineral distributions in estuarine sediments have been attributed to differential salt flocculation, but, given the likely minor role of this process, Eisma (1986) argues that they are more likely to be related to different sources, and to the mixing and dispersal patterns from various sources in the estuary and in the sea. Mixing of different continental and marine sources, rather than selective flocculation, now appears to be the consensus of most studies concerning the origin of estuarine clay mineral trends (Chamley 1989; Weaver 1989). A recent example is provided by the study of Algan et al. (1994) of the Solent region in southern England.

4.3.3.4 Properties of Aggregates and Flocs

Flocs and aggregates of particles are typically made from all of the diverse materials that occur in suspension, and as such they may have a wide variety of properties. However, almost all aggregates or flocs have substantial proportions of organic matter. Because of the aggregate structures, densities are lower than the component materials, while sizes may be up to one or two orders of magnitude larger than the components. The flocs found in estuaries are typically between a few to a hundred microns in size, although larger more loosely bonded aggregates known as macroflocs (Eisma 1986) have sizes up to 3–4 mm. In the oceanic environment the largest aggregates are known as "marine snow" and these also may have sizes of up several millimetres in diameter. The larger flocs often trap silt and fine sand grains between clay particles, increasing their settling velocities.

For platy clay-mineral particles, flocs may be formed by several modes of particle interaction (van Olphen 1977). Adhesion between the oxygen planes of basal surfaces is known as face-to-face (FF) association; adhesion between broken-bond surfaces at the sides of the plates is known as edge-to-edge (EE) association; and adhesion between broken bonds surfaces at the edges of the plates and the oxygen planes of the basal surfaces is known as edge-to-face (EF) association (Fig. 4.12). The ease of formation of the different types of associations depends on the balance of repulsive and attractive forces which are controlled by the chemistry of the solution and the van de Waals interac-

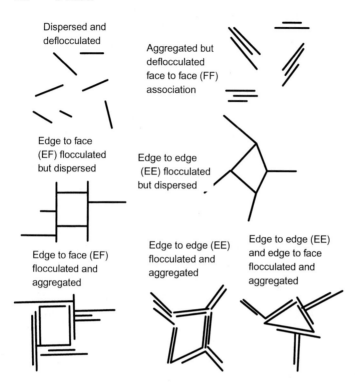

Fig. 4.12. Modes of platy clay particle associations in suspension and clay flocs according to the terminology of van Olphen (1977). (Modified from van Olphen 1977)

tion energy of the particles. The van der Waals interaction energy is different for the three types of association being highest for FF. Associations of the FF type give thicker particles whereas associations of EF and EE lead to much lower density, larger volume aggregates (Fig. 4.12). For the phyllosilicate clay minerals, F (face) surfaces carry permanent net negative charges due to substitutions within the mineral structure, whereas charges at edge sites are due to broken bonds and are pH dependent. Such charges become increasingly positive at low pH due to adsorption of H^+ ions and more negative at high pH due to adsorption of OH^- ions. EF interactions between particles are therefore more probable at low pH. The FF associations are sometimes referred to as tactoids. However, it should be pointed out that in natural materials, modes of association may be more closely controlled by any surface coatings on the particles rather than by the intrinsic properties of the particles themselves.

4.3.4 Erosion, Transport and Deposition by Wind

Although not as important globally as water, in certain areas of the world wind is an important agent of erosion, transport and deposition of clay minerals, and can contribute the dominant source of material to some deep-sea sediments.

Erosion and entrainment of clay particles by wind occurs principally in the worlds deserts and semi-arid areas where soil moisture content is most deficient. However, regions in more temperate areas can also be affected, especially where agricultural practices expose large areas of soil for part of the year. For example, Coote et al. (1981) have estimated that 160×10^6 tons of soil is lost annually by wind erosion of the Canadian Prairies, compared to an annual lost of 117×10^6 tons by water erosion. Apart from the degradation of soils, other environmental effects include the health hazards of dust and deterioration of air quality; in many parts of the world dust storms reduce visibility to less than 1000 m for 20 to 30 days a year (Nickling 1994). On the other hand, vast areas of the world are covered by wind-blown clay and silt, known as loess (Catt 1988), and the soils formed on these deposits can be very fertile.

Fine-grained aeolian sediments do not form important deposits in desert areas because the desert conditions ensure that they act principally as a source of material to other environments. Other types of terrain that may be important sources include glacial outwash plains, dried up lake beds, wadies, and alluvial fans.

The direct entrainment by wind, of particles smaller than about 80 μm in diameter is inherently difficult (Fig. 4.13) because at these small grain sizes sediment surfaces become aerodynamically smooth and interparticle forces increase in importance. As a result, most small particles, including clay minerals, are usually entrained by the impact of larger sand-sized grains and also by abrasion and break-up of larger particles during transport (Nickling 1994). Particles >20 μm in diameter are usually transported only within a few meters of the surface, but smaller particles may be transported to great heights (Fig. 4.14) and great distances. In arid regions dust storms typically transport material up to 3000 m height, and cover areas up to 500000 km² (Goudie 1983).

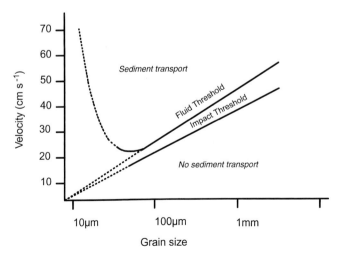

Fig. 4.13. Wind velocity required to entrain particles of various sizes. (Redrawn from Bagnold 1941)

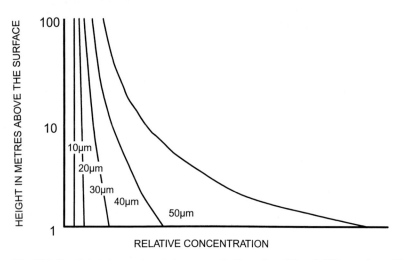

Fig. 4.14. Predicted changes in relative concentration of particles of different sizes with height in a severe wind storm. (Redrawn from Tsoar and Pye 1987)

Long-distance transport depends on many related factors including the texture of the eroding surface, ejection rate into the air stream, particle size and shape, and the turbulent structure of the wind (Nickling 1994). The maximum distance that particles of various sizes could be transported in suspension during a moderate storm was calculated by Tsoar and Pye (1987). For clay-size material, their calculations show that even moderate storms are capable of transporting particles over thousands of kilometres, whereas particles >20 µm in diameter are not likely to travel more than 30 km.

To a large extent the deposition of aeolian dusts depends on dampness and roughness of surfaces (Nickling 1994). Deposition increases over damp surfaces because of cohesion, and over open water there is no longer any further source of particles for entrainment. Changes in roughness, such as from unvegetated to vegetated ground, cause trapping of particles. Just as in sedimentation from water, fine-grained clay particles suspended in the atmosphere may become aggregated which increases their settling velocities. In the atmosphere, aggregation by scavenging and by shear become more efficient than aggregation due to Brownian motion, for particles >3 µm in size (Lerman 1979). The potential for long-distance aeolian transport is nowhere more evident than in the sediments of the deep ocean basins where, because of the paucity of other sediment sources, aeolian supply may become the dominant source of sediment. According to Windom (1969, 1975) dust transported by wind from deserts contributes from 10 up to 75% of the non-biogenic fraction of deep ocean sediments. The great variety of anthropogenic aerosol also suggests that there is likely to be a growing number of anthropogenic components in deep-sea sediments.

The major source areas for wind blown dust and dust trajectories are shown in Fig. 4.15. The most important meteorological conditions for dust

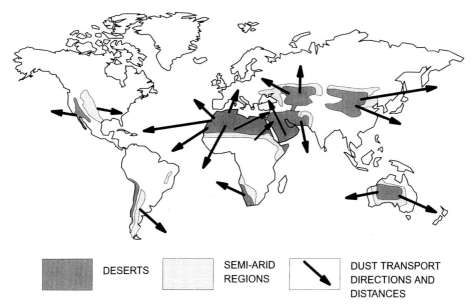

Fig. 4.15. Aeolian dust trajectories and transport distances from the world's desert areas. Long-distance transport involves particles with grain sizes of <20 µm. (Modified from diagrams in Péwé 1981; Pye 1987)

transport occur when low pressure fronts cross these regions, although locally many other weather patterns are capable of producing winds strong enough to entrain large amounts of dust. The composition of aeolian dusts varies greatly because it is determined by the minerals in the source area, although the influence of local sources diminishes with distance.

4.3.5 Erosion, Transport and Deposition by Ice

Ice can be an important agent of erosion, transport and deposition of clay minerals in two settings; glacial environments, and high-latitude oceans where the climate is seasonally cold enough for sea ice to develop (Fig. 4.16). At the present time glaciers cover about 10% of the Earth's surface, but during the Quaternary, maximum coverage was about 30% (Edwards 1986). As a result direct glacial deposits known as "tills", together with various sediments formed or fed by glacial outwash, form important deposits in many present-day temperate regions. There are many kinds of tills (Edwards 1986), but all are typically poorly sorted and often contain large amounts of fine-grained sediment forming a matrix for the coarser-grained material. Much of the fine-grained material is formed as "rock flour" generated by glacial grinding of rock against rock. Glacial meltwater or outwash also forms many kinds of sedimentary deposits (Edwards 1986), but as far as fine-grained sediments are

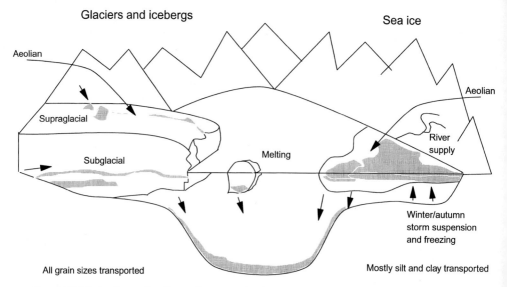

Fig. 4.16. Mechanisms of sediment incorporation into, and transport by, ice. One of the most significant processes for clay minerals is suspension freezing in shallow-shelf waters. Sediments transported by sea ice may be distinguished from those transported by icebergs by their grain-size distribution

concerned, the contributions of outwash deposits to wind blown loess (Catt 1988) and to glacio-lacustrine and glacio-marine sediments are probably the most important. Strong seasonal variations often result in changes in sediment load of glacial meltwater, resulting in the deposition of laminated deposits of silt and clay, which, when they are composed of annual couplets of silt and clay-rich layers, are known as varves.

Generally, temperate and subpolar glaciers produce more meltwater and release the largest quantities of suspended sediment which may form underflows or overflows depending on whether the receiving water body is marine or fresh, and on the concentration of material in suspension.

In addition to acting as fine-grained sediment sources to lacustrine and marine environments via meltwater, ice may accomplish direct sediment entrainment and transport in lakes and in the ocean by the process known as ice rafting. Where glaciers are concerned, ice rafting by icebergs is best known for its ability to transport and deposit coarse sediments in regions that normally receive only fine-grained pelagic sediment. As the ice drifts away from its source and begins to melt, coarse material incorporated within it is released and drops into the sediments below forming so called "drop stones". However, large quantities of fine-grained sediment may also be transported in this way by icebergs, and in sea ice it is typically the major sedimentary component (Barnes et al. 1982; Nürnberg et al. 1994).

Sediment distribution in sea ice is very variable but tends to be concentrated near the surface or in the interior of the ice in distinct layers, in some instances up to 5 m thick (Nürnberg et al. 1994). Sediment concentrations are also very variable. Typically, they range from a few mg l^{-1} to several hundred mg l^{-1} but they may be as high as several thousand mg l^{-1}. Near surface accumulations may become more concentrated by cycles of melting and freezing of surface ice. The processes by which sediment may be incorporated into sea ice include aeolian transport to the ice, bottom freezing in shallows, slumping of cliffs, flooding by sediment laden river water, and bottom dragging by ice keels, but all to these processes are of minor importance compared to suspension freezing (Fig. 4.16; Barnes et al. 1982; Nürnberg et al. 1994). Suspension freezing occurs when strong winds and intense turbulence in combination with extreme subfreezing temperatures, cause shallow-shelf waters to become slightly supercooled, resulting in the underwater growth of ice crystals which interlock with clay and silt sediment particles suspended in the water by the turbulence due to the storm. With time, the crystals accumulate at the surface as a layer of slush ice which gradually congeals into a continuous layer of turbid ice. The process of suspension freezing is confined to shallow shelves, and then only occurs seasonally during the autumn and winter freeze up. It is also most effective on the widest shelves where the progressive seaward extension of freezing is most prolonged, and where the prevailing winds blow offshore, promoting the continuous advection of newly formed ice seawards. By these processes sea ice and icebergs may transport clay minerals over hundreds of kilometres (Ehrmann et al. 1992; Kuhlemann et al. 1993).

For both icebergs and sea ice, cycles of freezing and melting can cause sediment to become aggregated into pellets which remain intact and are deposited as pellets onto the sea bed in areas where the ice melts (Goldschmidt et al. 1992). Pellets from icebergs appear to be larger, up to 2 cm diameter, than those from sea ice, which are at most a few millimetres in diameter. Sea-ice pellets can also be distinguished from those of iceberg origin by the finer grain size of the component grains.

4.3.6 Modifications and Transformations During Transport and Deposition

During transport, and during cycles of transport and temporary deposition, by any number of the processes discussed in the previous sections, clay minerals may undergo both physical and chemical modifications and transformations. Principal among these modifications are those that occur by the processes of ion exchange and fixation, indeed many modern-day environmental pollution problems are determined by the way in which pollutants interact with the surfaces of clay-sized materials. A useful summary of the kinds of interaction that may occur is given by Yariv and Cross (1979).

4.3.6.1 Ion Exchange and Fixation

One important process that affects clay minerals during transport is ion exchange. Most clay minerals are negatively charged so that most ion exchange

Table 4.1. Typical ranges of cation exchange capacity (mEq/100 g) of various clay minerals and river-suspended sediment. Values for clay minerals are from Grim (1968), values for suspended sediment are from Kennedy (1965)

Kaolinite	3–15
Chlorite	10–40
Illite	10–40
Smectite	80–150
USA river-suspended sediment	6–48

involves cations which balance these negative charges. The amount of exchangeable cations required to balance the net negative charge is known as the cation exchange capacity (CEC). CEC is traditionally quoted in units of milliequivalents per 100 g of dry clay. An equivalent of an ion is its molecular weight divided by its valence. The net negative charge of the clay mineral may originate from substitutions within the mineral structure, and from surface reactions such as those with "broken" bonds. The proportions of charges of different origin vary from one mineral to another, for example broken bonds at the edges of crystals are the major cause of CEC for kaolinite, whereas substitution in the mineral structure is the major cause of CEC for smectites. Surface related charge is pH dependent, and by convention the CEC is normally measured at neutral pH. Typical CECs of the common clay minerals are given in Table 4.1. along with a range of CECs for suspended river sediments.

Suspended sediments usually contain a range of clay minerals along with various oxides, hydroxides and organic matter derived from soils, all of which may have important CECs. Bulk CEC measurements of such materials obviously depend on both the proportions and the properties of the various different components.

The kinds and proportions of different exchangeable ions associated with a clay-mineral particle depend on the kinds and proportions of different ions in the surrounding solution. If the composition of the solution changes, the exchangeable cations also change in a related fashion. Such cation exchange reactions are effectively instantaneous. Hence, changes in river water chemistry, either spatially or through time, will cause concomitant changes in the exchangeable ions associated with clay particles. Such changes in water chemistry may occur as a result of significant changes in the geology, soils, and weathering in different parts of a river's catchment, or they may occur due to human influences such as the drastic chemical changes that characterize acid mine drainage. However, the most obvious and dramatic change in water chemistry occurs when river waters eventually meet and mix with the ocean.

Typically, the exchangeable cation suites of clay minerals in fresh waters are dominated by Ca^{2+} ions. On entering sea water Ca^{2+} ions are exchanged principally for Na^+ ions (Sayles and Manglesdorf 1977). To a lesser extent there is usually some increase in exchangeable Mg^{2+} and K^+ ions. Data from many early studies often showed that clay minerals took up Mg^{2+} from seawater in preference to Na^+, and this was sometimes suggested as indicating the formation of interlayer brucite and hence chlorite layers. However, these data are now known to be erroneous due to a flaw in the experimental procedure, namely the rinsing of samples with distilled water (Sayles and Manglesdorf 1977). Essentially, the dilution effect of rinsing sea water from the sample leads to the selective uptake of cations of higher valence as predicted qualitatively by Donnan equilibrium concepts. Experiments by Sayles and Manglesdorf (1977) showed that clay minerals in equilibrium with sea water contain a suite of exchangeable cations consisting of about 50% Na^+, 20–40% Mg^{2+}, up to 20% Ca^{2+} and around 5% K^+. The main variation appears to be in the relative proportions of the divalent cations Ca^{2+} and Mg^{2+}, but together their sum is usually constant at around 40%. The dominant change from Ca^{2+} to Na^+ is related to the change in relative abundance of these ions in fresh waters and sea water, respectively. However, the exchangeable cation populations of different clay minerals are not identical, because changes in concentrations of cations in solution are not the only factor involved in cation exchange. Certain cations may be selectively sorbed at various different exchange sites.

There are some data that suggest that the CEC of certain clay minerals may decreases on passing from river to marine waters. Processes such as K^+ fixation by vermiculitic clay minerals formed in soil environments may often be responsible for such observations (Weaver 1989). In this respect, Weaver (1989) notes that vermiculitic clay minerals are common in East and Gulf Coast rivers in the United States, but are not observed in the estuaries and marine bays. Weaver (1989) suggests that K^+ fixation may have resulted in much vermiculite reverting back to a more micaceous or illitic nature. Potassium fixation, and the resulting formation of illite layers, may also occur for smectites, particularly when they are subjected to cycles of wetting and drying (Srodon and Eberl 1984; Eberl et al. 1986). This process is most effective for smectites with high layer charge and may result in the formation of randomly interstratified illite/smectite with up to 50% of illite layers.

4.3.6.2 Pollutant Transport and Regulation

As a result of the same properties which give rise to cation exchange and fixation, clay minerals may also act as important transporters and regulators of environmental pollutants (Malle 1990; Förstner et al. 1990; Kühnel 1992). Many pollutants or contaminants such as heavy metals, like lead, zinc, cadmium and copper, tend to be quite insoluble in most surface waters and are strongly sorbed to surfaces. Consequently, both the amounts of many contaminants in the environment and their transport pathways, cycling, and ultimate removal, are not controlled by solubility, but by sorption–desorption reactions on sedimentary particulate matter (Schindler 1991).

The nature of sedimentary particulate matter is diverse, but in general it is the finest grain-size material, including clay minerals, with which most pollutants become associated (Baker 1980; Helios Rybicka 1992). This is because the finest size fractions have the largest surface areas and therefore dominate the surface properties of most sediments. It follows that the dispersal of pollutants in the environment depends on the processes which control the dispersal and accumulation of fine-grained sediments. In addition, the potentially toxic effects of pollutants and their bio-availability, depend on how the pollutants are sorbed to the clay-sized material, and how this may be altered by changes in the chemical environment such as pH and Eh, either during transport, or after deposition. The transport of pollutants by clay materials is not only important in surface environments, but also in the subsurface through the porous media of soils and aquifers by the process known as colloidal transport (McCarthy and Degueldre 1993).

An impressive example of the potential importance of clay minerals in regulating the effects of contamination is provided by the fate of radioactive caesium in the type of upland soils of Scotland that were contaminated following fallout from the Chernobyl nuclear disaster in 1986 (Cheshire and Shand 1991; Shand et al. 1994). Most upland soils in Scotland are high in organic matter, but immobilisation of caesium is related to the small amounts of mineral matter present within them, particularly the micaceous clay minerals because of their ability to fix caesium. This is an example of selective fixation. Fixation of caesium occurs in much the same manner as potassium, so that one might speculate that cycles of wetting and drying of the clay material, as might occur under natural conditions depending on the weather, may also be of some importance in governing the effect that the clay minerals have on the fate of caesium in particular environments.

Pollutants may be introduced to the environment from both point or diffuse sources. An example of a diffuse source, where its wider dispersal and later concentration is probably controlled by clays, is phosphorus applied to agricultural lands as fertilizer (Oglesby and Bouldin 1984). Movement of phosphorous in the environment may have disastrous ecological consequences by promoting the eutrophication of lakes and estuaries. The principal mechanism of phosphorus movement is by erosion and transport of clay particles to which phosphorus is sorbed (Froelich 1988).

4.4 Authigenic (in situ) Formation of Clay Minerals in Sediments

In many different sedimentary environments the new formation of clay minerals at surface or near surface temperatures is common place. Such in situ formation of a clay mineral in any environment, be it soil, sedimentary, diagenetic, hydrothermal or metamorphic, is known as authigenic formation.

Typically, in sedimentary environments such authigenic clay minerals are not abundant, but they have often been studied extensively because they contain much information about specific geochemical processes that are occurring there. Among the most common authigenic clays are various minerals of the smectite group, such as nontronite and stevensite, minerals of the mica group including glauconite and celadonite, various minerals related to the chlorites, and two fibrous clay minerals sepiolite and palygorskite. As a very general rule, in continental and evaporitic environments the most common clay minerals that form are usually Mg-rich, whereas in shallow marine environments they are Fe-rich, and in deep marine environments both Fe and Mg-rich varieties are common. This is illustrated in Fig. 4.17. Clearly, the geochemistry of both Mg and Fe must play a major role in the formation of authigenic clay minerals in surface and near surface environments.

4.4.1 Continental Authigenic Smectite

In continental environments, smectite formation by authigenesis and transformation processes is undoubtedly dominated by smectites formed in the soils that are an integral part of many sedimentary systems. Such smectite formation is dealt with in Chapter 3. Here, we deal only with authigenic smectites that form in continental aqueous sedimentary environments, either during or shortly after deposition, as a direct result of the chemistry of the depositional waters. Such smectite authigenesis is essentially restricted to saline alkaline lakes (Fig. 4.17), and is related to the unusual water chemistries that may develop in these settings. However, it should be emphasised that the conditions that give rise to clay mineral formation in alkaline lakes appear to be the exception, rather than the rule (Chamley 1989), although without detailed chemical and mineralogical data (e.g. Jones and Weir 1983) minor amounts of authigenic smectites might be difficult to detect amongst large amounts of detrital clays. In fact, the clay-mineral assemblages found in most saline, alkaline lakes are detrital (Droste 1961). Furthermore, where volcanic inputs coincide with high alkalinity, zeolite formation is often the prevalent type of silicate authigenesis.

The most commonly reported authigenic smectites in alkaline lakes are trioctahedral Mg-smectites, including stevensite, hectorite, and saponite (Dyni 1976; Trauth 1977; Jones 1986). In addition stevensite may occur interstratified with layers of kerolite, a disordered relative of the mineral talc (Eberl et al. 1982; Martin de Vidales et al. 1991). These clays often occur in carbonate lithologies disseminated throughout the sediment, but stevensite has also been reported in oolitic form from the Green River Formation (Tettenhorst and Moore 1978). Occurrences of peloidal and oolitic nontronite, a dioctahedral smectite, rich in Fe and Si, and poor in Al, are also known from recent sediments in lake Chad (Lemoalle and Dupont 1973) and lake Malawi (Müller and Förstner 1973).

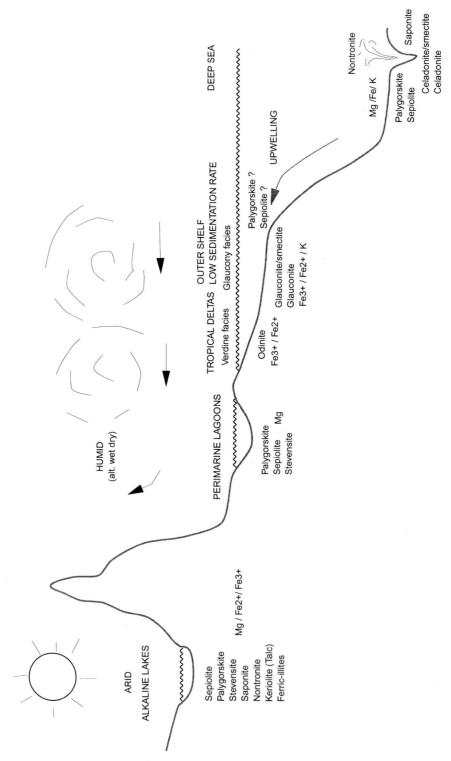

Fig. 4.17. Environmental situations in which many authigenic Mg-rich and Fe-rich clay minerals form. (Modified from Weaver 1989)

Often it appears that there is a transition from dioctahederal Al and Fe-rich smectites in lake marginal sediments, to more Mg-rich trioctahedral varieties in the lake-centre sediments (Trauth 1977; Darragi and Tardy 1987). Such sequences represent transitions from detrital-dominated sequences, supplied to the lake from the surrounding catchments, to authigenic dominated sequences in the basin centre. Because of the potential for mixing of detrital and authigenic smectites, the identification of authigenic varieties can be ambiguous and requires careful mineralogical characterisation. The problem is complicated because detrital inputs of reactive colloidal material, including smectite, probably serve as nuclei or templates for authigenic clay formation by some form of transformation mechanism, as emphasised by Jones (1986). Several of the most extensive deposits of lacustrine Mg-smectites are characterised by thick beds of these clays, often in close association with deposits of the fibrous clay minerals sepiolite and palygorskite. Such occurrences have much in common with those in peri-marine lagoonal environments in coastal locations (Fig. 4.17), where Mg is supplied periodically from sea water (Weaver and Beck 1977). According to Jones (1986), smectites such as stevensite precipitate under conditions of higher salinities and lower aqueous silica contents compared to the conditions that favor sepiolite and kerolite formation.

In many ancient continental evaporite basins, especially of Permo-Triassic age, Mg-rich chlorites and corrensite are very common clay minerals. Although many workers have interpreted these minerals as syn-sedimentary, they do not occur in any modern evaporitic environments, marine or non-marine. It seems more likely that they are the result of diagenetic alteration, either of the kinds of Mg-rich smectites described above, or of other early formed Mg-rich minerals, originally present in such sediments (Hillier 1993).

4.4.2 Marine Authigenic Smectites

In the marine environment, the formation of authigenic smectites can be divided essentially into three different categories. The first includes smectites associated with the alteration of volcanics and associated material, especially the hydrous alteration of volcanic glass known as palagonitisation. The second includes smectites formed in the mixing zones of ocean waters and plumes of hydrothermal fluids issuing from vents in the ocean floor. And the third category includes the so-called hydrogenous formation of smectite on the sea floor, by mechanisms that do not depend directly on volcanic and hydrothermal input but involve halmyrolysis (submarine weathering) and direct precipitation. The identification of authigenic smectites in marine settings is often a difficult task because of the potential for mixing with very similar detrital material of both allochthonous and autochthonous origin. This is particularly the case for the category referred to hydrogenous formation where the amount of authigenic smectite is often small. In most cases, where the authigenic formation of smectite is unequivocal, the smectites formed are nontronites. This contrasts with detrital terrigenous smectites which usually belong to the aluminous montmorillonite–beidellite series.

The palagonitisation of basaltic volcanic glass occurs by hydration and, among other minerals, results in the formation of nontronitic smectites. The process begins when the basalts are still hot, and this probably explains why many of the early formed smectites are trioctahedral Mg-saponites similar to those found where true hydrothermal alteration of basalts occurs deeper in the ocean crust. Later formed smectites are usually dioctahedral nontronitic varieties. In fact, considerable amounts of potassium are added in the process and many of the authigenic clay minerals are probably mixed-layer celadonite/ smectites (Weaver 1989).

Fine-grained glassy volcanic material is also dispersed across large areas of the ocean floor, but there is no general agreement on whether or not it is an essential prerequisite for smectite formation in this environment. Weaver (1989) discusses numerous studies that have shown that there are a lot of fresh volcanic ash layers in buried Tertiary sediments, which have not altered to smectite. Such ash layers are the precursors of bentonite beds and indicate that most marine bentonites must be diagenetic in origin, rather than sedimentary.

One of the best-documented and spectacular sites of authigenic smectite formation are the hydrothermal discharge deposits found at many locations above the mid-ocean ridge system. Well-known examples are from the Galapagos spreading centre (McMurtry et al. 1983) and from the Red Sea. The authigenic clay minerals are typically true nontronites, and precipitate during the mixing and cooling of hydrothermal brines with ocean water. However, with time many appear to have altered to celadonite/smectite or glauconite/ smectite by fixation of K^+ ions from sea water. Significant deposits of clay minerals are usually found only very close to the vents in the local depressions of the ridge system. However, dispersal of smectites from the plumes and transportation by ocean currents has been suggested to explain some occurrences of authigenic smectite in areas of the south-east Pacific and Indian Oceans that are far removed from the ridge systems. Indeed, hydrous-oxides of Fe and Mn are known to disperse laterally away from the ocean ridge vents to distances up to 2000 km, and there is every reason to believe that smectites will be similarly dispersed. Nonetheless, the distinction between transported hydrothermal clays and so-called hydrogenous clays is often obscure because it is difficult to find criteria with which to distinguish them.

Mechanisms envisaged for the hydrogenous formation of smectites include the halmyrolysis (submarine weathering) of volcanic glass, reactions between biogenic silica and Fe-oxy-hydroxide precursors, or possibly the ageing of amorphous ferrous-silicic complexes. Although there is plenty of evidence that the alteration of volcanic glass is not the only, or indeed the dominant, mechanism of smectite formation, the potential contribution of smectites from allochthonous sources often makes it difficult to identify mechanisms and origins with certainty. In this respect, smectites attributed to an hydrogenous formation often appear to be appreciably richer in Al and Mg compared with clearly authigenic nontronites, such as those at hydrothermal discharge sites: in many cases, this may be due to mixing of authigenic smectite with allochthonous Al-rich smectites derived from the continents.

4.4.3 Marine Glauconite and Glauconite/Smectite

One of the most extensively studied groups of authigenic clay minerals is that of glauconite and glauconite/smectite (Odom 1984; Odin 1988). These names refer to specific minerals and because of past confusion between the use of the term glauconite to refer both to a mineral and a facies, the term glaucony was recommended (Odin and Matter 1981) for use when reference is made to the aggregate green-grains and to uncharacterised minerals belonging to this group. This usage is followed here.

Typically, glaucony occurs as bright green rounded to subrounded pellets in the silt to sand size range. The pellets form near the sea water sediment interface by replacement of the fine-grained sediment/biogenic aggregates that comprise faecal pellets or, more commonly, the carbonate tests and contents of foraminifera and other microfossils. In detail, many types of substrates such as quartz, feldspar, phosphate or mica grains may also be replaced or serve as templates for glaucony growth, but faecal pellets and especially various carbonate grains are the most common. Glaucony may also occur as a film coating various types of substrate.

The distribution of glaucony in modern sediments is largely concentrated in continental shelf areas at water depths between 100 and 300 m (Fig. 4.18). Tropical latitudes appear to be especially favourable areas for glaucony development, but unlike the verdine facies (see below) glaucony is not restricted to tropical latitudes and is known to occur from latitudes 50°S to 65°N (Fig. 4.18). Deep-water occurrences down to 3000m have also been documented, but it is not clear to what extent these may represent relict occurrences that formed originally at much shallower water depths. Generally, glaucony is characteristic of shallow, fully marine, shelf settings, especially where sedimentation rates are very low, such as towards the shelf edge away from areas of terrigenous input and active sediment transport. Indeed, slow sedimentation rates appear to be one of the factors necessary for glaucony formation. Often during periods of sea-level transgression successive substrates of biogenic and terrigenous origin, each characteristic of different water depths, may be glauconitised in turn.

The predominate granular form of glaucony as a replacement of grains, the preferential glauconitisation of silt and sand sized material rather than smaller particles, and the more advanced glauconitisation of the interiors of grains compared to their peripheries, all indicate that the genesis of glaucony

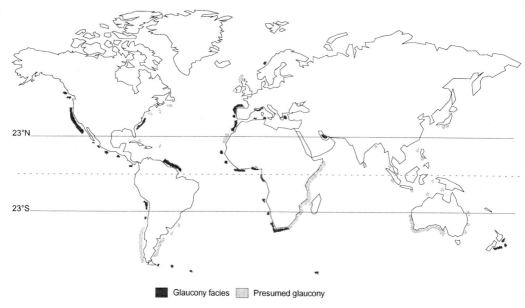

Fig. 4.18. Distribution of glaucony on the modern shelf. (Modified from Odin 1988)

is favoured by a semi-confined micro-environment (Odin and Matter 1981). Semi-confined means the environment is partially isolated from sea water via the microporous substrate. Initially, the first phase to form in the glauconitisation process is an Fe-rich smectite with about 20 wt% Fe_2O_3. With time, the potassium content increases and the structure of the mineral changes from smectite, through mixed-layer glauconite/smectite, towards the micaceous end member glauconite. This process occurs at the sea floor with most of the K^+ ions apparently supplied from sea water, because if the grains are buried the process is stopped at whatever stage it may have reached, presumably because the ready supply of K^+ ions is cut off. X-ray diffraction patterns of glaucony grains from the present-day continental shelf, illustrated by Odin (1988), demonstrate that the process of glauconitisation may proceed all the way to the micaceous end member glauconite at the sea floor. The enrichment in Fe occurs before K, and the amount of Fe shows little or no change from smectite to glauconite. Odin and Matter (1981) described the continuous progress of the glauconitisation process in terms of four stages (Fig. 4.19). The first (nascent) stage is characterised by the relatively rapid alteration of the support material, and the establishment of the microporous semi-confined environment. At the second (slightly evolved) stage the porosity allows ionic exchange with both sea water and the interstitial water of the underlying sediment. The grain becomes green and begins to loose most of the structures of the original support material. In the third (evolved stage) the structure of the support has disappeared entirely and preferential clay growth in the centre of the grain causes the grain to grow bigger, and the exterior of the grain to develop cracks. By this stage K_2O content has reached 6–8%. The final (highly evolved) stage corresponds to grains that have been completely replaced by minerals near the glauconite end member, and the surface cracks have been filled by less-evolved glaucony to form a more rounded outline. Potassium contents of glauconitic minerals are directly related to the proportion of expandable layers present. Other chemical changes which accompany the obvious increase in K^+ are more difficult to discern because of the heterogeneous nature of most glaucony samples (Odom 1984).

Various theories regarding the origin of glaucony have been advanced over the years (Odin 1988). Currently, most workers favor the neoformation theory. Much of the evidence that supports this theory has been gained by use of the scanning electron microscope to demonstrate the obvious neoformed morphologies of glaucony. Other pertinent observations that support the neoformation theory are: (a) glauconitization of biotites occurs by growth between sheets; (b) non-micaceous substrates may be entirely replaced; and (c) glaucony growth may be displacive, leading to the break up of shell substrates.

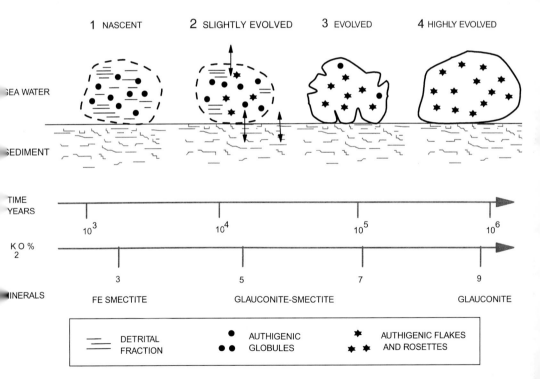

Fig. 4.19. The four stages in the development of pelloidal glaucony

4.4.4 Celadonite and Celadonite/Smectite

Celadonite is a mineral very similar to glauconite in composition, especially in its Fe content, the main chemical differences being that celadonite is usually richer in Si and Mg and poorer in Al than glauconite (Odom 1984; Odin 1988). The principal difference concerns their occurrences, in that unlike glauconite, celadonite never occurs in granular form and is characteristically a product of the submarine weathering of volcanics. However, mode of origin is not a criterion for the identification of glauconite or celadonite. Typically, celadonite forms by the alteration of basalts under oxidising conditions, probably at temperatures above those of ambient ocean waters. Mg and K are provided from sea water and Fe and Al from the rock.

Morphologically, celadonite usually occurs as well-shaped laths, in contrast to the more anhedral globules, box-work and irregular lamellar structures of glaucony. These differences may be related to the fact that a lot of glaucony is glauconite/smectite, whereas celadonite/smectite is less frequently reported with most material being close to end member celadonite. This may be due in part to the higher temperatures associated with celadonite formation, the more concentrated solutions involved, and the greater availability of pore space allowing the growth of euhedral crystals.

4.4.5 Non-Marine Glauconite and Ferric Illite

Glauconite-like clay minerals and clay minerals known as ferric illites have been reported from a number of lacustrine environments. Early data on these minerals were summarised by Porrenga (1968). Characteristically, these minerals are found in green clays often associated with carbonates and gypsum, and are thought to form by the alteration of smectite. Their contents of ferric iron (around 10 wt% Fe_2O_3) are intermediate between marine glaucony, and ordinary illites which are also more aluminous. Like marine glauconite, they are believed to form at surface temperatures. More recently, they have been reported by Norrish and Pickering (1983) from Oligocene sediments of lake Eyre and from desert soils in Australia, and by Deconinck et al. (1988) from the Purbeckian of the Swiss and French Jura. Deconinck et al. (1988) suggested that the Purbeck illites may have formed by cycles of wetting and drying, which is certain to be a common process in many lacustrine settings because of frequent changes in lake level. However, experimental work on wetting and drying cycles has shown that this process can only form randomly interstratified illite/smectite with up to 50% illite layers (Eberl et al. 1986). Typically, ferric illites and/or non-marine glauconites, including the Purbeck examples, are much more illitic than this, so that another process such as neoformation, as documented for marine glaucony, may be involved. More recently, Jeans et al. (1994) has shown that much of the illite in the Permo-

Triassic continental facies of western Europe has characteristics like those of ferric illite, and it is suggested that it may have formed in coeval desert soils before being eroded into the adjacent sedimentary basins.

4.4.6 Minerals Related to Chlorites and the Verdine Facies

The formation of a true chlorite (14 Å, 2:1 + 1 mineral) in recent sediments has never been demonstrated. Early reports in the literature of sedimentary chlorite formation, which usually refer to only minor amounts of 'chlorite', are often equivocal in their identifications, or use the term 'chlorite' in a different sense to that which is understood today. However, Fe-rich minerals with a 7 Å basal spacing, closely related in composition to chlorites, are relatively common components of green-grains in recent sediments at water depths of 10 to 60 m off the coast of many deltas in tropical regions, such as the Niger delta. Formerly, many of these occurrences where described as chamosite (e.g. Porrenga 1967), and then subsequently as berthierine (e.g. Odin and Matter 1981) when the name chamosite was reserved for a true (14 Å) Fe-rich chlorite. Like chlorites, the Fe in berthierine is mainly Fe^{2+} and it is a trioctahedral mineral. However, extensive study of these recent and subrecent green-grains by Odin (1988) showed that in almost all cases the 7 Å mineral in these grains contains predominantly Fe^{3+}, and is richer in Si and Mg, and poorer in Al compared to the berthierines described from sedimentary rocks. This led Odin (1988) to suggest that this mineral was a new type of phyllosilicate to which he gave the provisional name phyllite V. Subsequently, Bailey (1988) proposed the new name odinite. Odin (1988) has studied the minerals in these green-grains very extensively and has shown that in addition to odinite, the green-grains are sometimes dominated by a chemically similar mineral which more closely resembles chlorite in structure but contains a swelling component. In older green-grains, of 1000 to 20 000 years, the chemistry in still similar, but minerals with other structures are present including a 7–14 Å mixed layer, a ferric chlorite, and a pyrophyllite-like clay mineral. Collectively, all of these minerals comprise the verdine facies.

As mentioned above, the verdine facies occurs in relatively shallow marine waters and appears to be confined to tropical latitudes (Fig. 4.20), in contrast to the glaucony facies which occurs at greater water depths and is not confined to the tropics (Fig. 4.18). However, like glaucony, the substrates that are replaced are similar, the minerals of the verdine facies forming in the semi-confined micro-environments of faecal pellets, foraminifera tests, and porous mineral grains.

The latitudinal restriction of the verdine facies is curious. Offshore from the Niger delta, Porrenga (1967) noted that the seaward boundary to the verdine (chamosite) facies coincided with the thermocline. From this he inferred that a water temperature >20 °C was necessary for verdine (chamosite)

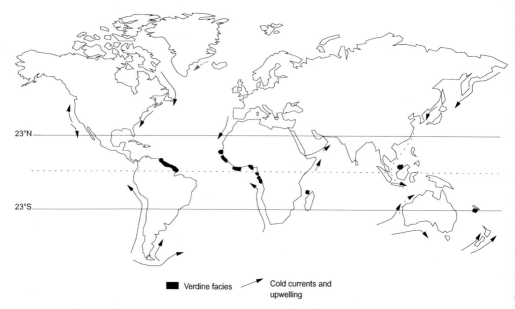

Fig. 4.20. The distribution of the verdine facies on the modern shelf. Modified from Odin and Matter (1981) and Odin (1988). The verdine facies appears to be confined to tropical areas mostly in association with fluvial inputs (which may provide a source of reactive colloidal Fe) and appears to be absent from shelf areas crossed by cold currents associated with upwelling, suggesting some temperature control

to form. An effect of temperature is also suggested by the observation of Odin and Matter (1981) that verdine (berthierine) is absent on continental shelves crossed by the cold oceanic currents that are frequently associated with areas of upwelling (Fig. 4.20). Another important factor may be the supply of Fe. Rivers draining tropical regions frequently carry high loads of dissolved and colloidal forms of Fe that would be reactive and available for clay formation in near shore sediments.

Although the verdine facies does not contain iron-oolites, many ancient marine sedimentary oolitic ironstones contain minerals which may have formed under similar conditions to those of the verdine facies. This hypothesis supposes that the minerals berthierine and chamosite found in oolitic ironstones, and which contain dominantly ferrous iron, formed from minerals such as odinite with dominantly ferric iron, as a result of burial diagenetic processes. Its attraction is that it is no longer necessary to try to explain the formation of minerals which require both reducing conditions and agitated waters at one and the same time. Instead, the minerals can be formed in essentially oxidizing conditions, and the Fe present reduced after burial along with other structural modification to produce berthierine and, with increasing temperature, chlorite (Hillier 1994).

The relationship of the verdine facies to other authigenic Fe-rich minerals in the marine environment is shown schematically in Fig. 4.21. In the deep ocean basins, various types of true chlorites, corrensite, and mixed-layer chlorites are commonly described as alteration products of submarine basalts, but the fluids involved in the formation of these minerals are undoubtedly hydrothermal.

4.4.7 Sepiolite and Palygorskite

The fibrous clay minerals palygorskite and sepiolite are relatively rare minerals, but locally, in certain lacustrine and peri-marine basins, they may occur in abundance. Palygorskite may also be found in many deep-sea deposits, while sepiolite is more rarely reported and usually less abundant. Both minerals are Mg-rich, especially sepiolite, but they are also very high in Si, and so their formation is favoured in environments where amorphous Si is abundant. There is some question as to whether they are formed by transformation of pre-existing phyllosilicates, or neoformed by direct precipitation from solution. According to Jones and Galan (1988) both sepiolite and palygorskite commonly form by a dissolution–precipitation mechanism which incorporates components of pre-existing detrital material.

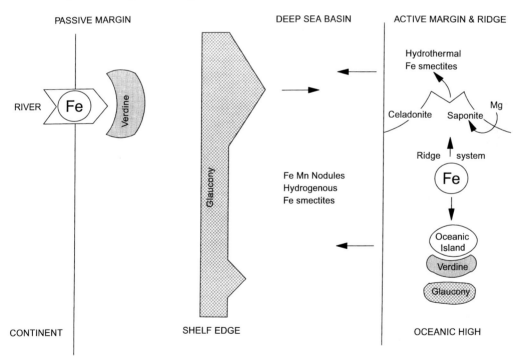

Fig. 4.21. The distribution of the verdine facies in relation to the glaucony facies and other iron-bearing marine clays. Fe is supplied to the marine environment from two main sources; rivers, and hydrothermal processes at active margin and ridge systems. (Modified from Odin 1988)

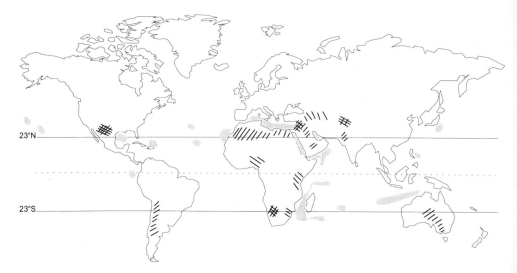

Fig. 4.22. The distribution of both continental and marine occurrences of palygorskites and sepiolite for the late Pliocene to Holocene. Oceanic occurrences are shown by the *dotted shading* and continental soil data by *diagonal shading*. *Cross-hatching* indicates occurrences where soils are superposed on sedimentary basins with palygorskite. (Modified from Callen 1984)

The distribution of both continental and marine occurrences of palygorskites and sepiolite for the late Pliocene to Holocene is shown in Fig. 4.22. Shallow water lacustrine and peri-marine occurrences are found only in arid to semi-arid climatic settings adjacent to areas of intensive chemical weathering and subjected to strong evaporation but also characterised by frequent mixing with supplies of fresh water. A supply of fresh water is necessary to maintain the brackish conditions and pH range 8 to 9 favourable for these minerals. Restricted detrital terrigenous input is also a characteristic feature, and various carbonates and cherts are typical associated deposits. Often there is a zonation of minerals from sepiolite in the basin centre to palygorskite in more marginal locations where the influence of detrital materials entering the basin becomes evident. This may reflect the fact that palygorskite formation requires a supply of Al, a fact which seems to be corroborated by observations that sepiolite is usually favoured over palygorskite in cases where fresh water supplies are replenished by ground waters rather than surface run off that is certain to be richer in aluminous colloidal material.

In deep sea deposits there is general agreement that some palygorskite and sepiolite may be formed in situ as a result of hydrothermal activity combined with the influence of sea water. Authigenic origins have been suggested for the more widely dispersed occurrences over large areas of the ocean floor, but in their reviews of these occurrences both Weaver (1989) and Chamley (1989) favor an origin by inheritance from sources on the continents via aeolian and marine transport, although they do not rule out authigenic occurrences. Indeed, it is now clear that the delicate fibres of these minerals tolerate transport more easily than was once thought to be the case (Chamley 1989). Observations in support of an inherited origin for many occurrences include the following; present and past occurrences in the deep sea correspond with the distribution of tropical to subtropical climatic belts suggesting a supply from contemporaneous continental deposits confined to these climatic zones (Fig. 4.22); past concentrations in the deep sea correspond to times of greatest abundance (potential source) on land; there is no correlation of occurrences to any particular sediment type, as there should be given that particular geochemical conditions are required; and electron micrographs show the mineral fibres to be short and broken giving the appearance of having been transported and redeposited. Cases where delicate fibre morphologies are preserved are often attributed to transport by wind. Note that, as well as lacustrine and peri-marine deposits, many calcareous soils in arid zones represent potential sources of palygorskite and sepiolite; the dry arid conditions often promoting aeolian transport of clay material.

However, the question of the more widespread authigenic formation of palygorskite and sepiolite in deep marine sediments is still unresolved. For example, in Miocene marine sediments from the North Carolina continental margin, palygorskite and sepiolite appear to be demonstrably authigenic because of their pore-bridging arrangement (Allison and Riggs 1994). They occur in association with authigenic dolomite, abundant siliceous microfossils, and phosphate and organic-rich sediments. This association is believed to be related to upwelling of nutrient-rich waters. Dolomite and Mg-rich clays are suggested to have formed from the early interstitial pore waters under anaerobic conditions. An association of palygorskite and sepiolite with phosphorites has been mentioned previously by Bentor (1980) and it is interesting to note that several of the deep sea occurrences shown in Fig. 4.22 coincide with areas of upwelling as shown in Fig. 4.20. As pointed out by Jones and Galan (1988), opinions on the origin of deep sea occurrences of the fibrous clays have ranged between the extremes of asserting that all are detrital, to asserting that all are authigenic. According to Chamley (1993), like smectites, certain occurrences of palygorskite and sepiolite can pose difficult problems in determining which side of the autochthonous–allochthonous boundary they lie on. Thus it appears that more investigation of marine occurrences is required.

4.5 Mineralogical Patterns in the World Ocean

Several workers have compiled data on the distribution of the major clay minerals in the oceans. Notable contributions include the work of Biscaye (1965), Griffin et al. (1968) and Rateev et al. (1969); the most recent compilation is that of Windom (1976). Such compilations show that the largest-scale feature of the distribution is a broad latitudinal zonation of the different clay minerals (Figs. 4.23, 4.24, 4.25, 4.26). This feature is related to the different latitudinal climatic zones that cross the continents and the consequent weathering regimes, and is testimony to the dominant detrital origin of oceanic clay-mineral assemblages from terrigenous sources. This dominant detrital origin is also attested by many other observations including; the absence of isotopic readjustments with respect to the composition of sea water; the radiometric age of the clay minerals which typically corresponds to the average age of rocks on the adjacent continents, and which is often older than the age of sedimentation or even the ocean basin itself; the flux of terrigenous clay particles to the ocean

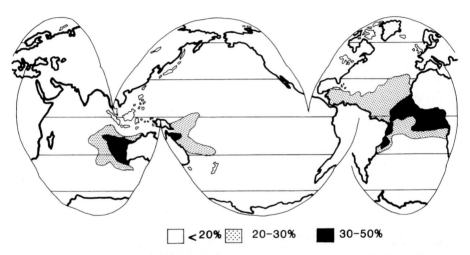

Fig. 4.23. The distribution of kaolinite in the world ocean. (Modified from Windom 1976)

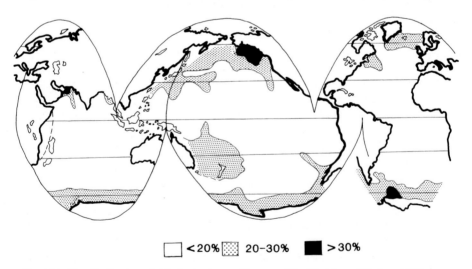

Fig. 4.24. The distribution of chlorite in the world ocean. (Modified from Windom 1976)

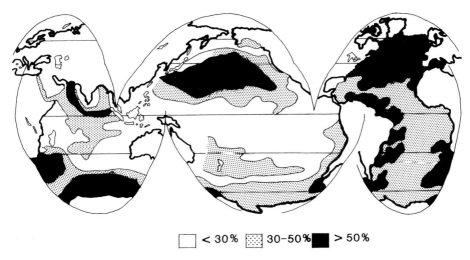

Fig. 4.25. The distribution of illite in the world ocean. (Modified from Windom 1976)

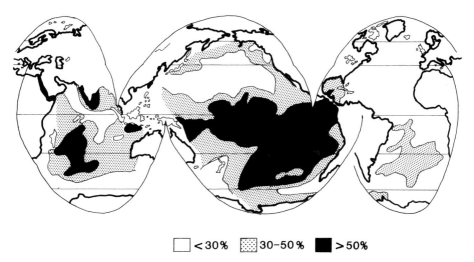

Fig. 4.26. The distribution of smectite in the world ocean. (Modified from Windom 1976)

(Fig. 4.5) which is much greater than the flux of dissolved material; the absence or extremely weak character of mineralogical transformations at the continent ocean boundary; the essentially physical nature of sorting and sedimentation during marine transport; and the constant similarity in any particular region between the clay mineral composition of rocks and soils and that of fluvial sediments, marine suspensions, aeolian dusts and the sediments from the ocean bottoms (Chamley 1989, 1993; Weaver 1989).

The two minerals that illustrate this latitudinal zonation best are kaolinite and chlorite. The distribution of kaolinite (Fig. 4.23) shows that it is concentrated in low latitudes where it is largely derived from kaolinite-rich soils developed by chemical weathering in humid climates. In contrast, chlorite is one of the most easily weatherable phyllosilicates and is one of the first minerals to be destroyed where soils are well-developed and chemical weathering intense. Consequently, it is usually minor or absent in low latitudes, but abundant and concentrated at high latitudes where physical weathering predominates (Fig. 4.24). In the southern oceans the increase in chlorite abundance below 50 °S coincides almost exactly with the northern limit of material transported by ice from Antarctica (Fig. 4.24)

The distributions of illite (Fig. 4.25) and smectite (Fig. 4.26) also show broad latitudinal patterns, but these are not as well defined as those of kaolinite and chlorite. Like chlorite, illite tends to increase towards the polar regions due to the decrease in chemical weathering, but it is also more abundant in northern than in southern oceans. This appears to be related to the concentration of the continental landmasses, and therefore terrigenous sources, in northern latitudes. The largest-scale features of the distribution of illite are similar to those for quartz. The grain-size characteristics of quartz indicate a dominantly aeolian supply and the illite is also thought to be dominantly aeolian, but with river inputs becoming more important toward the continental margins. (Chamley 1989; Weaver 1989).

Smectite abundance (Fig. 4.26) also shows a distribution that is quasi-parallel to the continental climatic zones, but it is much more abundant in the southern oceans than in northern oceans. Again this may reflect the distribution of the continents in that large areas of the southern oceans receive comparatively little terrigenous supply. Volcanic contributions, both subaerial and submarine, which may alter to smectite, are likely to be proportionally more important, as are smectites formed by other authigenic processes at the sea bed.

The average clay mineral composition of the world ocean has been estimated by Windom (1976). Illite and smectite are the most abundant minerals accounting for about 35% each, whilst chlorite and kaolinite account for about 15% each. Weaver (1989) estimates that about half of the smectite is authigenic. In detail, much of what is termed smectite or illite is often intermediate mixed-layer material (Weaver 1989). Other, but less common, clay minerals that may also be locally important include palygorskite and sepiolite which are characteristic of low latitudes (Fig. 4.22) and talc which is often thought to be entirely from anthropogenic sources, but may also have some authigenic input.

Still at a global scale, other features of the distribution of clay minerals in the oceans serve to emphasize that the latitudinal climatic control via weathering is only a very general pattern. These include: physical weathering dominated regimes related to mountain ranges at low latitudes; the minerals that are available where physical weathering is concerned; and the processes of transport and deposition that are responsible for the wider dispersal of the minerals in the oceanic environment. All of these can produce clay-mineral distribution patterns that are in total contradiction to the general latitudinal pattern described above. Furthermore, even at the scale of a depositional system like a delta, such variables may leave clay-mineral signatures that are totally unrelated to the contemporaneous latitudinal climatic zones. Examples, of large-scale discrepancies include abundant kaolinite and smectite in some polar regions simply due to their abundance in Mesozoic rocks of the sediment source area, and abundant kaolinite in the ocean sediments adjacent to modern-day desert regions because it has been stored for some considerable length of time in soils formed under wetter, more humid conditions (Singer 1984).

4.6 Environmental Interpretation of Clay Minerals

All environmental interpretations of sedimentary clay minerals are fundamentally based on determining the origin of the clay minerals present. Without doubt, this is a difficult and challenging task, because the origins of clay minerals are diverse, and because of the very nature of the sedimentary realm where clay minerals, of many different origins, are normally mixed together by various sedimentary processes. Nonetheless, clay minerals do contain information on many aspects of their origin, related to sources and provenance, dispersal patterns, depositional environment, climate, and even tectonics and eustasy (Chamley 1989). The previous sections of this chapter were devoted to the origins of clay minerals in sediments from the standpoint of the fundamental processes that determine them. Environmental interpretations are built upon our understanding of these processes and our ability to recognise the results and products of them in the sedimentary record. The purpose of this final section is to illustrate the kinds of studies that can be made and the sort of information that may be obtained, but with emphasis always on the problems of environmental interpretation that the careful investigator must try to tackle. More details of various aspects of environmental clay mineralogy, and compilations of many case studies, can be found in the detailed books of Chamley (1989) and Weaver (1989).

4.6.1 Sedimentary Environments and Provenance

Principally there are two ways in which studies of clay minerals may contribute to the analysis of sedimentary environments: firstly, the presence of certain authigenic clay minerals may be indicative of a particular depositional regime, and secondly, detrital clay minerals can be used as tracers of sediment transport processes, dispersal, and provenance. A classic example of the use of authigenic minerals is the inference that the presence of glaucony indicates a shallow marine environment of deposition. Another example is the presence of authigenic smectites or the fibrous clay minerals sepiolite and palygorskite, in lacustrine sediments from which inferences can be made concerning the chemistry of the lake waters from which they were deposited. Much of the background to these sorts of interpretations has been given previously in the section on authigenic clays. However, the point that must be made here is that there are exceptions of which the investigator must be aware, such as non-marine glaucony, for example. The formation of glaucony occurs under a specific set of geochemical conditions, and although these conditions are common in shallow marine environments, they are not restricted to them. Additionally, there is often more than one plausible explanation for the origin of a clay mineral, such as when palygorskite could form authigenically in lake sediments or form in arid soils surrounding the lake and subsequently be reworked into it by wind.

Both of these examples emphasise that there is often more than one possible origin for a clay mineral in a sediment. Although the examples concern authigenic clay minerals, this fact is even more true of detrital clays, which, as we have seen, are by far the most abundant. This important point is illustrated in Fig 4.27 which shows the potential for sedimentary smectite to be derived from six different sources (Chamley 1989), and to which can be added the authigenic formation of smectite in situ in the sedimentary basin. It is this diverse spectrum of origins which complicates the use of clay minerals as tracers of sediment sources and dispersal. Nevertheless, the use of clay minerals as tracers is the main sedimentological use to which they have been put.

The kinds of information that can be obtained can be illustrated by the study of Kolla et al. (1981) of the surface sediments of the Arabian Sea (Fig. 4.28). To the west and north the Arabian Sea is surrounded by arid lands, with no significant fluvial inputs. To the northeast the river Indus drains the Himalayas and forms the major fluvial source to the area; up until recent dam construction it delivered up to 400 million tons per year of suspended sediment to the region. In the east the main fluvial sources are the Namada and Tapai rivers which drain soils developed on the Deccan trap basalts. The Arabian Sea is an area where the distribution of clay minerals in the surface sediments reflects the nature of the surrounding land masses and a variety of different sediment transport processes. Smectite-rich clays, which extend all along the Indian margin, are related to their derivation from smectite-rich soils developed on Deccan trap basalts, and their subsequent southerly transport by the prevailing ocean surface currents. In the far south some smectite may also have been transported around the tip of India by surface currents from the Bay of Bengal. Over most of the rest of the region illite-rich sediments dominate and are derived from several sources. In the east the main source is the Indus river, from which illite-rich sediments are dispersed westwards and southwards by surface currents and turbidity currents across the Indus deep-sea sediment fan. Other sources of illite are from the arid lands of Iran, and

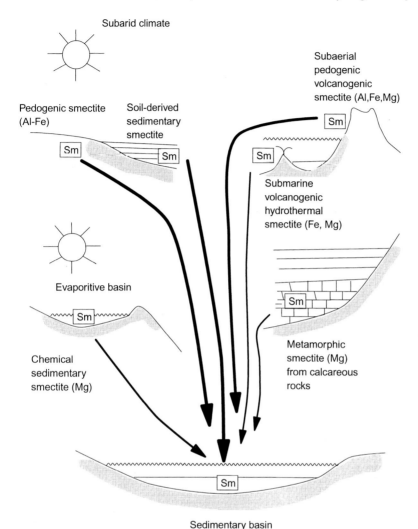

Fig. 4.27. The potential for the same type of clay mineral, in this case smectite, to be derived from very different sources. The *thickness of the arrows* is meant to be schematically proportional to the probable global importance of different sources of smectite to sediments, but in any one instance, one or other of these sources, including the more exotic ones such as metamorphic smectite, may be dominant. (Modified from Chamley 1989)

Makran in the north, and Arabia and Somalia in the west. The illite-rich sources from the west are associated with palygorskite and have been transported by wind as far as the Indian margin. Although some chlorite is introduced from the Indus, most of it appears to be derived by aeolian transport from northerly sources. The sediments of the deep-sea Indus fan appear to be a mixture of all of these sources. Kaolinite-rich sediments are restricted to a small area off southern India and to a belt across the very south of the region.

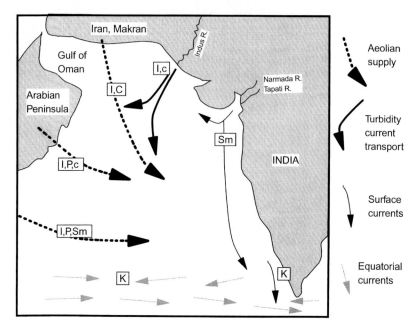

Fig. 4.28. Sedimentary dispersal processes and provenance of recent clay deposits in the Arabian Sea. Based on interpretation of Kolla et al. (1981) as illustrated by Chamley (1989)

It appears that they are derived from tropical soils in Africa and Madagascar and from southern India, their dispersal being related to the prevailing equatorial ocean circulation.

4.6.2 Palaeoclimatic Interpretation of Clay Minerals

The palaeoclimatic interpretation of sedimentary clay-mineral assemblages is based on the correlation of climate with weathering and soil formation, and hence with the clay minerals which these processes produce. This correlation is clearly shown by the changing clay mineralogy of the worlds zonal soils (Chap. 3; Sect. 3.2.3), and evidence of its affect on sedimentary clay-mineral assemblages is found in the broad latitudinal distribution of clay minerals in the world ocean. However, as emphasised by Singer (1984) the palaeoclimatic interpretation of clay minerals in sediments is anything but straight forward and needs to be undertaken with the utmost care. This is because in addition to climate there are many other factors which affect the clay mineralogy of sediments and, acting in isolation, or in combination, they may completely obliterate any climatic signal (Fig. 4.29).

As indicated in Chapter 3, weathering, soil formation, and the resulting clay minerals are not simply dependent on climate. The time that is available for weathering to proceed, parent materials, and topography, may all affect the kinds of soil profiles and clay minerals that may be produced (Fig. 4.29). The length of time over which weathering may occur, depends on tectonic and geomorphologic stability, and these periods may be shorter or longer than periods of climatic change. Furthermore, in certain climatic regimes weathering may be too weak or too strong to register certain climatic changes. Secondly, the clay minerals released to the sedimentary system are not only a function of weathering and soil formation, but of parent materials (Fig. 4.29). This is important in terms of the parent material's response to weathering, but even more so because of the possibility that the parent materials themselves may act as significant sources of clay minerals to the sedimentary system. Such clay minerals bear no relation to the prevailing climate. Thirdly, between the formation and the final deposition of clay minerals in a sediment, the processes of erosion and transport intervene (Fig. 4.29). Both erosion and transport may be selective, and furthermore transport distances may be so large that clay minerals are transported from one climatic regime to another. Lastly, after clay minerals have been deposited, they may be subjected to post-depositional changes, amongst the most important of which are those that occur during burial diagenesis. Indeed, burial diagenetic change may completely obliterate any palaeoclimatic signal. All of these aspects must be considered when trying to determine if clay-mineral assemblages preserve signals of palaeoclimate and/or palaeoclimatic change. Nevertheless, large sedimentary basins inte-

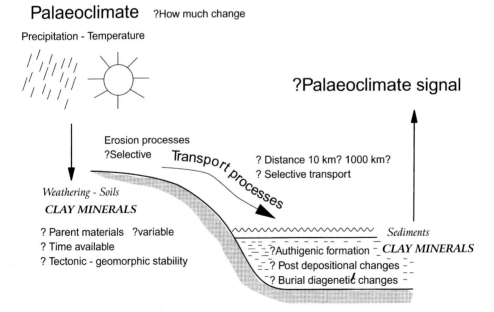

Fig. 4.29. The many processes and factors that must be considered when attempting to extract palaeoclimatic information from the analysis of clay minerals

grate the clay-mineral response of their source areas to climate change, so that in favourable cases climatic signals do appear to be preserved.

According to Chamley (1989), palaeoclimatic investigations in oceanic sediments are likely to be more favourable with increasing distance from land, and in sequences that have not been buried more than 2.5 km. These are favourable due to the integration effect of distance on different sources, and to the need to avoid the effects of burial diagenesis, respectively. Furthermore, the effects of climate change are typically more pronounced in mid-latitudes. To combat the problem of diverse origins of clay minerals, Chamley (1989) advocates the use of changes in relative abundance of clay minerals rather than absolute abundances. As well as direct effects of climate on clay assemblages, climate changes may also be recorded indirectly through the affects that changes in the intensity of rainfall, ice cover, sea level, and marine currents, all may have on clay-mineral source areas and dispersal. Clay minerals alone rarely provide primary evidence of climate change and the best approach is the interdisciplinary one where evidence from clays is compared and correlated with that from the likes of microfossils and geochemical data, such as oxygen isotopes (Singer 1984; Chamley 1989).

Clearly, the application of clay mineralogy to palaeoclimate analysis must be based on a thorough understanding of weathering and its clay-mineral products. So far, in practice, most studies use very simplified schemes wherein abundant illite, and chlorite are taken to relate to cold or arid climates, kaolinite to more humid climates, and smectite to seasonal climates with a pronounced dry season. Indeed, more detailed schemes may not be warranted given the many factors which may swamp any climatic information (Fig. 4.29). However, even if climatic information is obscured, clay minerals can be characterised in considerable detail by modern mineralogical techniques, and such characterisation may provide much more data concerning their origin than the above generalisations. An early attempt at such an approach is the use of crystallinity data in Quaternary hemipelagic muds of the Mediterranean sea by Chamley (1971). Cold dry periods corresponding to glacial intervals are characterised by relatively more abundant and better crystallised illite and chlorite. Warmer and wetter periods, corresponding to interglacials, are characterised by more abundant and better crystallised smectite and more abundant kaolinite. A more recent example is the work of Fagel et al. (1992) where the palygorskite/illite ratio in sediments from the Arabian Sea has been investigated by various signal processing methods and related to periods in the Earth's orbital cycles. Increasingly, there have also been many attempts to better characterise minerals such as smectites which show a wide range of chemical composition that may be related to specific origins. However, work on oceanic smectites shows that there can be many causes of apparently significant compositional variation, including what might be dismissed by some as 'minor' contamination, and it is clear that a careful approach to characterisation of these kinds of materials is necessary (Vali et al. 1993).

Various studies of Cainozoic sediments from the western Atlantic region appear to provide an impressive example of the widespread nature of climate induced changes in clay-mineral assemblages. Studies from the North Sea (Karlsson et al. 1978), the northern Bay of Biscay (Chamley 1979; Debrabant et al. 1979), the Sierra Leone Rise (Robert 1982), and the Goban Spur (Chennaux et al. 1985), although different in detail, all show the same general trends in their clay-mineral assemblages. Essentially, Palaeocene and Eocene sediments are dominated by smectite, while chlorite appears in the Oligocene or Miocene and persists through Pliocene to recent sediments, accompanied by ever increasing amounts of illite. These trends are thought to correspond to the general cooling of world climate throughout the Cainozoic, and to increased relief in the source areas due to Alpine tectonics, whereby the importance of chemical weathering diminished and that of physical weathering increased. However, the changes may in fact reflect some combination of climatic control and of tectonics on oceanic circulation. From mid-Miocene times onwards, subsidence of the Iceland Faroe ridge allowed increased southerly flow of cold bottom water from the Norwegian Sea to the North Atlantic, and these waters may have been responsible for some of the observed increase in illite and chlorite (Chennaux et al. 1985). Nonetheless, comparable trends in Tertiary clay mineralogy are also observed in areas of other oceans, such as the Antarctic (Ehrmann et al. 1992), and most investigators seem to agree that the overriding control is climate change, even though it may frequently be expressed indirectly by related changes in sediment-transport processes.

Palaeoclimate appears also to have left its record in certain marine Mesozoic rocks. For example, generally speaking, Cretaceous sediments of the Atlantic and Tethyan domains are relatively rich in smectite. This is believed to be due to the warm seasonal climates and general tectonic stability of this period, smectite being supplied from the continents (Chamley 1989). However, an alternative hypothesis has been put forward by Thiry and Jaquin (1993) which involves transformation of other minerals into smectite at the sediment–water interface and during early diagenesis while there is still free exchange of pore water with sea water. In addition, Jurassic sediments from the same region are frequently enriched in kaolinite which is related to the more humid climatic conditions of this period (Chamley 1989). Palaeoclimatic interpretations of older rocks become increasing difficult because of the greater chance that they have been altered by burial diagenesis.

Besides climate, changes in palaeocirculation and tectonics are amongst the most common factors that may produce changes in the clay-mineral assemblages of marine sediments. According to Chamley (1989), changes in circulation of ocean currents affect clay-mineral assemblages to the same order of magnitude as those resulting from climate change. In contrast, changes due to tectonic effects on clay-mineral supply are potentially much more significant and longer lived than either the effects of climate or circulation. Tectonic rejuvenation causes soil sources of clay minerals to be replaced by rock sources. Therefore, the effects are most pronounced when tectonics affects regions where soils have developed under conditions of intense chemical weathering such as in humid tropical climates.

A detailed compilation of changes in clay mineralogy through geological time is given by Weaver (1989). The main trend is an increase in smectite and decrease in illite from Cambrian to Recent times. Weaver (1989) discusses how this trend may be related to changes in weathering through time, or to the increased chance that older sediments have been affected by burial diagenesis. Weaver (1989) also discusses the interpretation of the data for various geological periods in terms of changes in climate and global tectonics.

It is evident, as will be seen in the following chapter, that as sediment becomes rock during burial, mineral changes occur which have a tendency to obscure the chemical and environmental orgin of sedimentary clays. If diagenesis occurs, it must strongly affect the mineral assemblages of clays in sedimentary rocks. Hence, the signature of sedimentary provenance will be restricted to superfical and young areas of the sedimentary pile.

References

Algan O, Clayton T, Tranter M, Collins MB (1994) Estuarine mixing of clay minerals in the Solent region, southern England. Sediment Geol 92: 241–255

Allison MA, Riggs SR (1994) Clay-mineral suites in cyclic Miocene sediments: a model from deposition in a mixed silicicalstic-phosphatic-dolomitic-biogenic system. J Sediment Petrol A64: 386–395

Arnold PW (1978) Surface-electrolyte interactions. In: Greenland DJ, Hayes MHB (eds) The chemistry of soil constituents. John Wiley, New York, pp 355–404

Avoine J (1987) Sediment exchanges between the Seine estuary and its adjacent shelf. J Geol Soc Lond 144: 135–148

Bagnold RA (1941) The physics of blown sand and desert dunes. Menthuen, London, 265 pp

Bailey SW (1988) Odinite: a new dioctahedral-trioctahedral Fe^{3+}-rich 1:1 clay mineral. Clay Min 23: 237–247

Baker RA (1980) Contaminants and sediments vols 1, 2. Ann Arbor Science Publishers, Ann Arbor

Barnes PW, Reimnitz E, Fox D (1982) Ice rafting of fine-grained sediment, a sorting and transport mechanism, Beaufort Sea, Alaska. J Sediment Petrol 52: 493–502

Bentor YK (1980) Phosphorites–the unsolved problems. In: Bentor YK (ed) Marine phosphorites–geochemistry, occurrence, genesis. SEPM Spec Publ 29, Tulsa, DK, pp 3–18

Berger WH (1974) Deep-sea sedimentation. In: Burk CA, Drake CL (eds) The geology of continental margins. Springer, Berlin Heidelberg New York, pp 213–241

Biscaye PE (1965) Mineralogy and sedimentation of recent deep-sea clay in the Atlantic Ocean and adjacent seas and oceans. Geol Soc Am Bull 76: 803–831

Biscaye PE, Eittreim SL (1977) Suspended particulate loads and transports in the nepheloid layer of the abyssal Atlantic ocean. Mar Geol 23: 155–172

Blatt H, Middleton GV, Murray RC (1980) Origin of sedimentary rocks. Prentice Hallm Englewood Cliffs

Callen RA (1984) Clays of the palygorskite-sepiolite group. Depositional environment, age and distribution. In: Singer A, Galan E (eds) Palygorskite-sepiolite, occurrence, genesis and uses. Developments in Sedimentology 37, Elsevier, Amsterdam pp 1–37

Catt JA (1988) Loess-its formation transport and economic significance. In: Lerman A, Meybeck M (eds) NATO ASI Series C. Mathematical and physical sciences. V251, Kluwer, Dordrecht, pp 113-142

Chamley H (1971) Recherches sur la sédimentation argileuse en Méditerranée. Sci Géol Strasbourg Mém 35, 225 pp

Chamley H (1979) North Atlantic clay sedimentation and paleoenvironment since the late Jurassic. In: Talwani M, Hay W, Ryan WBF (eds) Deep drilling results in the Atlantic Ocean. Continental margins and paleoenvironment. Maurice Ewing Series, vol 3. American Geophysical Union, Washington, DC, pp 342-361

Chamley H (1989) Clay sedimentology. Springer, Berlin. Heidelberg New York, 623 pp

Chamley H (1993) La sédimentation marine des minéraux argileux. In: Paquet H, Clauer N (eds) Sédimentologie et géochimie de la surface. Colloque: la mémoire de George Millot. Les colloques de l' Académie des Science et du Cadas. Institut de France, Paris, pp 217-241

Chennaux G, Esquevin J, Jourdan A, Latouche C, Maillet N (1985) X-ray mineralogy and mineral geochemistry of Cenozoic strata (Leg 80) and petrographic study of associated pebbles. DSDP Leg 80. Initial reports of the Deep Sea Drilling Project 80. US Government Printing Office, Washington, pp 1019-1046

Cheshire MV, Shand C (1991) Translocation and plant availability of radio caesium in an organic soil. Plant Soil 134: 287-296

Coote DR, Dumanski J, Ramsey JF (1981) An assessment of the degradation of agricultural lands in Canada. Agricultural Land Resource Institute, Contribution 118

Curtis CD (1990) Aspects of the climatic influence on the clay mineralogy and geochemistry of soils, palaeosoils, and clastic sedimentary rocks. J Geol Soc Lond 147: 351-357

Darragi F, Tardy Y (1987) Authigenic trioctahedral smectites controlling pH, alkalinity, silica and magnesium concentrations in alkaline lakes. Chem Geol 63: 59-72

Davies TA, Gorsline DS (1976) Oceanic sediments and sedimentary processes. In: Riley JP, Chester R (eds) Chemical oceanography, vol 5, 2nd edn. Academic Press, London, pp 1-80

Debrabant P, Chamley H, Foulon J, Maillot H (1979) Mineralogy and geochemistry of upper Cretaceous and Cenozoic sediments from the North Biscay Bay and Rockall Plateau (eastern North Atlantic), DSDP Leg 48. Initial reports of the Deep Sea Drilling Project 48. US Government Printing Office, Washington, pp 703-725

Deconinck JF, Strasser A, Debrabant P (1988) Formation of illitic minerals at surface temperatures in Purbeckian sediments (Lower Berriasian, Swiss and French Jura). Clay Min 23: 91-103

Droste JB (1961) Clay minerals in the playa sediments of the Mojave desert, California. Special report 69, California Division of Mines. Ferry Building, San Francisco

Dyni JR (1976) Trioctahedral smectite in the Green River Formation, Duchesne County, Utah. US Geol Surv Prof Pap 967: 14

Eberl DD (1984) Clay mineral formation and transformation in rocks and soils. Philos Trans R Soc Lond A 311: 241-257

Eberl DD, Jones BF, Khoury HN (1982) Mixed-layer kerolite/stevensite from the Amargosa desert, Nevada. Clays Clay Min 30: 321-326

Eberl DD, Srodon J, Northrop HR (1986) Potassium fixation in smectite by wetting and drying. In: Davies JA, Hayes KF (eds) Geochemical processes at mineral surfaces. ACS Symp Ser 323/14: 296-326

Edwards M (1986) Glacial environments. In: Reading HG (ed) Sedimentary environments and facies. 2nd edn. Blackwell, Oxford, pp 445-470

Ehrmann WU, Melles M, Kuhn G, Grobe H (1992) Significance of clay mineral assemblages in the Antarctic Ocean. Mar Geol 107: 249-273

Eisma D (1986) Flocculation and deflocculation of suspended matter in estuaries. Neth J Sea Res 20: 183-199

Fagel N, Debrabant P, de Menocal P, Demoulin B (1992) Utilisation des minéraux sédimentaires argileux pour la reconstitution des variations paléoclimatiques a court terme en Mer d'Arabie. Oceanol Acta 15: 125-136

Faugères J-C, Stow DAV (1993) Bottom-current-controlled sedimentation: a synthesis of the contourite problem. Sediment Geol 82: 287–297

Förstner U, Ahlf W, Calmano W, Kersten M (1990) Sediment criteria development. Contributions from environmental geochemistry to water quality management. In: Heling D, Rothe P, Förstner U, Stoffers P (eds) Sediments and environmental geochemistry: selected aspects and case histories. Springer, Berlin Heidelberg, New York, pp 311–338

Froelich PN (1988) Kinetic control of dissolved phosphate in natural rivers and estuaries: a primer on the phosphate buffer mechanism. Limnol Oceanogr 33: 649–668

Garrels RM, Mackenzie FT (1971) Evolution of sedimentary rocks. WW Norton, New York, 397 pp

Gibbs RJ (1967) The geochemistry of the Amazon River system. Part I. The factors which control the salinity and the composition and concentration of the suspended solids. Geol Soc Am Bull 78: 1203–1232

Gibbs RJ (1977) Clay mineral segregation in the marine environment. J Sediment Petrol 47: 237–243

Gibbs RJ (1983) Coagulation rates of clay minerals and natural sediments. J Sediment Petrol 53: 1193–1203

Glasby GP (1991) Mineralogy geochemistry and origin of Pacific red clays: a review. N Z J Geol Geophys 34: 167–176

Goldschmidt PM, Pfirman SL, Wollenburg I, Henrich R (1992) Origin of sediment pellets from the Arctic seafloor: sea ice or icebergs? Deep Sea Res 39: 539–565

Gorsline DS (1984) A review of fine-grained sediment origins, characteristics, transport and deposition. In: Stow DAV, Piper DJW (eds) Fine grained sediments: deepwater processes and facies. Geological Society Special Publication No 15, London

Goudie AS (1983) Dust storms in space and time. Prog Phys Geog 7: 502–529

Griffin JJ, Windom H, Goldberg ED (1968) The distribution of clay minerals in the world ocean. Deep Sea Res 15: 433–459

Grim RE (1968) Clay mineralogy, 2nd edn. McGraw-Hill, New York, 596 pp

Hahn HH, Stumm W (1970) The role of coagulation in natural waters. Am J Sci 268: 354–368

Hathaway JC (1972) Regional clay mineral facies in the estuaries and continental margin of the United States east coast. In: Nelson BW (ed) Environmental framework of coastal-plain estuaries. Geol Soc Am Mem 133: 293–316

Helios Rybicka E (1992) Heavy metal partitioning in polluted river and sea sediments: clay mineral effects. Miner Petrogr Acta 25-A: 297–305

Hillier S (1993) Origin, diagenesis, and mineralogy of chlorite minerals in Devonian lacustrine mudrocks, Orcadian Basin, Scotland. Clays Clay Min 41: 240–259

Hillier S (1994) Pore-lining chlorites in siliciclastic reservoir sandstones: electron microprobe, SEM, and XRD data, and implications for their origin. Clay Min 29: 665–679

Irion G, Wunderlich F, Scheldhelm E (1987) Transport of clay minerals and anthropogenic compounds into the German Bight and the provenance of fine grained sediments SE of Helgoland. J Geol Soc Lond 144: 153–160

Jannson MB (1988) A global survey of sediment yield. Geogr Ann 70: 81–98

Jeans CV, Mitchell JG, Scherer M, Fischer MJ (1994) Origin of the Permo-Triassic clay mica assemblage. Clay Min 29: 575–589

Jones BF (1986) Clay mineral diagenesis in lacustrine sediments. In: Mumpton FA (ed) Studies in diagenesis. US Geol Surv Bull 1578: 291–300

Jones BF, Galan E (1988) Sepiolite and palygorskite. In: Bailey SW (ed) Hydrous phyllosilicates (exclusive of micas). Reviews in mineralogy, 19. Mineralogical Society of America, Washington, DC

Jones BF, Weir AH (1983) Clay minerals of Lake Abert, an alkaline saline lake. Clays Clay Min 31: 161–172

Karlsson W, Vollset J, Bjørlykke, Jørgensen P (1978) Changes in mineralogical composition of Tertiary sediments from the North Sea. In: Mortland MM, Farmer VC (eds) International Clay Conference 1978. Developments in sedimentology 27. Elsevier, Amsterdam, pp 281–289

Kennedy VC (1965) Mineralogy and cation exchange capacity of sediments from selected streams. US Geol Surv Prof Pap 443-D

Kirby R (1987) Sediment exchanges across the coastal margins of NW Europe. J Geol Soc Lond 144: 121–126

Kolla V, Kostecki JA, Robinson F, Biscaye PE, Ray PK (1981) Distribution and origins of clay minerals and quartz in surface sediments of the Arabian Sea. J Sediment Petrol 51: 563–569

Kuhlemann J, Lange H, Paetsch (1993) Implications of a connection between clay mineral variations and coarse grained debris and lithology in the central Norwegian-Greenland Sea. Mar Geol 114: 1–11

Kühnel RA (1992) Clays and clay minerals in environmental research. Miner Petrogr Acta 25: 1–11

Lemoalle J, Dupont B (1973) Iron bearing oolites and the present conditions of iron sedimentation in Lake Chad (Africa). In: Amstutz G, Bernard AJ (eds) Ores in sediments. Springer, Berlin Heidelberg New York, pp 167–178

Lerman A (1979) Geochemical processes water and sediment environments. John Wiley, New York, 481 pp

Malle K-G (1990) The pollution of the river Rhine with heavy metals. In: Heling D, Rothe P, Förstner U, Stoffers P (eds) Sediments and environmental geochemistry; selected aspects and case histories. Springer, Berlin Heidelberg New York, pp 279–290

Martin de Vidales JL, Pozo M, Alia JM, Garcia-Navarro F, Rull F (1991) Kerolite-stevensite mixed-layers from the Madrid Basin, central Spain. Clay Min 26: 329–342

McCarthy JF, Degueldre C (1993) Sampling and characterisation of colloids and particles in groundwater for studying their role in contaminant transport. In: Buffle J, van Leeuwen HP (eds) Environmental particles, vol 2. IUPAC, Lewis, Boca Raton

McCave IN (1972) Transport and escape of fine-grained sediment from shelf areas. In: Swift DJP, Duane DB, Pilkey OH (eds) Shelf sediment transport. Dowden, Hutchinson and Ross, Stroudsboug, pp 225–248

McCave IN (1984) Erosion transport and deposition of fine grained marine sediments. In: Stow DAV, Piper DJW (eds) Fine-grained sediments deep water processes and facies. Geol Soc Lond Spec Publ 15: 69

McCave IN (1985) Recent shelf clastic sediments. In: Brenchly PJ, Williams BPJ (eds) Sedimentology: recent developments and applied aspects. Blackwell, Oxford, pp 49–65

McCave IN (1986) Local and global aspects of the bottom nepheloid layers in the world ocean. Neth J Sea Res 20: 167–181

McMurtry GM, Wang CH, Yeh HW (1983) Chemical and isotopic investigations into the origin of clay minerals from the Galapagos hydrothermal mounds field. Geochim Cosmochim Acta 47: 475–489

Meade RH (1972) Transport and deposition of sediments in estuaries. Geol Soc Am Mem 133: 91–120

Meade RH (1988) Movement and storage of sediment in river systems. In: Lerman A, Meybeck M (eds) Physical and chemical weathering in geochemical cycles. NATO ASI series C. Mathematical and physical sciences, vol 251, Kluwer, Dordrecht, pp 165–180

Meade RH, Parker RS (1985) Sediment in rivers of the United States. US Geol Surv Water Supply Pap 2275: 49–60

Meybeck M (1987) Global chemical weathering of surficial rocks estimated from river dissolved loads. Am J Sci 287: 401–428

Milliman JD, Meade RH (1983) World wide delivery of river sediments to the oceans. J Geol 91: 1–21

Millot G. (1970) The geology of clays. Masson, Paris.

Millot (1978) Clay genesis. In: Fairbridge RH, Bourgeois J (eds) The encyclopedia of sedimentology, vol 6. Dowden, Hutchinson and Ross, Stroudsburg, pp 152–155

Müller G, Förstner U (1973) Recent iron ore formation in Lake Malawi Africa. Miner Depos 8: 278–290

Nickling WG (1994) Aeolian sediment transport and deposition. In: Pye K (ed) Sediment transport and depositional processes. Blackwell, Oxford, pp 293–350

Norrish K, Pickering JG (1983) Clay minerals. In: Soils – an Australian viewpoint. Division of Soils CSIRO, Australia, pp 281–308

Nürnberg D, Wollenberg I, Dethle D, Eicken H, Kassens H, Letzig T, Reimnitz E, Thiede J (1994) Sediments in Arctic sea ice: implications for entrainment, transport and release. Mar Geol 119: 185–214

Odin GS (ed) (1988) Green marine clays. Developments in sedimentology, vol 45. Elsevier, Amsterdam, 445 pp

Odin GS (1990) Clay mineral formation at the continent ocean boundary: the verdine facies. Clay Min 25: 477–483

Odin GS, Matter A (1981) De glauconarium origine. Sedimentology 28: 611–641

Odom IE (1984) Glauconite and celadonite minerals. In: Bailey SW (ed) Micas. Reviews in mineralogy 13. Mineralogical Society of America, Washington, DC, pp 545–571

Oglesby RT, Bouldin DR (1984) Phosphorous in the environment. In: Nriagu JO, Moore PB (eds) Phosphate minerals. Springer, Berlin Heidelberg New York, pp 400–423

O'Melia CR, Tiller CL (1993) Physicochemical aggregation and deposition in aquatic environments. In: Buffle J, van Leeuwen HP (eds) Environmental particles. Environmental analytical and physical chemistry series, vol 2, IUPAC. Lewis, Boca Raton, pp 353–386

Peaver DR (1972) Sources of nearshore marine clays, southeastern United States. In: Nelson BW (ed) Environmental framework of coastal-plain estuaries. Geol Soc Am Mem 133: 317–335

Péwé TL (1981) Desert dust: an overview. Geol Soc Am Spec Pap 186: 1–10

Porrenga DH (1967) Clay mineralogy and geochemistry of recent marine sediments in tropical areas. Publ Fysisch-Geographisch Lab Univ Dort Stolk Amsterdam, 9, 145 pp

Porrenga DH (1968) Non-marine glauconitic illite in the lower Oligocene of Aardebrug, Belgium. Clay Min 7: 421–430

Potter PE, Maynard JB, Pryor WA (1980) Sedimentology of shale: study guide and reference source. Springer, Berlin Heidelberg New York

Prospero JM (1981) Eolian transport to the world ocean, Chap 21. In: Emiliani C (ed) The sea, vol 7. Academic Press, New York, pp 801–874

Pye K (1987) Aeolian dust and dust deposits. Academic Press, London, 334 pp

Rateev MA, Gorbunova ZN, Lisitzyn AP, Nosov GL (1969) The distribution of clay minerals in the oceans. Sedimentology 13: 21–43

Reid I, Frostick LE (1994) Fluvial sediment transport and deposition. In: Pye K (ed) Sediment transport and depositional processes. Blackwell, Oxford, pp 89–155

Robert C (1982) Modalité de la sédimentation argileuse en relation avec l'histoire géologique de l'Atlantic Sud. Thesis, Univ Aix-Marseille II, 141 pp

Sayles FL, Manngglesdorf PC Jr (1977) The equilibration of clay minerals with seawater: exchange reactions. Geochim Cosmochim Acta 41: 951–960

Schindler PW (1991) The regulation of heavy metal concentration in natural aquatic systems. In: Vernet J-P (ed) Heavy metals in the environment. Elsevier, Amsterdam, pp 95–124

Selby MJ (1994) Hillslope sediment transport and deposition. In: Pye K (ed) Sediment transport and depositional processes. Blackwell, London, pp 61–87

Shand CA, Cheshire MV, Smith S, Vidal M, Rauret G (1994) Distribution of radiocaesium in organic soils. J Environ Radioactivity 23: 285–302

Singer A (1984) The palaeoclimatic interpretation of clay minerals in sediments – a review. Earth Sci Rev 21: 251–293

Srodon J, Eberl DD (1984) Illite. In: Bailey SW (ed) Micas. Reviews in mineralogy 13. Mineralogical Society of America, Washington, DC, pp 495–544

Stoddart DR (1971) World erosion and sedimentation. In: Chorley RJ (ed) Introduction to fluvial processes. Methuen, London, pp 8–29

Stow DAV (1994) Deep sea sediment transport. In: Pye K (ed) Sediment transport and depositional processes. Blackwell, Oxford, pp 257–291

Strakhov NM (1967) Principles of lithogenesis. Oliver and Boyd Edingburgh

Stumm W (1992) The chemistry of the solid water interface: processes at the mineral water and particle water interface. John Wiley, New York, 440 pp

Tettenhorst R, Morre GE Jr (1978) Stevensite ooolites from the Green River Formation of central Utah. J Sediment Petrol 48: 587–594
Thiry M, Jacquin T (1993) Clay mineral distribution related to rift activity, sea level changes an paleogeography in the Cretaceous of the Atlantic Ocean. Clay Min 28: 61–84
Trauth N (1977) Argiles évaporitic dans la sédimentation carbonatée continental et épicontinentale tertiaire. Bassins de Paris, de Mormoiron et de Salinelles (France), Jbel Ghassoul (Maroc). Sci Géol Strasbourg Mem 49, 203 pp
Tsoar H, Pye K (1987) Dust transport and the question of desert loess formation. Sedimentology 34: 139–153
Vali H, Martin RF, Amarantidis G, Morteani G (1993) Smectite-group minerals in deep-sea sediments: monomineralic solid-solutions or multiphase mixtures? Am Mineral 78: 1217-1229
van Olphen H (1977) An introduction to clay colloidal chemistry. John Wiley, New York, 318 pp
Weaver CE (1989) Clays, muds and shales. Developments in sedimentology 44. Elsevier, Amsterdam, 819 pp
Weaver CE, Beck KC (1977) Miocene of the S.E., United States: a model for chemical sedimentation in a peri-marine environment. Sediment Geol 17: 234
Wells JT, Coleman JM (1981) Physical processes and fine-grained sediment dynamics, coast of Surinam, South America. J Sediment Petrol 51: 1053–1068
Windom HL (1969) Atmospheric dust in permanent snow fields; implications to marine sedimentation. Bull Geol Soc Am 80: 761–82
Windom HL (1975) Eolian contributions to marine sediments. J Sediment Petrol 45: 520–529
Windom HL (1976) Lithogenous material in marine sediments. In: Riley JP, Chester R (eds) Chemical oceanography, vol 5. Academic Press, New York, pp 103–135
Yariv S, Cross A (1979) Geochemistry of colloid systems for earth scientists. Springer, Berlin Heidelberg New York, 450 pp

5 Compaction and Diagenesis

B. Velde

This chapter follows logically that dealing with the origins of clay in the sedimentation environment. Sedimentation followed by more sedimentation leads to diagenesis, burial diagenesis which is the most important diagenesis for silicates. What is the nature of the change in state which brings about new minerals upon their burial? As sediments accumulate in a basin (the basic concept of a sedimentary basin being a recipient that has a bottom which continues to grow deeper with time, a sort of un-filling cup; otherwise the recipient would fill up and the story would end) they are subjected to two major changes in their environment. First as burial proceeds, sediment temperature increases. As any miner knows, deep mines are hotter than shallow ones, and in fact the Earth is hotter inside than at its skin (i.e. the solid gas interface known as the Earth's surface). So as sediments get buried in basins, or on the edges in continents, they get hotter.

The second effect of burial is the expulsion of water. Sedimentary deposition includes much of the sedimentation environment, water. The portion of water included in sands is almost 30% but that of clay-rich sediment is about 80%! This water has to be eliminated if one is to produce a rock, a *sedimentary* rock. The water will escape, upwards for the most part, to join its milieu of origin, leaving behind the solids as sediments and the salts in a concentrated form in interstitial water.

Therefore two major changes are accomplished with the burial of sediments; temperature changes and water is eliminated. In the case of clay-rich sediments this produces a silicate dominated system. Instead of a water-dominated system one finds a clay or silicate-dominated one. The chemical potentials which govern the changes in mineral stability are dictated by the silicates present, not the water or fluids. The rock becomes the master rather than water. The sequence of decreasing water influence is from weathering, to sedimentation, to burial diagenesis.

There are limits to the change in temperature with depth that should be considered in terms of burial diagenesis. The normal geothermal gradients of sedimentary basins are in the range of 20–50 °C/km depth. Above these values one should consider using the terms metamorphism or geothermal conditions. This rate of heating is greater than that found in most sedimentary basins over the period of their development, which is on the scale of millions of years. Higher gradients can be experienced in basins, but they are likely to be of shorter duration. If a heating period is of less than one or perhaps two million years duration, it can probably be considered as a short episode, one of "geothermal" heating. These periods can have an effect on clays if they reach high enough temperatures, above 100 °C, but they appear to affect clays less when they operate for short periods. In contrast, organic material will react at a much greater rate under short heating periods and where these materials are present they can give an idea of maximum temperatures attained, even over very short geological periods. Clay minerals then generally give an indication of an integration of time and temperature.

5.1 The Geologic Structure of Diagenesis

The geologic situation where diagenesis occurs is as stated above, in basins which are continuously deepening and filling with sediments. Not all ocean or sea bodies are sites of active deposition, even though they present one of the characteristics necessary, i.e. depth. For example, in the Mediterranean Sea, on one side of the Italian peninsula there is an active diagenetic system (the Adriatic Sea) where recent sediments from the Alps fill a deep trough of many kilometers depth. The sea is shallow, less than 1000 m depth for the most part. However on the western side of the same peninsula, the ocean floor is at 5000 m depth, but at present sedimentation is low and burial of sediments does not produce thick sequences of sediments which turn into rocks. Hence, it is not enough to have a recipient for the sediments in order to form a sedimentary sequence, one must also have a source of sediment. The Po river in northern Italy furnishes much of the eroded material from the Alps to the Adriatic Sea basin, whereas the Rhone river in France does not have such a source area. In the same way, not all of the off-shore continental shelf areas with deep and subsiding floors receive sediment to make a sedimentary basin. Sedimentary rocks need both a place to be deposited and sediments.

The situation of sedimentation and burial is one of tectonic movement. The continent is uplifted to provide sediment and the sea or ocean is depressed to receive the sediment as it is produced, or perhaps not. Very schematically, one can represent these effects as in Fig. 5.1. This shows the typical passive continental margin setting where the foreland is produced by essentially vertical movement, uplift, and the basin by block displacement. Zones of sedimentation, burial and compaction are found progressively further off-shore. One can change the relative size of blocks and their movements to give different dimensions to represent continental basins and graben situations. These variations are also shown in Fig. 5.2 which indicate the common types of structure where active sedimentation and diagenesis occurs. The mountain or upland areas provide eroded material from basement or old consolidated rocks, the foreshore areas provide sedimentary material from recently com-

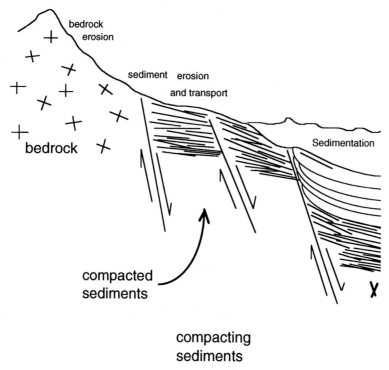

Fig. 5.1. Drawing of a typical passive continental margin sedimentation setting where source area and sedimentation are directly linked

pacted sedimentary rocks, and the sea, or basin, is an accretive area of sedimentation and compaction.

5.1.1 Compaction and Porosity

Clays have a very special behaviour in water dominated solutions, and this is the key to their physical state upon sedimentation. This behaviour is very well explained in detail by R.M. Meade (Geological Professional Paper 467-B, 1964). The clay platelet organisation is the key to the capture and eventual expulsion of water, and the key to the ultimate behaviour and changes in clays during diagenesis (Sect. 4.3.3.2).

Initially, if clays are put into a pure-water medium, they tend to remain in suspension due to their small grain size which holds them aloft by Brownian motion, i.e. thermal agitation. The clays attract water to their surfaces, as do all particles. But, given their large surface area, due to their lath or sheet shaped form, their surface is great. Normally, the small residual charges on the surfaces of clays are concentrated on their grain edges. If clays are observed as they aggregate, they are seen to form loosely linked sponge-like masses con-

Compaction and Diagenesis 223

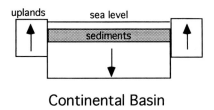

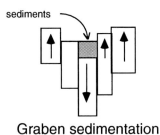

Fig. 5.2. Block diagrams of situations giving rise to different geometries of sedimentary basins: **a** the continental basin configuration; **b** the graben configuration; **c** the continental passive margin configuration. *Arrows* indicate the relative movements of continental blocks

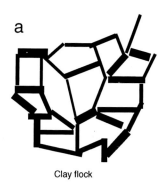

Fig. 5.3. a Clay flock in aqueous suspension; **b** clay particles oriented by sedimentation pressures

nected by their edges. A schematic representation in two dimensions is given in Fig. 5.3. As one can see at a glance, there is not much clay and there is a lot of space which in the case of clays in water is taken up by water. Most of the material in the sedimentation of clays is water, up to 80 or even 90%. Compaction causes the clay platelets to lie flat and, when in contact, they do not retain much water.

The reason that the clays tend to aggregate by their ends, is that there are apparently higher charges there, due to broken or incomplete bonds leaving negative or positive charges at these points. The frequency of encounter at edges is greater than at the parallel surfaces of the plates where the charge is less. Once the distance barrier is overcome, there is a link of negative-charge–cation–negative-charge between the two plates, being bridged by a cation, or more rarely an anion. This has been discussed in detail in Section 4.3.3.2.

As the sediment is buried, the open structure of the flocculate is gradually flattened to form a more or less, parallel oriented clay sediment, as indicated in Fig. 5.3b. Compaction brings about loss of water content and increases the density of the clay-rich sediment. The change in density is large, as one can imagine. The porosity, i.e. the portion of the sediment that is not mineral, changes drastically. While clay-rich sediments can have up to 80% porosity or non-solid content, the porosity of sands is generally below 50% percent upon their initial deposition. Therefore, clay sediments tend to lose more of their volume upon compaction under sedimentary burial conditions than do other sediment types. Sand grains compact in a different way, leaving in the end much more pore space between the grains than do clay sediments. Figure 5.4 shows sand compaction compared to Fig. 5.3 for that of clay compaction.

The result of these differences in geometry of sediment particles is shown by the typical sedimentation curves, shown as porosity (percent of non-solids) of sediments against depth of burial, which are shown in Fig. 5.5. The initial stages of compaction of clay-rich sediments occur in the first kilometer of burial. This is also the zone where large changes of water chemistry take place and certain adjustments of clay mineralogy are seen.

In a system where 80% of the volume is water, or more precisely aqueous solution, the chemical dominance will come from the solution. The activity of ions such as Na, K, Ca will be determined to a large extent by the values of the initial sea-water solution. However, as the solutions decrease in proportion due to compaction and de-watering the solids become more important in mass compared to the solutions, and they will impose their chemical activity on the solution-solid mixture of the sediment by an effect of mass balance. Hence, the

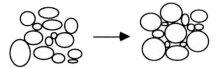

Fig. 5.4. Compaction of sand grains

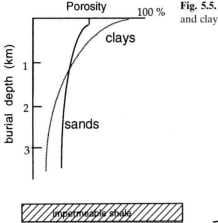

Fig. 5.5. Typical compaction (porosity) curves for sands and clay-rich (shales) sediments

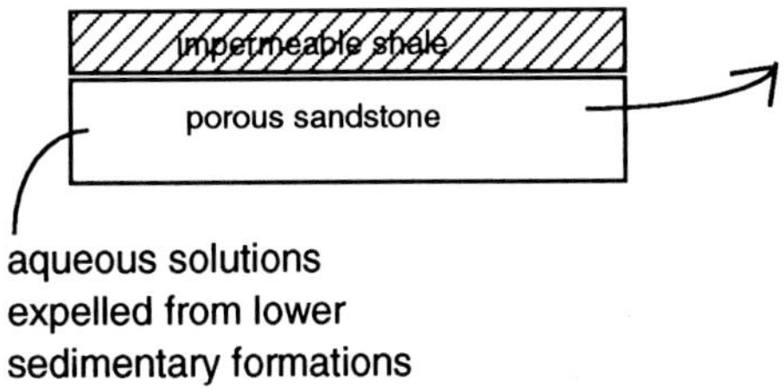

Fig. 5.6. Effect of compaction on aqueous flow in contrasted sand and clay-rich sediment layers

sediments become, more and more, a closed system in which the chemical activity of the solutions is determined by the bulk chemistry of the solids. In clay-rich sediments, the permeability becomes low, porosity decreases to only 10–20%, and the sediments do not "see" the solutions that, for example, circulate in sands. One can find for example, sandstones at say 2 km depth with a porosity of 30%, whereas claystones which are adjacent will have 10–12% porosity. This situation dominates the mineralogical changes which claystones experience. Figure 5.6 illustrates this situation. The difference in compactibility gives very different properties to the rocks with respect to fluid flow. The porous sands allow fluids to migrate within them, whereas the more impermeable clay-rich rocks act as barriers to fluids. The different proportions of fluids means that the minerals present in a clay-rich layer will change phase and composition according to the bulk composition of the sediment. Each layer will behave in an independent way. Little outside influence will affect the minerals and hence only the variables of time and temperature will govern the mineralogy that we see in a given sample. In all sediments, the initial input is, to a certain extent, heterogeneous and many crystals are not in chemical or mineralogical equilibrium with their neighbours upon deposition. Some old micas and biotites, for example, will remain present when they should be a trioctahedral smectite or vermiculite. Some plagioclase will be present when it should be either a zeolite or an albite. These mineral species inconsistencies will be sorted out as the temperature of the sediments increases.

A second, and more important, aspect of the difference in porosity and permeability is related to petroleum migration and accumulation. Fluids migrate upward because they are less dense than rocks. They will migrate until they reach a barrier, such as a clay-rich rock. If there is sufficient sealing capacity in three dimensions, hydrocarbons will migrate and accumulate. Most often clay layers are the barrier to further migration and the principal cause of hydrocarbon accumulation.

5.1.2 Temperature

In different geological settings, i.e. continent edges, subduction zones, pull-apart basins, etc., the thermal regime is different. The thermal input into the sedimentary pile can vary to a great extent, affecting different temperatures at the same burial depths. This is the thermal or temperature variable which affects diagenetic reactions. In continental-margin basins, the thermal input commonly produces a thermal gradient of about 25–35 °C/km. In basins formed in mid-continent regions one can often find gradients of 30–40 °C/km, whereas basins on continent edges have lower gradients. In rift areas, where block faulting is an important manifestation of the tectonic forces, gradients can be in the 40–60 °C/km range. Local magmatic activity can effect even higher gradients until one reaches geothermal areas where gradients can be above 200 °C/km. Thus sedimentary basins can have very different thermal conditions which effect the same diagenetic processes.

5.1.3 Sedimentation Rate

Upon burial, the sediment layers reach greater depths and their temperature increases. However, the rate at which the sediments increase in temperature is different in each basin of sedimentation. This is due to two effects, one the temperature gradient in the sedimentary pile and another the burial rate. In the Los Angeles, California, area for example, one can find sediments of only 2 million years of age at depths of up to 5 km. In the Paris Basin (France), the basin basement at 2.8 km depth has sediments of 250 million years of age. In both basins, the increase in temperature is similar to depth, about 30 °C/km. Hence, in California, the deepest sediments have been at 180 °C for less than 1 million years, whereas those in the Paris basin have been at or near the maximum of 80 °C for more than 80 million years (Fig. 5.7).

5.1.4 The Kinetics of Clay Transformations

In nature, and the laboratory which is an extension of nature, when physical or chemical conditions change, a disequilibrium is established. The nature of

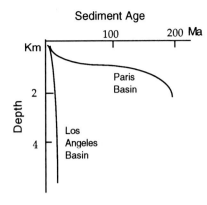

Fig. 5.7. Different types of sedimentation history comparing the Los Angeles Basin (a young and deep basin) and the Paris Basin (an old and shallow basin)

nature is to establish another equilibrium. This creates a need for chemical reaction. The description of the rate at which things try to attain equilibrium is called kinetics.

Normally, chemical reactions, especially silicate reactions, are affected by three factors. These factors are time (t), temperature (T) and composition (x). According to the law of mass action, the higher the concentration of a reactant, the faster the reaction will go to completion. Thus, for a given concentration, there will be a specific rate of change in the reacting phases. An increase in temperature tends to hasten reactions, it brings the reacting phases into equilibrium faster. If enough time is allowed, any reacting assemblage will find another equilibrium assemblage.

The basis for such transformations is often expressed as:

$dC/dt = -kC$, which shows the time (t) relationship, and

$k = A \exp^{(-E/RT)}$, which shows the temperature (T) relationship.

C is the concentration of the phase which is changing to another, A is a constant generally independent of temperature, and k is the reaction constant which relates the time to the temperature formulation.

Clay mineralogy is often controlled by kinetics. In burial diagenesis, it has been found that most pelitic clay-rich rocks show a high concentration of smectite in surface layers and a high concentration of illite in the deeper portions of a sedimentary series. This has been described as the smectite to illite reaction series. The intermediate stages of the reaction are expressed by the appearance of mixed-layer minerals of illite/smectite compositions. The reaction progress can be described by the illite content of the mixed layer mineral. In Fig. 5.8 reaction progress is shown as it has been observed in many sedimentary basins. The figure indicates the relations of the two basic parameters of burial diagenesis, those of temperature and time, as they affect clay-mineral transformation during diagenesis. In the basins which were used to gather the data (Tertiary in Japan, Gulf Coast and Los Angeles in the USA, Paris basin in France), the increase in temperature with depth is, today, be-

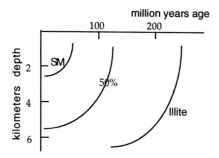

Fig. 5.8. Range of depth–time conditions (thermal gradient is near 30 °C/km) where one finds the critical stages in the smectite to illite conversion in sedimentary basins. *SM* Smectite starting point of the reaction; *50%* illite/smectite mixed layer minerals with half smectite and half illite

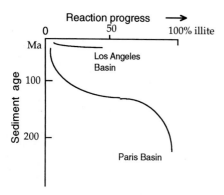

Fig. 5.9. Representation of the clay composition of the smectite to illite reaction (reaction progress is towards illite composition) and the age of sediments in basins with similar thermal gradients but different sedimentation rates, i.e. the same increase in temperature with depth but where the time over which the sedimentation takes place is very different

tween 25 and 35 °C/km. Thus, Fig. 5.8 can be used to show the influence of both time and temperature on the clays in the smectite to illite reaction under comparable thermal conditions. Here time is the most important variable in forming the clay minerals in the reaction. The clay reactions follow the differences in time under the conditions of burial diagenesis.

However, if the thermal gradient in the basin is similar, the time factor will be the most important factor in forming illite. In Fig. 5.9 the smectite to illite reaction has been followed in two basins of very different age but similar geothermal gradient. In the young basin the reaction has not progressed very far, while in the older basin it has gone to near completion. In the young basin (California) the sediments are nevertheless buried to near 5 km depth, at a present-day temperature of near 180 °C. The deepest part of the Paris basin is only buried to 2.5 km depth at a temperature of only 85 °C. Hence, for the clays, at low to moderate temperatures, the reaction progress is highly dependent on time.

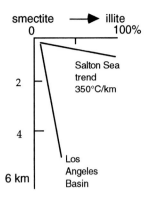

Fig. 5.10. Reaction progress in two basins of similar ages and similar burial rates but different thermal gradients

The same reaction can be followed in young sediments, where the thermal gradient has been very different (Fig. 5.10). In these two examples the sediments have experienced the thermal gradients indicated for periods of less than 2 million years (Ma), and the temperature was the main agent in mineral change. The reaction has progressed to completion in the well with a high thermal gradient (Salton Sea) whereas the low temperatures of the other well (Los Angeles Basin) did not allow the reaction to progress nearly as much.

Figures 5.9 and 5.10 illustrate the importance of trajectory in geologic space. Clay reactions in closed systems, where the solids, clays and other silicates dominate the chemistry, proceed at low rates under most basin sedimentation conditions. In metamorphic rocks one can often find in the field instances where a change of mineral assemblage takes place over a distance of less than 1 m. A rock changes its mineral assemblage rapidly in geologic space. However, in sedimentary-basin rocks mineral reactions can be followed over distances of up to 6 km! Not all clay-mineral reactions take place at the same rates and hence some mineralogies change over short burial distances while others in the same rocks change over kilometers.

A special point should be made concerning the burial history of sediments. The rate of sedimentation in a given basin often changes during its history. Some basins start out having a high sedimentation rate and vice versa. This has an effect on the mineralogy of each layer in the rocks of the basin. Each layer has a history in time and temperature space. Figure 5.11 illustrates this point. Two different burial and sedimentation histories are imagined for sediments each having been deposited about 150 Ma (million years) ago, as shown as the area in the box of the figure. The thermal gradient in both basins is supposed the same over their entire history. The history of the zone has a very different thermal history in the two series, A and B. In series A, there is a period of 70 Ma of initial slow burial. The sediment investigated had a history of slowly increasing temperature followed by rapid burial and a short period, about 20 Ma, of higher temperatures. The history of the layer in series B of the same sedimentary age is quite different. It experienced a rapid initial increase in

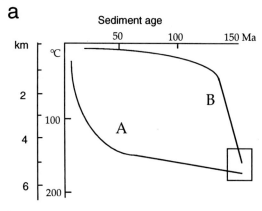

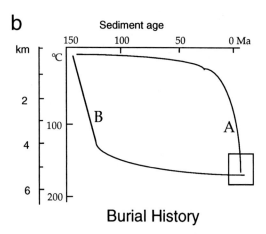

Fig. 5.11. Time–temperature relations of sediments in different basins with the same thermal gradient and having the same age of sedimentation but different burial histories

temperature over a 20-Ma interval with great burial depth to 4 km, followed by a very long period of high temperatures. The total time that layer B spent at 150 °C is nearly 100 Ma while layer A in the other basin has a much shorter high temperature history. Both layers are at the same present day temperature of near 150 °C. One can expect, therefore, that the clay minerals in layer B will be much more mature, i.e. have a greater stage of reaction advancement, than those of layer A, even though they both have the same age and are at the same present depth in sedimentary basins with the same geothermal gradient and hence at near the same present day temperatures. This shows the importance of burial history.

5.1.5 Chemically Driven Reaction in Clays

5.1.5.1 Solution Transport

The change in clay mineral assemblage can be traced as a function of time and temperature as outlined in the previous section. However, chemical potentials (the law of mass action) can also affect clay mineralogy. The most simple case is that of sandstones, rocks which are permeable to migrating solutions and which can contain a certain amount of material susceptible to being changed into clays, or clays susceptible to being transformed under a change of chemical potential.

A first caution to the reader is to remember that the sandstone remains essentially sand, only about 30% of its volume can be filled with fluids. These fluids have a certain capacity to dissolve substances, and hence the mass of ions in the fluids is limited to an absolute amount. On the other hand, the sandstone has a large capacity of exchange (80% of the rock) which can react and which will affect the chemistry of the solutions through exchange processes. This is demonstrated by the relations in Fig. 5.12 in which the amount of reacting material (solids) is shown for the different parts of a porous sediment. The diagram shows the rock as solids (shaded) and pores. The pore space is occupied by fluids in which a small amount of solids are present as dissolved

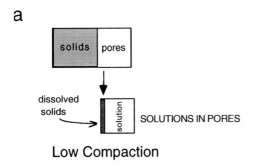

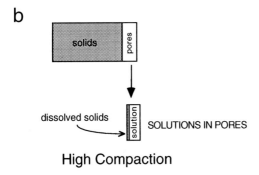

Fig. 5.12. The effect of compaction increases the amount of ions in solution. However, in a given rock the solids represent an increasing proportion of material (*shaded area*) compared to the solutions (*pore space*) and their dissolved solids (*shaded area* in solution portion). Solution diagram *rectangle* has the same size as the pore space

matter (shaded area in pore-fluid space). In both examples, the dissolved material is much less than the solids in the rock. In both the low compaction example (Fig. 5.12a) and the high compaction state (Fig. 5.12b), the potentially reactant ions dissolved in solution are a minority compared to the silicate solids. Hence, in order to create a significant silicate assemblage through interaction of solids and dissolved elements in solution, fluid-solid interaction, it is necessary to transport material via the fluid phase. Two possibilities are available (1) transport by convection, i.e. movement of solutions in which dissolved phases are present, (2) transport by diffusion. In the instance of convection, a large amount of solution is necessary for reactions. For example, when alumina is involved in a reaction its transport is limited by its extremely low solubility in aqueous solution, on the order of several parts per million, or about 0.0001% of the solution. As the solution represents a small portion of the rock, if one intends to precipitate, for example, kaolinite in a sandstone, very large amounts of water are necessary in order to transport even several percent of kaolinite, in the order of 1000 times the volume of the rock.

A more simple method of transport is available, although it operates over a limited range. This vector is one of diffusion and transfer in the solution. If a mineral reaction consumes alumina, for example, to form kaolinite, as soon as the kaolinite is precipitated, the solution will be undersaturated with respect to this element and some new alumina will be called into solution from another source by dissolution. If an adjacent layer of rock, say an alumina-rich clay-bearing layer, is available and in fluid connection with the rock precipitating kaolinite, then the reaction can be accomplished without moving the water at all. In this case, the rock moves and not the solution.

In either case, chemical reaction can be accomplished by a change in chemical potential of the elements in the rock–solution environment. If a strong chemical imbalance is created by the arrival of a new solution with a strong concentration of an element, reaction will be rapid.

5.1.5.2 Chemical Equilibrium Among Clay Particles

If clay-mineral reaction, i.e. the formation of new phases, is a function of chemical disequilibrium, the composition of clay particles themselves is also guided by chemical forces. When a new mineral is produced, this is the result of the overall discord between the mineral forms present and new chemical (x) and physical (temperature) conditions. However, when a new mineral is produced it can remain stable as a phase although its composition can be out of equilibrium as diagenesis proceeds. Figure 5.13 indicates possible phase relations in chemical space as conditions of diagenesis change (essentially temperature). Here, the phases present are in fact the same, but the change in temperature changes the compositions of the phases which will coexist at

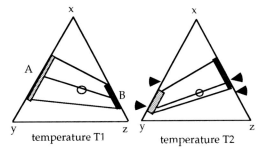

Fig. 5.13. Schematic phase relations in triangular chemical space with the variables *xyz*. Phases A and B are both stable together at the two temperatures, *T1* and *T2*. However, the range of compositions of the two phases changes with temperature and those which will co-exist in a rock of the composition denoted by the *circle* will change with temperature. In the *right-hand triangle*, the *arrowheads* indicate the different compositions at different temperatures

equilibrium. Let us suppose that the rock composition lies at the point of the circle in the two triangles in compositional space xyz. The two coexisting phases for this composition are A and B of compositions at either end of the intermediate line passing through the circle representing the bulk compositor of the rock and joining the phases A and B of a given composition. The line joining the two phases of different compositions is called a tie-line. Phase A has a composition of about 40% x at temperature T1, but at temperature T2 it has only 10% of component x. The new temperature phase has changed the compositions of the phases A and B which can coexist. The tie-line has changed inclination in the compositional triangle. If all goes according to equilibrium theory, as the rock changes temperature from T1 to T2, the compositions of phase A and B will change according to the relations shown in the figure by the tie-lines.

However, the changes of composition are not instantaneous. In fact the changes of composition are very slow. The problem is that the means of attaining chemical equilibrium in a crystal are through internal diffusion. This is a much slower process than the dissolution–recrystallization process which effects mineral-phase change in diagenesis. For example, a chlorite can take up to 500Ma to change composition in the burial-temperature range of 20–120 °C, whereas dissolution–recrystallization processes in the slow smectite to illite transformation will occur in the time range of about 80Ma. Other mineral reactions during diagenesis are much more rapid than this. Thus phases can come into equilibrium faster than their individual compositions.

5.2 Major Progressive Clay Mineral Reactions During Burial Diagenesis

5.2.1 Mixed Layer Mineral Series

Clay mineralogy becomes simplified during burial of sediments, the highly variable forms of expanding minerals such as vermiculites and smectites are quickly replaced by a smaller number of minerals. The detrital clays also become unstable and react to be replaced by the smaller number of phases. The simplified mineralogy eventually becomes one of kaolinite, illite (fine-grained potassic low charge mica), and chlorite in aluminous sediments. In sediments formed from basic rocks, one finds chlorite and mixed-layered chlorite–smectites in the more mature parts of the sedimentary column. In a very simplified manner one can write the relations

Smectites → illite, chlorite.

The less abundant or infrequent phases such as vermiculite, sepiolite and palygorskite are incorporated into the general change of phases. However, as indicated above, things do not occur in one single, abrupt step in depth – temperature space. The reactions occur over a range of depths in basins as a function of temperature gradients and sediment age or the length of time a sediment has spent at aggregate depths. Two particular reactions can run over the same range of geologic conditions; rarely, but still at times, they can be found together in the same rock. They are

Dioctahedral, aluminous smectite → illite/smectite minerals → illite;

and,

Trioctahedral smectite → smectite/ chlorite minerals → chlorite.

Both reactions show an intermediate phase which is in fact a metastable form, well-identified by X-ray diffraction, but which is a phase that constantly changes its composition as time and temperature push the reaction forward. The smectite to chlorite reaction typically occurs at slightly higher temperatures, perhaps 50 °C, than the smectite–illite reaction.

For example, in the smectite to illite conversion the initial stages of reaction show an illite–smectite mineral with an 100% smectite composition, while in the middle of the sequence it will have an aggregate composition of 50% illite and 50% smectite layers in the mixed-layer minerals. In the latter stages of the reaction, the crystallites of illite–smectite will contain on average 10% smectite component. The end point of this reaction will occur when all of the illite–smectite minerals are converted into illite. It is thus possible to measure the reaction progress by measuring the smectite or illite content of the intermediate mixed-layered mineral phase present in a sedimentary rock. The same is roughly true for the trioctahedral smectite to chlorite conversion, but with a time – temperature lag behind the smectite – illite series.

In the smectite–illite reactions, other minerals can and must take part in the conversion process by their destruction which furnishes needed elements for the new minerals. In the case of smectite to illite conversion, one needs potassium to complete the reaction as illite is a mica, and only potassic micas are found to form in sedimentary rocks. Smectites have predominantly calcic or magnesian interlayer ions, i.e. the interlayer ion population as it occurs in sedimentary rocks is predominantly calcic or magnesian. Thus, one can envision a reaction such as

$$\text{smectite} \rightarrow \text{illite},$$

i.e. $Ca + 2K^+ \rightarrow 2K + Ca^{2+}$.

The charged species found in the aqueous solution is the exchange matrix of the reactions. In most natural solutions, calcium and magnesium are more abundant than potassium. The source of aqueous potassium must be less stable, usually detrital, minerals which dissolve to give the needed potassium ions to produce the new, stable phase. These minerals do not, as such, figure in the reaction, despite the fact that they furnish ions to the aqueous solution. The normal candidates for potassium sources are detrital potassic feldspar or detrital, high-temperature mica, usually muscovite. Both minerals contain more potassium per aluminium ion in their structures than does illite. Feldspar has a 1:1 aluminium to potassium ratio and it should be remembered that illite has about 0.85 potassium ions per $O_{10}(OH)_2$ units whereas muscovite has 1.0 ions. Both muscovite and feldspar are therefore less stable than illite when in contact with a solution undersaturated in potassium with respect to illite. In dissolving (due to their metastable compositions), both minerals release aluminum and silicon ions into solution, as well as the needed potassium ions.

In these relations, one sees that the resistant, detrital minerals in a sediment take a significant role in reactions, as do those minerals formed at or near the sedimentation surface as they are sedimented into the sedimentary mineral matrix. There are in fact two parts of every sedimentary rock, those that are highly reactive and those that are more inert. Of course nothing is entirely and forever inert, but there are cases that come close to this state. Zircons and garnets tend to react very, very slowly in sedimentary rocks, but they react all the same. The micas and feldspars of high-temperature metamorphic or igneous rocks are in an intermediate category. They play a role of constant, but slow, reaction components which are critical to the reactions observed in burial diagenesis and early metamorphism, those reactions which operate under low temperatures over long periods of time.

5.2.2 Silica Polymorph Change During Diagenesis

The activity of silica in solution is "anchored" by the prevalent chemically active form of silica which is present in a rock. In low-temperature environments, those of sedimentation, stability of this material is limited but it nevertheless is the phase which permits the most silica to be present in solution. It is necessary to have more than 150 ppm of silica in solution in order to form a free silica phase. By contrast, the stable form of silica, quartz, can accept only 5 ppm of silica in solution at the same conditions. However, quartz is slow to crystallize and amorphous silica is the phase which determines the maximum silica concentration in solution. The form of silica found in sedimentary rocks is a function, again, of time and temperature. As both increase, the form found has a lower tolerance of silica in solution. The sequence is amorphous silica, cristobalite, quartz.

Initially one finds amorphous silica. As temperature or time increases cristobalite forms and the maximum amount of silica in solution decreases to several tens of ppm in solution. This fact excludes certain other silicate phases from forming. As temperatures increase yet further, or the rocks become older, the stable mineral is quartez.

One can draw a parallel between the stabilities of pure silica and the clay minerals. Smectites are the most silica-rich forms, and they occur at lowest temperatures in the youngest rocks. Mixed-layer minerals and vermiculites contain less silica occur at intermediate temperatures and ages. Micas, illite, chlorite, and kaolinite are the most stable clay minerals and contain the least silica.

5.2.3 Zeolite Mineralogy During Diagenesis

In surface environments zeolites are silica-rich, however, with burial, the silica content becomes reduced because temperature, and perhaps time, are important factors in their diagenetic history. The alkali zeolites are stable at the lowest temperatures, the calcic forms appear to persist to higher temperatures. Some rocks contain more than one form (see Chap. 7, on hydrothermal alteration) but generally the forms in sedimentary rocks are sodium and potassium-rich. The most siliceous forms are found in sediments and they are found at some depth. The alkali zeolites such as phillipsite, erionite, stilbite and clinoptilolite contain different proportions of Na and K ions as well as some Ca ions. Only natrolite and analcime are purely sodic. Thus, at low temperatures one can have a maximum of three alkali zeolites present in equilibrium. As temperatures and time increase, the field of zeolite solid solution (range of substitution of K, Na, Ca) decreases. These alkali zeolites are replaced by sodic, less siliceous forms upon deeper diagenesis. Analcime becomes the only zeolite stable in deeper and older rocks. Again, it appears that kinetics is a factor in the distribution of minerals in the clay fraction, in older Paleozoic rocks zeolites are a rare occurrence. The transition of alkali zeolite to analcime and calcic zeolites (heulandite and those forms stable at higher temperatures) seems to occur at temperatures between 80–95 °C in the

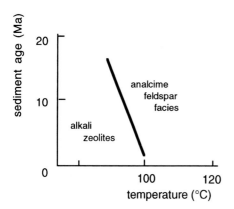

Fig. 5.14. Representation in time–temperature space where the alkali zeolites can exist compared to the analcime facies

age range of 2–18 Ma. Figure 5.14, adapted form Velde and Iijima (1988), indicates these relations. As alkali zeolites are eliminated from the pelitic facies, albite and eventually potassium feldspar is seen. The intermediate assemblage of analcime–albite is considered as a distinctive mineral facies. Eventually, all alkali and calcic zeolites are replaced by sodic or potassic feldspar.

As mentioned above it appears that zeolites are sensitive to kinetic factors, those of time, as well as temperature. However, the rather abrupt changes in mineral facies (alkali, sodic and feldspar) indicate that the reaction is highly temperature-dependent. The zone where several zeolites are present is not very large. The transition takes place over a relatively short time–temperature span. This indicates that the activation energy for the transition is relatively high, probably more than 100 kJ, and the pre-exponential factor is high. In short, temperature is more important than time in the conversion. This contrasts with clay minerals, at least the smectitic types, where the mineral conversion occurs over a range of 100 °C and over the time spans of tens to hundreds of millions of years.

5.2.4 Changes in Organic Matter

The evolution of organic matter, though not strictly in the realm of clay mineralogy, is all the same important to geology and students of the environment. Concentrations of organic matter in sediments, preserved from biological and oxidation effects, can produce petroleum through the maturation of the unstable, complex and heavy organic molecules that form organic matter. The activation energy of these processes is rather high, hundreds of kilojoules per mole, and hence the reactions are very temperature-dependent. The evolution of this material to produce petroleum molecules occurs in the range of 80–110 °C. This is the petroleum window. A further increase in temperature

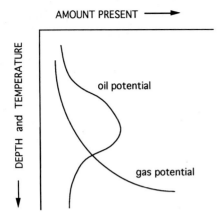

Fig. 5.15. Relations between the generation of oil and gas (potential) as a function of depth (temperature)

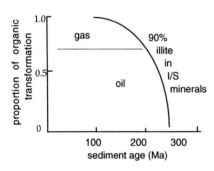

Fig. 5.16. Relation between clay maturity and organic matter maturity. The *curve* shows the end of the smectite to illite clay transformation as plotted against the maturation of organic matter and the age of the rocks in which both clays and organic matter were found

produces smaller molecules including gases, e.g. pentane, methane, etc. These are produced in the gas window. The windows overlap to a certain extent and can be diagramatically represented as in Fig. 5.15. There are three major types of organic material which are found to survive surface biological and oxidation effects to be preserved in sediments. Under diagenesis they produce petroleum reserves. These types of material are of marine and terrestrial origin. The terrestrial material can also produce coal due to a higher concentration of cellulose material in its structure. Two types of marine organic material have been identified; they are differentiated to a certain extent on the existence, or not, of cellulose. The major difference lies in the chain structure of the C–C bonds called aliphatic, or cyclic C–C aromatic structures. The former is present in type I and II organic residues and the latter dominates in the type III terrestrial material.

The occurrence of liquid and gaseous hydrocarbons seems to be, to a certain extent, controlled by kinetics; the older sedimentary basins have petroleum reserves at lower present-day temperatures than do the younger basins. There can be a temperature difference of up to 50 °C between Paleozoic and Cenozoic sedimentary basins. The difference in the kinetics of clay and those of hydrocarbons is very marked. Figure 5.16 indicates the relations between the end of the illite–smectite conversion sequence (90% illite) and petroleum maturity in several basins. The older the basin, the further the clays have evolved compared to the hydrocarbons.

The importance of clay minerals in petroleum geology is that most source rocks, i.e. sediments containing more than 0.5% organic carbon, are clay-rich. The clay minerals seem to have a role in protecting and preserving the organic material from ultimate destruction, oxidation to carbon dioxide. Hence shales are investigated with regard to their organic matter contents and their sealing capacities, i.e. their ability to form traps and reservoirs for petroleum accumulation. Shales are almost never considered as reservoirs. Their permeability is too low to allow a sufficient flow of material to be removed from depth. Their porosity is much too low for hydrocarbons to become concentrated.

Hence, in a simple minded way, shales protect and mature organic matter to produce hydrocarbons in liquid and gaseous form, but the hydrocarbons must leave the shale to accumulate in some trap, often created by a shale layer. Displaced, trapped hydrocarbons form reservoirs useful to mankind.

The precise role of clay–organic interactions during the period of hydrocarbon maturation is at present the subject of much debate. There is no clear relation which can be demonstrated at the moment. However, viscous organic material leaves very low permeability clay-rich rocks to accumulate in porous rocks, either carbonates or sandstones. This occurs at depths usually greater than 3 km, very often at much greater depths. The exact depth of the present-day accumulation is influenced by tectonic movement, such as uplift and erosion, migration and trapping. Some resources are still at great depth (in Texas and Oklahoma, wells have been drilled to more than 10 km), while others are found in shallower sequences due to erosion of the upper portion of the sedimentary rocks which were formerly above them. The La Brea tar pits, in Los Angeles, occur at the surface where they have entrapped prehistoric animals to their death.

5.3 Sequential Mineralogical Changes During Burial Diagenesis

5.3.1 The First Kilometer

In burial diagenesis, the change of sedimentary materials under the influence of normal sedimentary basin burial, the thermal gradients are normally not very great (25–40 °C/km) and hence the temperature of the material buried does not greatly increase during its first kilometer trajectory in time–temperature space. The difference between 15 and 45 °C is not very important in the kinetics of silicate transformation. The major change in the chemical life of the sediment is in the change in the ratio of water to solids in the immediate environment of the solids. The solids have the upper hand in the balance of chemical influence, they determine the activity of species in solution. The initiation of chemical change has nevertheless begun. The solids which determine the activity of species in solution will control the fate of those minerals that are not in equilibrium with their chemical environment at near surface temperatures (<45 °C). Therefore, in the first kilometer of deposition, it appears that few significant changes in mineralogy occur among the silicates. Since clays are silicates and are thus directly influenced by these mechanisms, one can expect that there will be few perceptible changes in clay mineralogy in the first kilometer of burial. In the second kilometer of burial, the temperature increases enough (at a 30 °C/km gradient one reaches 50 °C or so), and some reaction is evident on a large scale. Here the most reactive minerals begin to change and interact with others in the sediment.

5.3.2 The Stability of Detrital Minerals

In mineral assemblages in sediments, the majority of the silicate minerals present are not in chemical or thermal equilibrium with their chemical and mineralogical surroundings. This is normal, in that the sediments came from various geological situations; minerals from rocks subjected to rapid erosion exhibit strong chemical disequilibrium, minerals of soil formation approach the chemical equilibrium of surface conditions, rocks formed by aerial transport, such as volcanic ash, contain minerals which are eminently in disequilibrium at surface conditions. All of these minerals will react at their own rate to attain a sort of chemical and mineralogical harmony, depending upon the rate of burial and the time over which it occurs.

One can describe families of minerals and phases which will react more rapidly, and others which will be slower to attain chemical equilibrium.

Family A, the *rapid reaction family* is as follows:
   volcanic ash, smectites formed under surface conditions, amorphous materials, amorphous silica or cristobalite, hydrous alumino-silicate minerals from soils (imogolite), halloysite (hydrated kaolinite), alkali zeolites, sepiolite and palygorskite, and soil vermiculites.

Family B, the *slowly reacting* silicate minerals, comprises:
   quartz, potassium feldspar, berthiérine (7 Å chlorite), kaolinite, olivine and pyroxenes, biotite and muscovite.

Family C contains examples of *persistent minerals* present over long periods of time and a range of temperatures, but which should ultimately be transformed:
   garnet, zircons, and iron oxides such as magnetite.

In general, the minerals in family A will change phase during the first 1–1.5 km of burial. The minerals in family B will change in the 2–4 km zone and the minerals in family C will often persist into the zone of initial metamorphism, beyond a depth of 10 km, where there are no longer clay minerals in the rocks.

These depths are those of normal burial, where the thermal gradient is near 30 °C/km, and hence the temperatures reach 200 °C at 7 km depth. If the gradients are higher, one can expect that the clays will have been transformed at shallower depths, but they will always be transformed above about 250 °C in younger basins, <200 Ma.

5.3.3 Clay Mineral Assemblages in the Second Kilometer of Burial

The second kilometer of burial diagenesis, depending upon the temperature and age of burial, shows in general a decided tendency to produce fully expandable smectite minerals, for the most part one assumes dioctahedral in composition. The various interstratified minerals inherited in the detrital material of soil origin are converted to a more homogeneous clay assemblage of smectite, kaolinite and illite or detrital mica. Other detrital minerals such as chlorites, sepiolite and palygorskite, and zeolites remain in the clay fraction, but may not be present in older rocks. Glauconites and sedimentary 7Å berthiérines (Sects. 4.4.5 and 4.4.6) seem to maintain their initial sedimentary composition.

In the upper parts of the second kilometer of the sedimentary column the silicates are silica-rich. The number of phases or mineral types decreases due to the destruction of very metastable weathering products such as irregularly-mixed-layer found in soils. The more incompatible minerals, kaolinite and palygorskite for example, are transformed into a more restricted number of clay-mineral types. In this zone the apparent randomness of sedimentary processes is "ironed out" and a bit of mineralogical order instated.

Zeolite facies show alkali, silica-rich mineral assemblages in the upper levels of the second kilometer, and a transformation to the analcime facies further down, depending upon the age and temperature gradient. The highly reactive amorphous silica is transformed into cristobalite and eventually into quartz. The pure silica minerals, such as amorphous silica, cristobalite and quartz, change in relative abundance in the 50–70 °C range in young sediments subjected to burial diagenesis. It is well-known that quartz is the stable phase of silica at conditions near the surface of the earth, but it is equally well-known that the precipitation and crystallization of quartz is difficult due to unexplained energy barriers. The end result is that the precipitation of free silica from aqueous solution is governed by that of the formation of amorphous silica, which occurs at several orders of magnitude greater silica concentrations in solution than that of quartz. This reaction to form the quartz phase which governs the lower activity of silica in solution is dependent on time and temperature. It is also probably pH-dependent. Hence, the stability of silica, or the precipitation from solution, is probably the driving force of other mineral stabilities which react to their "hostile" environment at different rates. The non-precipitation of quartz allows aqueous solutions to contain relatively high amounts of free, soluble silica which is the governing factor for the formation of other silicate phases, especially the phases rich in aluminum. The control of the silica content of solutions is extreme important to silicate mineralogy. Several species commonly found in sediments and sediments of shallow burial cannot coexist with quartz, they require more silica in solution. Other than cristobalite (a silica polymorph) one can cite the siliceous Na–K rich alkali zeolites (analcime, phillipsite, erionite, stilbite, heulandite, mordenite, clinoptilolite) and probably the smectites.

Smectites are the most silica-rich clay minerals, containing more silica per unit cell than micas. Micas are the most silica-rich clay-like minerals to survive into the higher temperature realm of metamorphism. Thus, it is likely that the stability of smectites is dependent on the non-stability, or activity of quartz in the mineral assemblages of sedimentation and the very early stages of diagenesis in the second kilometer. However, in most deep wells, one finds a fully expanding smectite occurring in the late first and early second kilometer of burial. Below this level the smectites, di- and trioctahedral, begin to change to the more stable non-expanding minerals, as temperature and time effect mineral change.

It is in the second kilometer that the diagenetic minerals form and commence their transformations. The quartz, analcime, and I/S interstratified minerals begin to dominate. Kaolinite and chlorite are present but do not seem to change in their proportions notably.

5.3.4 The Last Kilometers

The description of mineral reaction is governed by the relations between the amount of water and solids (compaction and porosity), by burial temperature, and by burial rate. Thus far, in the discussion of the first kilometers of burial we have dealt with the silicates and mainly the clay minerals. The majority of mineral transformations in the first kilometers occur in the silicate, fine-grained, and hence clay, mineral fraction of clay-rich sediments. There is little interaction between clays and the other minerals in the sediments in the first 1–2 km of burial in younger sediments. However, as time and temperature increase with burial depth, other minerals and materials begin to interact with clays in sediments. The influence of carbonates and the reduction of the oxidation state of iron oxides become, at times, major factors in diagenetic mineralogy. Also, in this zone, interaction between layers of sediment of different compositions seem to be very important, i.e. clay layers can give out fluids which influence sand-layer clay mineralogy.

In general, after the first kilometer of burial is achieved, the clay-mineral assemblages of mixed phases, due to the detrital mixtures of sediments, have been simplified. The various types of mixed-layer minerals, which were produced by rapid or incomplete reaction in soils, have been transformed to a smectite-rich mineral. The minerals smectite (or mixed-layer illite–smectite), illite and kaolinite dominate the terrigineous sediments. The sedimentary minerals glauconite, berthiérine, sepiolite and palygorskite tend to remain unchanged in the first and upper part of the second kilometer of burial. As described above, some minerals are more unstable than others and those with the greatest instability are the first to react.

In these reaction sequences, some material reacts at a given constant rate determined by the physical conditions of burial. In such reactions, the extent of the reaction is governed by the temperature and duration of reaction. In some reactions time is more important than temperature, and in some the reverse is true. Temperature-sensitive reactions are a good measurement device, a sort of geothermometer. One such material is organic matter. The transformation of complex molecules of carbon, hydrogen and oxygen into more simple ones, of higher carbon content is the key to some of nature's greater wonders, and not the least the production of liquid petroleum and gas. This process is largely conditioned by the temperature attained by the sediment holding the organic matter. The reactions become highly activated at about 80–160 °C. Our problem here is not one of organic geochemistry, but one of the possible influences of these reactions on the surrounding silicate mineralogy.

In this temperature–time range sepiolite and palygorskite disappear early on. The sedimentary chlorite berthiérine is transformed to a 14 Å chlorite, usually very rich in iron and aluminum. Glauconites seem to be little affected by burial diagenesis.

The change in the state of organic matter is important to silicate mineral stability. The essential process of the evolution of organic matter under thermal stress is the loss of hydrogen ions from the various compounds present in various forms such as gases and liquids, and to concentrate carbon atoms in the organic residue. This process prduces molecules which are lighter (fewer atoms per molecule) and heavier than the original material. The formation of gaseous and liquid hydrocarbons and a heavy residue, called kerogen, are the results of these thermally activated processes. The release of hydrogen-rich materials (gases) has an overall effect of depressing the activity of oxygen in the ambient enclosing silicate and oxide materials of the sediment. Schematically,

$$Fe_2O_3 + H_2 \rightarrow 2FeO + H_2O.$$

The reactions in sedimentary rocks are of course more complex, but the simple principle will serve our purposes. In the above situation, the oxidation state of iron is reduced from the very stable trivalent form to a divalent form. The mono-oxide of iron is very rare in nature. In fact, almost all divalent iron is found in silicates or in the mixed oxide magnetite in which iron is found in both oxidation states to form Fe_3O_4. This being the case, when iron is reduced by the transformation of organic matter, it is readily incorporated into the silicates through mineral and phase interaction.

Schematically, one can consider

$Fe^{3+} \rightarrow Fe^{2+}$

and then

$$2FeO + Al_2Si_2O_5(OH)_4 \rightarrow Fe_2AlSiAlO_5(OH)_4 + SiO_2$$
$$\text{kaolinite} \qquad \text{7 Å chlorite (berthiérine)}$$

as representing a typical iron reduction–silicate reaction in which iron is changed from a free oxide phase and incorporated into a silicate mineral. Such relations change the normal path of phase reactions for the silicates. The introduction of divalent iron into a clay assemblage can cause, for instance, kaolinite to disappear from a clay-mineral assemblage. Also, introduction of divalent iron can cause an aluminum-rich mineral such as illite–smectite to produce chlorite in the later stages of its reaction sequence.

Other interactions, such as carbonate–silicate, can occur in the burial diagenetic changes. For example, dolomite and kaolinite have been observed in some series to form magnesian chlorites. Siderite has been seen to be destabilized in the presence of kaolinite to form iron-rich chlorite. In these interactions kaolinite is the silicate source of aluminum needed to produce chlorite.

As a result, kaolinite is generally abundant in the early stages of diagenesis, but as depth and maturity of rocks increase the kaolinite becomes less abundant. In the highest grade diagenetic facies, one finds a typical assemblage of illite, chlorite and feldspars. These have formed at the expense of smectites, kaolinite and zeolites.

The most common clay-mineral assemblage in deeper sedimentary rocks is illite and chlorite. The variability of clay phases, commonly including smectites and other swelling minerals plus kaolinite, becomes restricted to a few forms.

5.4 Conclusions

The effects of burial diagenesis are those of simplifying the clay-mineral assemblage, either through rapid mineral reaction between phases at given temperatures (carbonate–kaolinite reactions for example) or through the gradual change in mineralogy as a function of time and temperature. This is the most important aspect of burial diagenesis. This is the kinetic aspect. The transformation of zeolites, silica polymorphs and smectite minerals are all kinetically driven. Thus a complex and changing assemblage of each phase type (silica polymorph, alkali zeolite assemblage, mixed-layer smectite and illite or chlorite) will be present at different depths in a burial sequence or in layers of different ages. It is important to consider the aspect of time, comparing sediments by their age as well as their depth of burial (indicating temperature).

During the first kilometers of burial, the most unstable minerals are changed forming the initial diagenetic assemblages, alkali zeolites and smectites. The deeper sediments show a simplification of the mineral assemblage and an increasing reaction rate as temperatures become high enough to affect mineral change. The limits of diagenesis are a function of thermal history and the age of the sediments. The upper limit of diagenesis is the beginning of metamorphism. This limit is of course a function of temperature and, to a lesser extent, time.

Suggested Reading

Because this chapter is a general review of diagenesis and a summary of other recent papers and books, we have chosen to give general or review works which can be of use for an overview of diagenesis. There are of course many papers published in specialized journals which deal with individual cases of diagenetic clay mineral change which are indicated below.

Books where one can follow the changes in diagenesis experienced by clays

Chilingarian GV, Wolf KH (1988) Diagenesis I. Elsevier, Amsterdam, 591 pp
Sudo T, Shimoda S (1978) Diagenesis in sediments and sedimentary rocks. Elsevier, Amsterdam
Velde B (1985) Clay minerals: a physico-chemical explanation of their occurrence. Elsevier, Amsterdam, 427 pp
Weaver CE (1989) Clays, muds and shales, developments in sedimentology 44. Elsevier, Amsterdam, 819 pp

Journals where one will find the greatest amount of information on diagenesis of clay-bearing rocks

American Association of Petroleum Geologists (American)
Clay Minerals (European)
Clays and Clay Minerals (American)
Journal of Sedimentary Petrology (American)
Journal of Sedimentology (American)
Petroleum Geology (European)
Sedimentology (European)

Recent special issues of journals with numerous articles on clay diagenesis

Clay Minerals (1991) vol 26, no 2
Clays and Clay Minerals (1993) vol 41, no 2

Zeolite minerals and silica polymorphs

Aoyagi K, Kazama T (1980) Transformatonal changes of clay minerals, zeolites and silica minerals during diagenesis. Sedimentology 27: 179–188
Barrer RM (1982) Hydrothermal chemistry of zeolites. Academic Press, London, 360 pp
Gottardi G, Galli E (1985) Natural zeolites. Springer, Berlin Heidelberg New York, 409 pp
Iijima A, Utada M (1966) Zeolites in sedimentary rocks, with reference to depositional environments and zonal distribution. Sedimentology 7: 327–357
Mitzutani S (1970) Silica minerals in the early stages of diagenesis. Sedimentology 15: 419–436

Mitzutani S (1977) Progressive ordering of cristobalitic silica in early stages of diagenesis. Contrib Mineral Petrol 61: 129–140
Velde B, Iijima A (1988) Comparison of clay and zeolite mineral occurrences in Neogene age sediments from several deep wells. Clays Clay Min 36: 337–342

Organic matter

Bjorlykke K (1989) Sedimentology and petroleum geology. Springer Berlin Heidelberg New York, 363 pp
England WA, Fleet AJ (eds) (1991) Petroleum migration. Geol Soc Lond Spec Publ 59: 280
Hunt JM (1979) Petroleum geochemistry and geology. Freeman, San Francisco, 617 pp
Selley RC (1989) Elements of petroleum geology. Freeman, San Francisco, 488 pp
Tissot G, Welt DE (1984) Petroleum formation and occurrence. Springer, Berlin Heidelberg, New York

6 Hydrothermal Alteration by Veins

A. MEUNIER

6.1 Introduction

Sedimentary series, as well as crystalline basements, are cross-cut by fracture networks along which fluids of different origins (meteoric, connate, metamorphic, magmatic) have percolated. Chemical interactions with rocks along these pathways produce secondary minerals among which phyllosilicates are usually the dominant species. The mineral deposit sealing the fracture and the altered wall rocks are described as hydrothermal veins. These veins have long been studied as guides for prospective ore deposits (Rose and Burt 1979). Pioneer studies have demonstrated that the altered wall rock is zoned, and that the zones develop simultaneously by outward growth of several alteration fronts (Sales and Meyer 1950). Compared to pervasive alteration systems (Chapter 7), the volume ratio of altered versus unaltered rock is several orders of magnitude lower in a vein system. Nevertheless, veins are of great importance in environmental problems because they concentrate clays in the zones of fluid circulation.

Clay minerals can be produced by vein hydrothermal alteration according to three different pathways: (1) transformation of pre-existing high-temperature silicates, (2) transformation of pre-existing clay minerals, (3) direct precipitation from fluids. Each path corresponds to typical geological occurrences which will be described here. Whatever the path, the crystallochemical properties of the resulting clay minerals depend on local temperature and chemical conditions. They can be non-expandable (mica and chlorite), expandable (smectites) or intermediate (mixed-layered minerals). Two complete transitional series of interstratified minerals exist in the dioctahedral and trioctahedral families between the non-expandable and the fully expandable end members: illite–smectite (I/S) and chlorite–smectite (C/S) series.

In this chapter, the principal kinds of veins encountered in hydrothermal systems are presented. Then, an injection-vein case study will be detailed in order to describe the physico-chemical mechanisms of hydrothermal alteration. Finally, the formation of clay minerals by alteration of pre-existing ones and direct precipitation from solutions will be considered.

6.2 Structure of the Hydrothermal–Wall Rock System

6.2.1 Central Deposit and Altered Wall Rocks

Hydrothermal fluid circulation in fractured rocks induces a series of mineral reactions, the results of which can be seen as a central deposit surrounded by altered wall rocks. In the literature the term "vein" is usually used to designate the central deposit as well as the wall rocks. In order to avoid any confusion, we will consider here that vein is the equivalent of the central deposit. The central deposit is made of secondary minerals which completely, or partially, seal the space between the two wall rocks (this space comprises the open volume of the initial fracture and the volume of dissolved fracture edges). The secondary mineral deposit can have two origins (Fig. 6.1).

1. Deposition after transportation by fluids (suspensions) produces coatings on the fracture edges. In such a case, the wall-rock alteration, if it exists, has no genetic relations with the clay deposit in the fracture. This is frequently observed in surface alteration processes (soils and weathered rocks).
2. Precipitation from the saturated fluids (solutions) causes mineral growth on the edges of the fractures. Such features, rare in surface alteration conditions, are systematically described in hydrothermal vein systems.

From a theoretical point of view, a fluid will flow in a fracture and penetrate inside a rock if a pressure gradient (ΔP) is established between the source region and the open spaces (fracture and rock porosity). Consequently, in open systems, two pressure gradients are established: (1) $\Delta P_1 = P_{ff} - P_{fr}$ (between the fracture and the unaltered rock); (2) $\Delta P_2 = P_{fs} - P_{ff}$ (between the source region and the fracture). P_{fs}, P_{fr} and P_{ff} represent the fluid pressure in the source region, the porous space in rock and the fracture respectively ($P_{fr} < \sigma 3$; Fyfe et al. 1978). Pressure gradients ΔP_1 and ΔP_2 are not fixed in time; they vary from a maximum value (ΔP_{1max} and ΔP_{2max}) to zero. If their values become negative, the direction of the flow is reversed.

Altered wall rocks are easily observed in outcrop, because their colour is different from that of the unaltered rock. Usually, their apparent thickness is at least one order of magnitude greater than that of the central deposit. Petrographic observations show that dissolution–recrystallization features extend a long way inside the apparently unaltered rock. The altered wall rocks more or less conserve the structure of the rock, in spite of the fact that the primary minerals can be strongly dissolved or replaced by secondary ones. The altering fluid can have three different sources giving three different types of hydrothermal veins:

1. Injection vein type: the source region is located in high-pressure–high-temperature zones; the fluid is injected into the fracture and the rock (ΔP_{1max} and $\Delta P_{2max} > 0$).

Hydrothermal Alteration by Veins

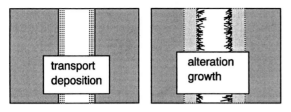

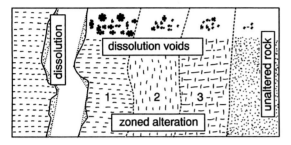

Fig. 6.1. *Above* the two ways for formation of the central deposit: deposition from a clay suspension flowing inside the fracture, alteration of wall rocks, if it occurs, is not genetically related to the formation of the central deposit; growth of secondary phases on the edges of the fractures; secondary phase growth while wall rocks are altered by a sequence of interdependent mineral reactions. *Below* a vein is composed of a central deposit surrounded by altered wall rocks. Three zones are distinguished here in altered wall rocks according to differences in mineralogical composition. Dissolution features of primary minerals decrease in number and size with distance to the fracture

2. Infiltration vein type: the source region is located in higher formations; fluids circulate by gravity (ΔP_{1max} and ΔP_{2max} are controlled by the hydrostatic pressure).
3. Drainage vein type: the source region is located in the rock itself; the fluid is drained from the rock towards the fracture until pressure in rock and fracture are equilibrated ($\Delta P_{2max} < 0$; ΔP_{2max} is controlled by the temperature during the cooling of the rock body).

6.2.2 Fluid Injection Vein Type

Hot springs and geysers in geothermal fields are typical of an injection regime: the hot fluid, coming from deep high P-T formations, rises up to the surface and permeates through the surrounding rocks (Fig. 6.2). A strong alteration producing zoned wall rocks is observable along the fractures. Different secondary mineral assemblages are formed as the distance to the fracture increases (Bonorino 1959; Beaufort and Meunier 1983). An example of zonation is given in Fig. 6.3. Generally, the initial microstructure is still recognizable

INJECTION VEIN TYPE

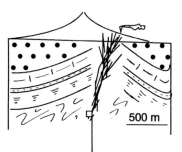

Fig. 6.2. Location of an injection vein in an active hydrothermal system. Fluids are injected inside the fracture network from deep hot zones and rise up to the surface where they escape to the atmosphere through hot springs or geysers. The wall rocks are zoned around the fracture

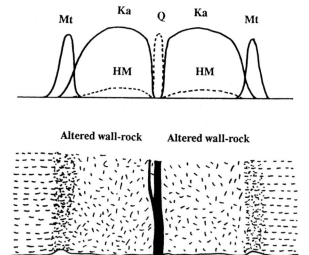

Fig. 6.3. Example of zoned alteration in hydrothermal veins from the Caribou Mine, Colorado (modified from Bonorino 1959). *Q* Quartz; *Ka* kaolinite; *Mt* montmorillonite; *HM* hydromica. This is a typical injection vein-type structure

even in the more altered zone; mineral reactions are volume conservative. They can be written as follows:

Primary mineral → secondary mineral(s) + dissolution voids.

The fluid which impregnates the altered wall rock cannot move in the same way that it does in the open space of fractures (the tortuosity of the pathways is much more important than that of the fracture). Here, most of the chemical transfers from rock to fracture are controlled by a diffusion rather than convection processes. The diffusion processes are activated by a thermal gradient.

6.2.3 Fluid Infiltration Vein Type

Fluids percolate downwards under gravity through different geological formations and precipitate clay minerals along their pathways. For instance, a sequence of C/S mixed layered minerals was observed in the crystalline basement of the Paris basin at Sancerre (GPF Scientific Drill-Hole). The formation of these minerals was obviously controlled by the composition of the surrounding rock which is dissolved in a thin layer on the fracture walls: saponite, corrensite and chlorite were observed in fractures crosscutting pyroxene-rich amphibolites, amphibolites and gneisses respectively (Fig. 6.4). All these rocks have reacted with diagenetic fluids originating from the sedimentary cover. These fluids were not renewed; their residence time was very long (Boulègue et al. 1990). Consequently, corrensite, saponite and C/S mixed-layer minerals are very well crystallized: the are of large size and display good crystal faces (Beaufort and Meunier 1994).

6.2.4 Fluid Drainage Vein Type

These veins originate in tension fractures which create local low pressure domains. The fluids trapped in the rocks or liberated by a crystallizing magma

INFILTRATION VEIN TYPE

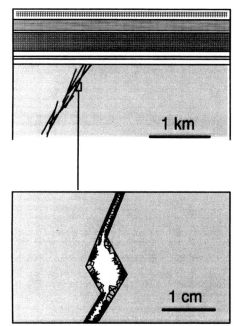

Fig. 6.4. Infiltration vein in the metamorphic basement (amphibolite) of the Paris sedimentary basin. The fractures are still open occasionally. Large corrensite crystals are observed on the fracture edges with albite (*below*). The stacking of chlorite and saponite layers is exceptionally regular in these conrrensite crystals

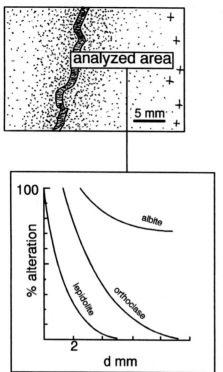

Fig. 6.5. Simplified representation of a drainage vein sealed by muscovite in an albite–lepidolite granite (Merceron 1988). The intensity of the alteration of primary phases by muscovite is shown by the decrease in altered lepidolite, orthoclase and albite with distance from the vein. The intensity of alteration is measured by the surface area of each primary mineral replaced by muscovite

(fluid pressure approaching rock pressure) are drained toward these domains. The secondary phases which precipitate inside the fractures are identical to those already present in the surrounding rocks. Monophased or polyphase assemblages are observed according to the chemical composition of the rocks which is the dominant parameter in such alteration processes.

The Echassières albite–lepidolite granitic body shows a sequence of hydrothermal vein deposits: muscovite, pyrophyllite, donbassite, tosudite and kaolinite were shown to result in reactions of the granite with a fluid of decreasing temperature which is expelled from the crystallizing magma. Isotopic data and mineralogical analysis showed that there is no contamination from the intruded micaschists (Merceron et al. 1992). Regarding muscovite veins as an example, no crystallochemical differences were observed between crystals inside the fracture and the altered surrounding rock (Merceron 1988). The quantities of muscovite were shown to decrease from the vein to the unaltered rock (Fig. 6.5). Drainage veins are common in the pervasive alteration areas such as propylitic zone around magmatic intrusions (Beaufort et al. 1992). In this case, the secondary assemblages which seal the veins are identical to that of the surrounding altered rock: chlorite + epidote + quartz for example (Fig. 6.6). These assemblages are polyphase because the chemical composition of the volcanic and plutonic rocks is the controlling

DRAINAGE VEIN TYPE

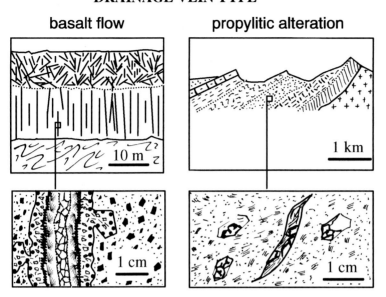

Fig. 6.6. Two examples of drainage veins are shown in a basalt flow and a propylitic zone around an intrusive granodiorite body. In both cases, fluids come from the surrounding rocks and are concentrated in open fractures (low pressure zones). Rythmic and simple zonations are observed in the basalt joints and propylitic veins respectively. In spite of this difference, both cases (*right* basalt flow vein and *left* propylitic rock alteration) show that the secondary parageneses is identical in veins and rocks

parameter (Beaufort et al. 1992). Such drainage veins are also active during the degassing stage of lava solidification. Indeed, fluids which were dissolved in the magma are expelled to the atmosphere when the lava flow cools. These fluids interact with the hot rock producing phyllosilicates, namely celadonite, saponite and nontronite, in association with cristobalite and zeolites (Goncalves et al. 1990).

6.3 Alteration Mechanisms

6.3.1 Zone Formation

The mechanisms necessary to produce zoning of hydrothermal alteration have been discussed for some time by petrographers. Indeed, Sales and Meyer (1950) concluded that the development of alteration zones was due to a metasomatic-like process. Nevertheless, Bonorino (1959) using such a model, has encountered some difficulties in explaining the formation of all the mineral zones as they are depicted in Fig. 6.3. All the secondary minerals were considered to crystallise at the same time in the wall rock in spite of the fact that they cannot crystallize under the same temperature conditions. Further, when each zone migrates away from the fissure, it was considered to grow at its outer edge and simultaneously recede at its inner edge because of the encroachment by the innermost zone. More recently, these questions have been considered in order to establish a general model of vein alteration. As wall rock zonation is easier to observe in high-temperature systems than in low-temperature ones, white mica veins were studied in detail at La Peyratte (Deux-Sèvres, France). It is assumed that the model obtained can be applied to low-temperature conditions in which clay minerals crystallize.

The vein system appears as white bands of different widths which crosscut a black, fine-grained biotite granite (Turpault et al. 1992a,b; Berger et al. 1992). The secondary phyllosilicates which crystallize in the altered wall rocks are chlorite and phengite. The alteration process was shown to depend on a sequence of interdependent mineral reactions. The first one, which is that observed further inside the rock, away from the vein fracture, is the dissolution of oligoclase. It is observed as etch pits whose size and density per surface unit decreases with distance from the vein. These relations will be described further.

The dissolved oligoclase components move towards the vein, the driving force being the stabilization of chemical gradients. These components are consumed by different mineral reactions inside the wall rock; chloritization of biotite and albitization of the remaining oligoclase. In their turn, these two new reactions liberate other chemical components which migrate towards the vein. A great part of them is consumed by the crystallization of white micas (mainly phengite) inside the vein or in the adjacent wall-rock zone (50–300 µm wide veins) or adularia and muscovite (20–50 µm wide veins).

Surprisingly, the chemical mass balance calculations shown that gains and losses of elements are not as important as they had been assumed to be in the past. Indeed, the mineral phases which precipitate inside the fracture-altered wall rock system consume almost entirely the chemical components liberated by the dissolution of biotite and oligoclase. These reactions can be written in a simplified manner as follows:

1. oligoclase $\rightarrow$ Na^+ + Si^{4+} + Ca^{2+} + Al^{3+}
2. oligoclase + Na^+ + Si^{4+} $\rightarrow$ albite + Al^{3+} + Ca^{2+}
3. biotite + Al^{3+} $\rightarrow$ chlorite + K^+
4a. Si^{4+} + Al^{3+} + K^+ + $(Fe,Mg)^{2+}$ $\rightarrow$ phengite (50–300 µm wide veins)
4b. Si^{4+} + Al^{3+} + K^+ $\rightarrow$ muscovite + adularia (20–50 µm wide veins).

The interdependent relations between the different mineral reactions operating in a 300 µm wide vein are shown in Fig. 6.7. Only water and small amounts of sulphur have been added to the hydrothermal alteration fluid which absorbs only the most soluble components; Si and Ca are partly leached out. If an oversimplification is allowed, the major effect of alteration seems to be reduced to the "transfer" of the open volume of the fracture to produce the

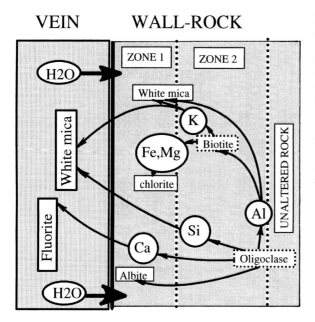

Fig. 6.7. Skecth showing the chemical transfer controlled by the interdependent mineral reaction sequence during the wall-rock alteration stage of an injection vein. Biotite and oligoclase are dissolved and replaced by white micas, chlorite and albite near the vein. White mica and fluorite precipitate inside the vein. (See Turpault et al. 1992)

dissolution voids inside the wall rocks. Illite–smectite mixed layered minerals have grown on the walls of these voids. The smectite percentage decreases from 70 to 40% with increasing distance from the vein. This is the last reaction which occurs during the final cooling stage.

The fluids trapped inside the dissolution voids have reached saturation with clay minerals which are the stable silicate phases in low-temperature conditions. Two processes can explain the regular increase in the amount of smectite in I/S mixed-layer minerals toward the vein:

1. If I/S are considered to have crystallized at the same time in all the altered wall rock, then we must admit the presence of an inverse temperature gradient established by cold fluid circulation in the porous altered wall rock after the sealing of the vein (Parneix 1992);
2. The I/S sequence could have been established if the I/S crystallization stage was more recent closer to the vein. In that case, the latest stage must be the cooler one. Whatever the interpretation chosen, the circulation of fluids in the altered wall rock was certainly very weak because of the low interconnectivity of the secondary porous spaces.

6.3.2 Zone Development

In granite, oligoclase and biotite are altered in equilibrium with quartz and microcline early in the sequence. The fluid can be considered to be a buffered solution. This means that the development of zoned wall rocks does not

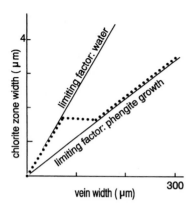

Fig. 6.8. Relations between the thickness of the chlorite zone and the width of the fracture. The limiting factor for narrow veins is the amount of water: alteration reactions stop when water is totally consumed. For wider veins, the limiting factor is the growth of white micas which are the end product of the sequence of mineral reaction. Growth stops because of lack of free space

depend on specific fluid compositions but is due to a self-evolution in a rock dominated system. The driving force of this self-evolution process is related to the stabilization of a pH gradient from acidic in the vein to more alkaline conditions in the wall rock. The dissolution–precipitation processes lead to a slight and roughly constant increase in the solid-phase volume. For a high water/rock ratio (W/R) in the widest (300 μm) veins, the volume variation of solids is negligible when compared to the solution volume and does not affect the rock porosity. In contrast, for low W/R (100–20 μm wide veins), porosity and mass of solution decrease. Consequently, the quantity of available fluid is a limiting factor for the alteration in narrow veins. In larger veins, (100 to 300 μm), the limiting factor is related to the crystallization of the end-member product of the reaction sequence, phengite. If the phengite growth stops (because of a lack of free space for instance) all the sequence stops. Taking the width of the chloritization zones as an indicator of alteration intensity, it was shown that it is dependent on the aperture of the fracture as indicated in Fig. 6.8.

6.3.3 Kinetics of Zonation

Summarizing the data obtained on the La Peyratte vein system, four successive stages can be distinguished in the alteration process from the opening to the sealing of the initial fracture (Figs. 6.9, 6.10). At time t_0, the fracture is opened. At time t_1, the oligoclase and quartz crystals dissolve at the fracture edges; fluids invade the wall rocks. At time t_2, the interdependent mineral reaction sequence begins to operate; the chloritization and albitization zones migrate from the fracture edge inside the rock. The development of zones in

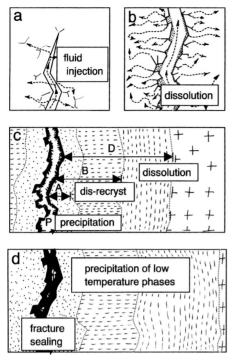

Fig. 6.9. Cartoon showing the four steps in the "life" of an injection vein from the opening of the fracture (birth) to the post-sealing state (death). **a** Opening of a fracture and injection of the hydrothermal fluid (*arrows*). If fluid pressure inside the fracture is greater than pore pressure in the rock, the fluid penetrates inside the wall rocks (*small arrows*). **b** The mineral species unstable with respect to the fluid are dissolved along the fracture edges (*stippled areas*); the fluid penetrates deeper into the wall rocks. **c** The sequence of interdependent mineral reactions is activated by chemical diffusion processes leading to the formation of a mineral zonation inside the altered wall rocks: *zones A* and *B* are characterized by the replacements of unstable primary phases by secondary ones (dissolution–precipitation). The dissolution of the most unstable phase extends largely out of *zones A* and *B*. The end point of the reaction sequence is the precipitation of a monophase mineral assemblage inside the fracture. **d** The fracture is completely sealed, the remaining fluid is trapped inside the dissolution voids of the most unstable phase. When cooling, the fluid becomes oversaturated with respect to low-temperature, hydrated secondary phases (clay minerals)

altered wall rocks does not necessitate the receding of their inner edge as was stated by Sales and Meyer 1950. They grow simultaneously but at different speeds. At time t_3, the limiting factor (available amount of water or phengite growth for narrow and large veins respectively) stops the reaction sequence. Generally speaking, the progress of a mineral reaction with time is recorded in some particular petrographic features, e.g. the size and shape of crystals. For instance, in thermal metamorphic environments, the size of wollastonite crystals was shown to be a good marker of the temperature versus time conditions (Joesten and Fischer 1988; Joesten 1991). Unfortunately, re-

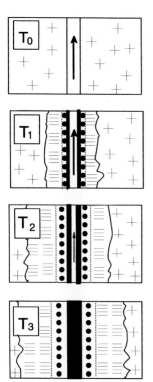

Fig. 6.10. Schematic representation of the kinetics of the altered zone growth in wall-rocks from the opening of a fracture (*To*) to its sealing (*T3*). The size of the *arrows* is roughly proportional to the quantities of hydrothermal fluid which have been injected into the fracture

presentative measures of crystal size are difficult to obtain in a vein system because of the very short distances between the fissure and the crystals to be analyzed. Another feature must be chosen, for example dissolution voids in a given primary mineral. Indeed, dissolution as well as growth is a kinetically controlled phenomenon and it is quite easy to measure the size and number of voids with a scanning electronic microscope (SEM). The kinetics of "nucleation" and growth of dissolution voids were summarized by Berner (1981) in weathering phenomena. The number and the size of voids depend on the process which controls the dissolution; transport control, surface reaction control or mixed transport and surface reaction control. The general law is expressed as follows:

$$d(rc)/dt = uD_s(C_v - C_{eq})/rc,$$

where:

- rc = average radius of voids;
- u = molar volume of the crystalline substance;
- D_s = coefficient of molecular diffusion in aqueous solution;
- C_v = concentration in solution away from the crystal surface;
- C_{eq} = equilibrium concentration adjacent to the crystal surface;
- t = time.

The variation in the number and shape parameters (width and length) was studied by the means of statistical analyses of dissolution voids in the altered wall rocks. As alteration progresses from a fluid-solid interface (edge of the fracture) into the surrounding rock, time is a direct function of distance to vein. The number of dissolution voids in oligoclase crystals in the La Peyratte vein system decreases exponentially with distance from the vein (Fig. 6.11a) but length and width of dissolution voids do not vary with distance according to the same simple exponential function (Fig. 6.11b). Such a difference is due to the fact that the formation of voids is much more rapid than their growth. Indeed, it is known that the chemical attack of crystal surfaces is so rapid in the case of transport-controlled dissolution, that etching occurs everywhere. On the contrary, void growth (dissolution) is a time-dependent process. The complicated shape of the curves in Fig. 6.11b is due to the fact that void

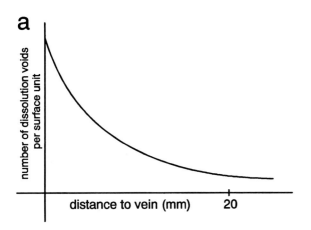

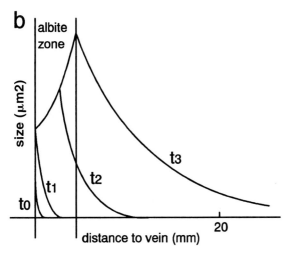

Fig. 6.11. Kinetics of zonation are recorded in the number and the size of dissolution voids inside oligolase crystals. **a** The number of voids decreases with distance to the fracture following an exponential relation; **b** the size (length, width or area) of voids varies with distance from the fracture as given by the curve $t3$. The increasing size portion is due to the competition of two processes, dissolution and albitization of oligoclase, then size decreases according to an exponential relation

growth is "poisoned" by the replacement of the unstable state of plagioclase (oligoclase) by the stable one (albite) in the vicinity of the fracture. In other words, voids can grow at a given place until oligoclase is replaced by albite. Then, dissolution stops and the voids are "frozen" at the state they have reached. Their size depends on the progress of the albitization front: the dissolution voids are "frozen" very early near the vein and later further into the rock.

6.3.4 Quantities of Fluids and Flow Regime in Fractures

The quantities of fluid involved in chemical reactions can be evaluated by the calculation of the mass balance. Indeed, some components are leached out of the system, e.g. Si and Ca. Silica has a temperature-dependent solubility (Fournier and Potter 1982; Michard 1985). For a given temperature, the loss of silica can be used to calculate the quantities (Qw) of solutions which have flowed in the system according to the following relations:

$$Qw = NSiO_2/N_T \times rw_T,$$

where:

$NSiO_2$ = number of moles of dissolved silica lost by the system;
N_T = number of moles of dissolved silica in water at temperature T;
rw_T = density of water at temperature T.

Such calculations have been done for a 40 dm³ reference volume of altered wall rock around a phyllic vein in the La Peyratte granite (Turpault et al. 1992). The fracture volume is 0.375 dm³; fluid temperature was estimated to be 300 °C by fluid inclusion analyses. The calculated solubility of silica is 9.6110^{-3} mol kg^{-1} of water (269 ppm of Si). The mass balance of alteration reactions indicates that 12 mol SiO_2 were leached out from the reference volume, thus $Qw = 1785$ l. This quantity represents 4762 times the volume of solution inside the vein the reference volume of rock. The problem now is how fluids are renewed inside the fracture. The answer can be found by studying the temperature reached in each point of the wall rock.

The study of secondary fluid inclusions trapped in quartz crystals inside the altered wall rock of a 300 µm wide vein showed that a 75 °C thermal gradient was established between the fracture and the unaltered rock (Fig. 6.12a). Such a temperature gradient is established in just a few seconds in the case where the flowing regime of hot fluids in the open fracture is strictly continuous (Grimaud et al. 1990). In such a case, the thermal gradient is not stabilized for long enough periods of time for mineral reactions to proceed. A thermal gradient is stabilized for periods exceeding few seconds only in the case of a pulsed regime.

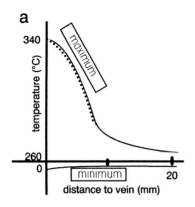

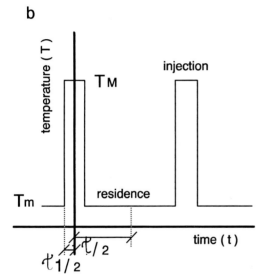

Fig. 6.12. The study of secondary fluid inclusions trapped inside quartz crystals of the altered wall rock shows that a thermal gradient was established for periods of time long enough for mineral reactions to proceed. Such local thermal gradients are established by a pulsed flow regime inside the fracture. *TM* Maximum temperature; *Tm* minimum temperature; τ period of oscillation; τ_1 time during which the temperature is at a maximum

Assuming that the pulsed regime is periodic, as it is more or less in geyser systems, Grimaud et al. (1990) established the equations describing the steady state of the thermal gradient.

The thermal gradient measured in the 300 µm wide vein is fitted with the following values: τ (oscillation period) = 1286 αs (about 20 min) and = 0.015. These values are comparable to that of geyser eruptions (Elder 1981). The minimum values for temperature are estimated to lie between 255 and 260 °C which is consistent with the precipitation of phengite or adularia (Fig. 6.12b). In fact, the oscillatory temperature conditions are known to favor the nucleation process (Baronnet 1980). The chemical compositions of fluids change during a pulse. Indeed, during the injection time, the chemical properties of hot fluids are controlled by the source region, but during the residence time fluid composition is controlled by mineral reactions in the altered zone.

6.3.5 Successive Fluid Circulation

The circulation of hydrothermal fluids into rocks depends on the geometric properties of the connected fracture network in a 3D space. When sealed, fractures can be reopened by mechanical stress due to tectonic events. This has been described as a crack-seal process (Ramsay 1980); the successive fluid flowing periods are generated by a "pumping effect" (Sibson et al. 1975). Reopening of old fractures and the opening of new ones modifies the geometrical properties of the network. The time relations between some of the successive flow episodes are recorded in crosscutting veins and zoned vein deposits.

The Sibert porphyry copper deposit (Burgundy, France) is a good example of a hydrothermal environment where the crosscutting relations between

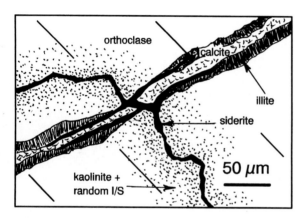

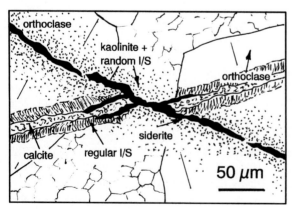

Fig. 6.13. Three steps of the cooling process in a cupromolybdenic porphyry at Sibert (France) are observed in crosscutting relations of hydrothermal veins: illite, regular I/S + calcite and kaolinite + random I/S + siderite. The temperature conditions for the precipitation of illite, ordered I/S and random I/S, are about 200, 160 and 100–80 °C respectively. *Above* The illite vein is crosscut by a vein containing ordered R = 1 I/S. *Below* An illite vein is crosscut by a vein with kaolinite + random R = 0 + Ca-Mg carbonates

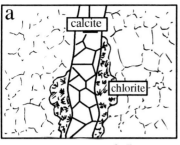
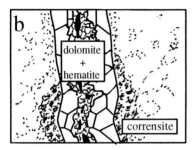

Fig. 6.14. Schematic representation of the superposition of two different hydrothermal alterations due to the reopening of a calcite vein in the "Les Crêtes" granites (Vosges, France). **a** Formation of a drainage calcite + chlorite + pyrite vein during the propylitic alteration event; **b** reopening of the vein and deposition of a corrensite + dolomite + hematite mineral assemblage (injection vein type)

argillic veins have recorded the different steps of the cooling process. Three types of vein were observed (Fig. 6.13): (type 1) illite + calcite; (type 2) regular I/S mixed-layer mineral (Rectorite) + calcite, (type 3) kaolinite + random I/S mixed-layer + Fe-Mg carbonates. Type 1 is crosscut by types 2 and 3, while type 2 is crosscut only by type 3. Some of the observed crosscut relations are shown in Fig. 6.13. The sequence illite–regular (ordered) I/S–random I/S is known to record decreasing temperature conditions (Velde 1985). Multiphase depositions are frequently observed in hydrothermal veins, because even if they are completely sealed they remain mechanically less resistant than the unfractured rock. Numerous examples have been described in the literature devoted to economic ore studies. Figure 6.14 shows the relations between two alteration episodes which were superimposed on a granitic rock from the Vosges massif France (Meunier et al. 1988): (1) chlorite + calcite + pyrite; (2) corrensite + dolomite + hematite.

6.4 Alteration of Pre-existing Clay Minerals

6.4.1 Layer-Charge Control of Clay Hydrothermal Reactions

The reactivity of clay minerals under hydrothermal conditions has been extensively studied in experiments (Eberl 1978; Eberl et al. 1978; Lahann and

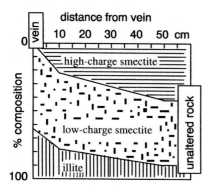

Fig. 6.15. Schematic representation of the reaction of a clay-rich material in the vicinity of an hydrothermal vein (Martinique). The high-charge smectite component is transformed into a low-charge smectite + illite assemblage. The intensity of the transformation increases toward the fracture

Robertson 1980; Howard and Roy 1985; Whitney and Northrop 1988 among others). However, hydrothermal alteration of clayey rocks in natural conditions is poorly documented because of the rarity of suitable outcrops. Some examples have been described in volcanic environments as natural analogs of engineered barriers for radioactive waste storage (Bouchet et al. 1988a,b). Compared to diagenetic mineral reactions, which are governed by the progressive increase of the time-temperature factor, the hydrothermal reactions are much shorter in duration and are controlled by the crystallochemical heterogeneity of the starting material.

Two reactions have been observed in hydrothermal veins which crosscut argillized ash deposits at "Les Trois Ilets, Martinique" (Bouchet et al. 1988a) which lead to an increase in the smectite content of the clay fraction. The first reaction is the transformation of the high-charge smectite layers into low-charge ones:

  illite–high-charge smectite mixed-layer + quartz + kaolinite → illite–low-charge smectite mixed layer.

The second reaction produces a totally expandable smectite of low-charge montmorillonite composition (Fig. 6.15):

  illite – low-charge smectite mixed layer + quartz → low-charge montmorillonite + K^+

As chemical systems in hydrothermal zones are open, dissolved components (noticeably K^+ ions) are leached out of the altered rock. This explains why illite cannot be formed. Similar reactions were observed in hydrothermal reactions occuring in porous sandstones of the Colorado River hydrothermal system (Jennings and Thompson 1987). In both cases, the temperature conditions did not exceed 100 °C. The layer charge is an important parameter in mineral reactions involving smectite-rich materials. The high to low-charge transformation observed in natural environments has been reproduced in experimental conditions. The alteration of bentonites in closed systems (Whitney and Northrop 1988) has shown that the first reaction to occur is the transformation of the low-charge montmorillonite into a high-charge one.

For instance, it was shown that the layers of low charge do not react in the same way as high-charge ones under hydrothermal conditions;

I/S mixed-layer high-charge smectite + Si^{4+} + K^+ →
I/S mixed-layer low-charge smectite + illite,

Si^{4+} and K^+ are supplied by the hot fluid. The reaction operates as long as the hydrothermal fluids can flow in the fracture and can be stopped before complete illitization (Fig. 6.15). In diagenetic series, the long time duration of mineral reactions allows more advanced transformation of smectite into I/S and then into illite.

6.4.2 Polyphase Clay Mineral Assemblages

In some cases, hydrothermal alteration leads to the formation of multiphase I/S assemblages with different expandabilities and structures. As the starting material tends to be homogeneous, the presence of different I/S species is mainly controlled by temperature. Contrary to diagenesis, it is possible to observe intermediate stages of the smectite-to-illite conversion in hydrothermal alteration. Each stage is characterized by the coexistence in the same sample of several I/S species. This is due to the fact that the smectite-to-illite reaction is complex. The transitional phases (more smectite-rich) are going to be replaced by more illitic ones with sufficient time. Each mechanism of the bulk reaction is characterized by its own I/S species (i.e. illite content and ordering) and kinetics.

6.5 Conclusion

There are two main reasons for interest in hydrothermal vein studies in environmental sciences: (1) veins concentrate clay minerals in the preferential pathways of solutions; (2) veins can be considered as natural analogues of alteration processes induced by deep nuclear waste storage. They are fascinating geological objects since they impose all the intermediate conditions between fluid dominated systems to rock dominated ones. Besides, the kinetics of the dissolution-crystallization processes can be studied by considering the distance to vein as a time-dependent parameter.

Phyllosilicates (particularly clay minerals) are formed in two different ways: direct precipitation from solutions in the vein and dissolution-crystallization processes on primary silicates inside the wall rocks. Their crystallochemical properties are heavily dependent on the time-temperature parameter. Short hydrothermal pulses favor the formation of low-charge expandable clays. Non-expandable species crystallize in systems in which fluids circulate during longer periods. Unfortunately, little has been known, until recently, regarding the kinetics of clay formation in active systems.

References

Baronnet A (1980) Polytypism in micas: a survey with emphasis on the crystal growth aspect. In: Kaldis E (ed) Current topics in material science, 5. North Holland, Amsterdam, pp 447–548

Beaufort D, Meunier A (1983) Petrographic characterization of an argillic hydrothermal alteration containing illite, K-rectorite, kaolinite and carbonates in a cupromolybdenic prophyry at Sibert (Rhône, France). Bull Miner 106: 535–551

Beaufort D, Meunier A (1994) Saponite, corrensite and chlorite-saponite mixed-layers in the Sancerre-Covy deep drill-hole (France). Clay Miner 29: 47–61

Beaufort D, Patrier P, Meunier A, Otaviani MM (1992) Chemical variations in assemblages including epidote and/or chlorite in fossil hydrothermal systems of Saint Martin (Lesser Antillies). J Volcanol Geotherm Res 51: 95–114

Berger G, Turpault MP, Meunier A (1992) Dissolution-precipitation processes induced by hot water in a fractured granite. Part 2: Modelling of water-rock interaction. Eur J Miner 4: 1477–1488

Berner RA (1981) Kinetics of weathering and diagenesis. Rev Mineral 8: 111–132

Bonorino FG (1959) Hydrothermal alteration in the front range mineral belt, Colorado. Bull Geol Soc Am 70: 53–90

Bouchet A, Meunier A, Velde B (1988a) Hydrothermal mineral assemblages containing two discrete illite/smectite minerals. Bull Miner 111: 587–599

Bouchet A, Proust D, Meunier A, Beaufort D (1988b) High-charge to low-charge smectite reaction in hydrothermal alteration processes. Clay Min 23: 133–146

Boulègue J, Benedetti M, Gauthier B, Bosch B (1990) Les fluides dans le socle du sondage GPF Sancerre-Covy. Bull Soc Geol Fr 8: 787–795

Ebertl DD (1978) The reaction of montmorillonite to mixed-layer clay: the effect of interlayer alkali and alkaline-earth cations. Geochim Cosmochim Acta 42: 1–7

Eberl DD, Whitney G, Khoury H (1978) Hydrothermal reactivity of smectite. Am Miner 63: 401–409

Elder J (1981) Geothermal systems. Academic Press, New York, 508 pp

Fournier RO, Potter RW (1982) A revised and expanded silica (quartz) geothermometer. Geotherm Resour Counc Bull 12: 3–9

Fyfe WS, Price NJ, Thompson AB (1978) Fluids in the Earth's curst. Elsevier, New York, 383 pp

Gonclves, NMN, Dudoignon P, Meunier A (1990) The hydrothermal alteration of continental basaltic flows in northern Parana basin (Ribeirao Preto, Sao Paulo state, Brazil). Proc 9th Int Clay Conf Sci Geol Mem 88: 153–162

Grimaud PO, Touchard G, Beaufort D, Meunier A (1990) Pulsated flow regime in fractures: a possible explanation of local temperature gradients in hydrothermal system. Geothermics 19: 329–339

Howard JJ, Roy DM (1985) Development of layer charge and kinetics of experimental smectite alteration. Clays Clay Min 33: 81–88

Jennings S, Thompson JR (1987) Diagenesis in the Plio-Pleistocene sediments of the Colorado River delta, southern California. J Sediment Petrol 56: 89–98

Joesten RL (1991) Kinetics of coarsening and diffusion-controlled mineral growth. Rev Mineral 26: 507–582

Joesten RL, Fischer G (1988) Kinetics of diffusion-controlled mineral growth in the Christmas mountains (Texas) contact aureoles. Geol Soc Am Bull 100: 714–732

Lahann RW, Robertson HE (1980) Dissolution of silica from montmorillonite: effect of solution chemistry. Geochim Cosmochim Acta 44: 1937–1943

Merceron T (1988) Les altérations hydrothermales de la coupole granitique d'Echassières et de son environnement (sondage G.P.F. ECHA N°1). Thèse Univ Poitiers, 167 pp

Merceron T, Vieillard P, Fouillac AM, Meunier A (1992) Hydrothermal alterations in the Echassières granitic cupola (Massif central, France). Contrib Mineral Petrol 112: 279–292

Meunier A, Clement JY, Bouchet A, Beaufort D (1988) Chlorite-calcite and corrensite crystallization during two superimposed events of hydrothermal alteration in the "Les Crêtes" granite, Vosges, France. Can Miner 26: 413–422

Michard G (1985) Equilibres entre minéraux et solutions géothermales. Bull Miner 108: 29–44

Parneix JC (1992) Effects of hydrothermal alteration on radioelement migration from a hypothetical disposal site for high level radioactive waste: example from the Auriat granite, France. Applied Geochim Suppl Issue 1: 253–258

Ramsay JG (1980) The crack-seal mechanism of rock deformation. Nature 281: 135–139

Rose AW, Burt DM (1979) Hydrothermal alteration. In: Barnes HL (ed) Geochemistry of hydrothermal ore deposits, 2nd edn. John Wiley, New York, pp 173–235

Sales RH, Meyer C (1950) Interpretation of wallrock alteration at Butte, Montana. Colo Sch Mines Quat 45(1B): 261–274

Sibson RH, Moore J, Rankin AH (1975) Seismic pumping – a hydrothermal fluid transport mechanism. J Geol Soc Lond 131: 653–659

Turpault MP, Berger G, Meunier A (1992a) Dissolution-precipitation processes induced by hot water in a fractured granite. Part 1: Wall-rock alteration and vein deposition processes. Eur J Miner 4: 1457–1475

Turpault MP, Meunier A, Guilhaumou N, Touchard G (1992b) Analysis of hot fluid infiltration in fractured granite by fluid inclusion study. Appl Geochem 1: 143–150

Velde B (1985) Clay minerals: a physico-chemical explanation of their occurrence. Elsevier, Amsterdam, 427 pp

Whitney G, Northrop HR (1988) Experimental investigation of the smectite to illite reaction: dual reaction mechanisms and oxygen isotope systematics. Am Miner 73: 77–90

7 Formation of Clay Minerals in Hydrothermal Environments

A. INOUE

7.1 Initial Statement

Formation of clay minerals under hydrothermal influence is the result of rock alteration by circulating hot water in the Earth's crust. A pre-existing rock-forming mineral assemblage is altered to a new set of minerals which are more stable under the hydrothermal conditions of temperature, pressure, and fluid composition. The interaction of hot water and rocks forms a spatially and temporally regular zonal pattern of new clay minerals, as the fluid with cooling temperature moves through the surrounding rock mass. This chapter discusses the formation of clay minerals in such dynamic processes of hydrothermal alteration. The approach is one of clay-mineral facies formed under conditions of massive alteration in the rocks. The chemical and mineralogical changes which occur on the scale of a rock or rock mass are considered to have been dealt with in the preceding chapter. The exact process of change via local, vein-influenced exchange processes is ignored for simplicity (see Chap. 6).

Hydrothermal alteration is often accompanied by hydrothermal ore deposits and active geothermal systems which are of economic importance. A huge amount of field and laboratory data on alteration mineralogy and geochemistry have been accumulated through exploration of hydrothermal ore deposits and active geothermal fields due to the economic interest of this geologic process. Additionally, the amount of theoretical information pertaining to the hydrothermal alteration has been rapidly increasing in the last decade. These data have been reviewed from various aspects of interest, e.g. in connection with the geneses of porphyry copper deposits (Meyer and Hemley 1969; Rose and Burt 1979; Beane 1982), epithermal ore deposits (Hayba et al. 1985; Henley 1985; Heald et al. 1987; Shikazono 1988), geothermal systems associated with recent volcanism (Ellis and Mahon 1977; Browne 1978; Utada 1980; Henley and Ellis 1983), Kuroko deposits (Utada 1988), and hydrothermal systems at sea floor spreading centers (Rona et al. 1983, for a summary). In the present review, particular attention has been paid to the formation of clay minerals under these different hydrothermal environments.

Initially, the general backgrounds of hydrothermal alteration will be briefly reviewed (Sect. 7.1). Section 7.2 is concerned with the terminology and definition of hydrothermal alteration. Sections 7.3 and 7.4 introduce the geologic settings and the physicochemical nature of hydrothermal systems. The basic concepts pertaining to the formation of alteration minerals and their zoning on a scale of tens of hundreds of meters, are discussed in Section 7.5. In Section 7.6, the great variety of hydrothermal alterations which appear in nature are classified in terms of two variables, temperature and solution composition, following the classification scheme of Utada (1980). Section 7.7 deals with the morphology of alteration zones. Based on these general backgrounds, the formation processes of clay minerals in some individual cases of hydrothermal systems are explained in Section 7.8. Section 7.9 gives a brief summary of hydrothermal alteration as a function of rock type. Detailed mineralogical aspects of selected clay minerals typical of hydrothermal alteration will be discussed in Section 7.10. The interstratified illite/smectite, chlorite/smectite, sericite, chlorite, and smectite minerals are considered.

7.2 Definition of Hydrothermal Alteration

Hydrothermal solution refers to all types of hot water that exist in the Earth's crust. The temperature is appreciably warmer (e.g., 5°C or more) than the surrounding environment (White 1957). Hydrothermal alteration is defined as the interaction of rocks with such hot waters. Despite such a simple definition there appears to be confusion in the literature in using the terms of hydrothermal alteration and the related hydrothermal solution. The definition of Utada (1980) is used throughout this review. He defined hydrothermal reactions as follows: hydrothermal alteration is a rock alteration where solution higher in temperature than that expected from the regional geothermal gradient in a given area interacts locally with the surrounding rocks. Accordingly, the water that was stored in rocks for geologically long periods and heated in situ, as a fluid within a formation is considered not to be a hydrothermal solution. Rock alteration by such hot water in equilibrium with the surrounding rock is usually categorized as diagenesis or metamorphism.

The different rock alterations involving the formation of clays are generally known as weathering, diagenesis, and hydrothermal alteration. This classification is based on the predominant factors operating on each rock alteration. The important factor operating in weathering is the composition of the solution. The rock alteration occurs under a constant temperature–pressure condition at the surface of the earth. In the course of burial diagenesis, sediments with interstitial solutions may be regarded as a nearly closed system in terms of solution movement. The rock alteration is controlled essentially by increase in temperature and pressure of both the solids and the solutions associated with them. Hydrothermal alteration is alteration controlled by variations of all three factors of solution composition, temperature, and pressure. As the number of variables increases a greater variety of clay minerals can be expected to form under hydrothermal environments.

7.3 Geologic Settings of Hydrothermal Systems

Hydrothermal systems occur in a wide variety of geologic settings and rock types. The presence of a heat source and rock fracture is essential for a hydrothermal system to be stably established in a given area. A subjacent heat source gives a part of the thermal energy to descending cool water. The heated water has a lower density and is able to rise along the fractures. The circulating water involves meteoric water and seawater. In some case, fluids emanated from a magmatic heat source can contribute to a considerable extent to the circulating hot water.

Ellis and Mahon (1977) classified active geothermal systems into two types, cyclic and storage systems. The hot fluid in the storage systems is derived from water which was heated in situ at depths as formation water, and thus it is excluded from the present criteria of hydrothermal solution as defined above. If a static storage system is disturbed by a certain tectonic activity, the hot water stored in the deep formation can ascend along the generated faults and fractures toward the surface. In this case, as hot water ascends it becomes a hydrothermal solution, according to the present definition, because its temperature is discordant with the regional geothermal gradient. For simplicity, however, we do not deal with the hydrothermal alteration of storage systems here.

In the cyclic systems, the hot water is mainly meteoric water which has passed through a cycle of deep descent, heating, and subsequent rising. Minor additions of other water, such as magmatic fluids, can also exist. Convective forces are important in these systems. The cyclic systems (convective) were subdivided by Ellis and Mahon (1977) as follows: (1) high-temperature systems associated with recent volcanism, (2) high-temperature systems in non-volcanic zones of Cenozoic tectonic activity, and (3) warm water systems in near-normal heat flow zones. High-temperature systems associated with recent volcanism occur in a variety of situations. For example the Circum Pacific Orogenic Belt presents zones of predominantly calc-alkaline rhyolitic or andesitic volcanism. In contrast the hydrothermal activity in Iceland, which is a hot spot located along the Atlantic Mid-Ocean Ridge, occurs in extensively fractured and predominantly basaltic rocks. Examples of the high-temperature systems in non-volcanic zones of Cenozoic tectonic activity are the Larderello region, Italy, the Kizildere field, Turkey, and the Pannonian basin field, Hungary. The Larderello region, for instance, is in a region of Paleozoic metamorphic rocks, Mesozoic limestones, and shales. Examples of the warm water systems in near-normal heat flow zones occur at Carlsbad in Czech

Republic and other sites. The Carlsbad spring is situated in Bohemia in the western end of the Krusne Hory graben. In this area, the recurrent faulting and fracturing allow water to circulate to great depths. The temperature and total dissolved salt concentrations are usually low in the warm water part of the systems.

The above examples are geothermal systems which are active today. On the other hand, recent geochemical studies of fossil (or extinct) geothermal systems, which are also hydrothermal ore deposit systems, indicate that the majority of hydrothermal ore deposits hosted by volcanic rocks or their subjacent plutonic suites were formed within geothermal systems of similar size, chemistry and behavior, as well as similar geologic settings, to those we see active today (Henley and Ellis 1983). For instance, porphyry copper deposits are abundant in the late Cretaceous-Tertiary volcanic belts of the Circum-Pacific and Alpine zones. Kuroko-type ore deposits are situated typically in a region of back-arc basin where rhyolitic volcanism occurred on the seafloor during the middle Miocene. The genesis of Kuroko-type ore deposits may be related to that of sea floor massive deposits which occur at sea floor spreading axes at present associated with basalt volcanism. Thus fossil hydrothermal systems can most often be identified with modern analogues.

7.4 Physico-chemical Nature of Hydrothermal Systems

7.4.1 Temperature and Pressure

From the above definition of hydrothermal alteration, the temperature of hydrothermal solutions can range from several tens of degrees close to the surface to several hundred degrees, for example those of magmatic fluids. Recent geochemical studies of hydrothermal fluids indicate the important contribution of magmatic fluids to hydrothermal solutions circulating within geothermal systems (Fournier 1987; Hedenquist 1987; Shinohara 1991).

Water has a critical point at approximately 374 °C and 22 MPa (Fig. 7.1), values above which, in these hydrothermal fluids, there is no distinction between vapor and liquid. The amount of interaction of rocks with such a supercritical fluid may be insignificant in hydrothermal systems because the escape of gaseous phases is rapid. Natural hot water is not pure water, but is an aqueous solution containing various dissolved materials such as salts, gases, etc. The critical point values of these aqueous solutions are generally greater than those of pure water (Fig. 7.1). For instance, seawater (3.2% NaCl) shows the critical point at approximately 405 °C and 30 MPa (Bischoff and Rosenbauer 1985). It has been known that salinities of hydrothermal solutions in common geothermal systems including Kuroko deposits and epithermal deposits, estimated from the melting temperature of fluid inclusions, range

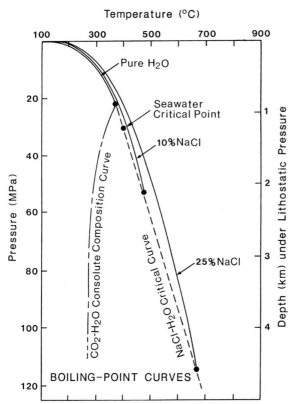

Fig. 7.1. Pressure–temperature diagram showing approximate positions of boiling-point curves and critical points *dots* for pure water, seawater (3.2 wt%), 10, and 25 wt% NaCl solutions, and CO_2-bearing water. (Modified from Fournier 1987)

from zero to several percent (wt%) in equivalent NaCl concentration (Roedder 1984). The presence of more saline brine is recognized, exceptionally, in some specific types of hydrothermal systems such as porphyry copper deposits and geothermal systems associating with subjacent evaporite beds. Accordingly, it is likely that the critical temperature of hydrothermal solution with a common salt concentration does not greatly exceed 400 °C. Fournier (1987) states that the convection systems of groundwater in geothermal fields are able to exist stably at temperatures below 400 °C. Because deformation in most rocks changes from frictional (brittle fracture) to plastic flow at about 400 °C the permeability reduction becomes increasingly more important. Consequently, the most likely upper limit of hydrothermal solution pertaining to rock alteration discussed here is around 400 °C. Rock alteration which has occurred at temperatures as high as 400 °C can be identified in porphyry copper deposits where biotite is extensively formed in the highest-grade zone. Biotite is not considered to be a member of the clay-mineral group because its grain size is relatively coarse. In the lower-grade zones of porphyry copper deposit systems, biotite is replaced by chlorite and/or sericite which are clay minerals. If we are concerned with the formation of clay minerals, the maximum temperature of the related hydrothermal solution may be around 300 °C instead of 400 °C, as deduced from information on the thermal stabilities of many clay minerals.

The minimum temperature of hydrothermal solution may be somewhat arbitrary. It is in fact limited by kinetic effects of rock–water interaction. As the temperature is lower, any reactions between rock and solution proceed more slowly, and therefore significant production of secondary minerals cannot always be recognized in the rocks during short periods of interaction. Consequently the minimum temperature observable may be around several tens of degrees, which must be essentially higher than the surrounding environment according to the definition of hydrothermal alteration.

The average thermal gradient in the earth's crust is about 30 °C/km. In hydrothermal systems, however, the thermal gradient is necessarily greater than the average value and often attains in excess of 200 °C/km.

The pressure of a fluid which has emanated from a magma source is initially close to the lithostatic pressure at its depth. It decreases upon cooling after separation from the magma. Convecting ground water can descend to depths of 2–3 km (Fournier 1987). At 3 km depth, lithostatic and hydrostatic pressures are 74 MPa and 29 MPa, using an average rock density of 2.5 kg/m^3 and water density of 1.0 kg/m^3, respectively, It is obvious therefore that the pressure of fluid related to hydrothermal alteration does not exceed 100 MPa. In general, both pressure and temperature in convective geothermal systems, excepting conductive geothermal systems associating with geopressured zone, follow the boiling point curve of the hydrothermal solution (Henley et al. 1984).

As a summary, Fig. 7.2 shows the most likely conditions of temperature and pressure for some types of hydrothermal systems discussed here, together with conceptual flow directions of convecting water and evolved magmatic fluid. Among the hydrothermal systems, porphyry copper deposits and Kuroko and sea floor massive sulfide deposits are representative of a high-temperature hydrothermal system, with magmatic fluids and a hydrothermal system with seawater, respectively.

7.4.2 Fluid Compositions

The composition of magmatic fluids depends upon those of parent magmas (e.g., felsic or basaltic magma) because the solubility of volatile components is a function of magma composition. Felsic magma is generally more rich in volatile components such as water, fluorine, and chlorine compared to basaltic magma. This was shown by analyses of their quenched glass compositions (Devine et al. 1984; Anderson et al. 1989; Bacon et al. 1992). The initial composition of volatiles in the hydrothermal fluid coming from a magma may be deduced from the composition of high-temperature volcanic gases which were carefully collected from fumaroles and craters (Table 7.1). The major

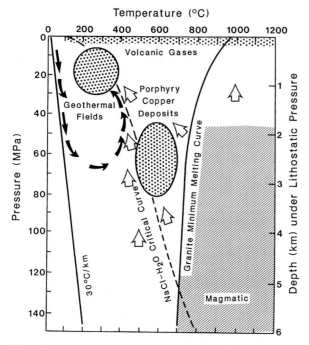

Fig. 7.2. Pressure–temperature diagram showing the likely conditions for various types of hydrothermal systems (modified from Shinohara 1991). The P-T fields for Kuroko, epithermal, sea floor massive sulfide deposits overlap the field of "geothermal fields". *Arrows* represent possible fluid flows of circulating meteoric water and ascending magmatic fluids

Table 7.1. Analyses of volcanic gases from Kilauea, Erta'Ale, and White Island (mol%)

	Kilauea[a]			Erta'Ale[a]			White Island[b]
H_2O	36.18	61.56	67.52	84.8	69.9		79.6
CO_2	47.68	20.93	16.96	7.0	15.8		13.6
CO	1.46	0.59	0.58	0.27	0.68		
COS				0.001	0.01		
SO_2	11.15	11.42	7.91	5.1	10.2		4.8
SO_3	0.42	0.55	2.46				
S_2	0.04	0.25	0.09	0.4	0.0		
HCl	0.08	0.00	0.20	1.28	1.22		0.17
HF							0.2
H_2	0.48	0.32	0.96	0.85	2.11		0.16
N_2	2.41	4.13	3.35	0.10	0.25		0.018
Ar	0.14	0.31	0.66	0.001	0.001		

[a] Kilauea and Erta'Ale (Krauskopf 1979).
[b] White Island (Ellis and Mahon 1977).

constituents of volcanic gas are H_2O, CO_2, and SO_2. The HCl and HF contents are generally small, although this may be due to the fact that part of them has been fractionated preferentially into the liquid phase after vapor-liquid exsolution at deeper levels (Bischoff and Rosenbauer 1985). The presence of these volatile components, other than H_2O, in magmatic fluid contributes to the generation of acidity in the evolved hydrothermal solution.

Stable isotopic studies, ^{18}O and D, of aqueous solutions in many active and fossil geothermal systems (Craig 1961; Taylor 1974) suggest that this water is mostly of meteoric origin. However, this does not exclude the contribution of some magmatic fluids to the hydrothermal solution. Most hydrothermal solutions may actually be a mixture of magmatic fluid and meteoric water in various proportions. In mixing a magmatic fluid with meteoric water, the nature of the resultant solution will diversify through interaction so as to differ from the initial nature of ascending magmatic fluid in porportion to the amount of mixing. For example SO_2 in magmatic fluid is disproportioned to form H_2S and SO_4^{2-} by the following reaction:

$$4SO_2 + 4H_2O = 3H_2SO_4 + H_2S. \qquad (1)$$

Reaction (1) generates the acidity in hydrothermal solutions.

A magmatic fluid which rises adiabatically, boils as it encounters the boiling point curve. By boiling, the volatile components will be fractionated between the liquid and gas phases. The volatiles tend to concentrate in the gas phase. The subsurface boiling, with removal of the evolved gas, is a possible mechanism that can generate high pH in a hydrothermal system. For instance, carbonate ions dissolved in fluids will be decomposed by boiling,

$$HCO_3^- = CO_2 + (OH)^-. \qquad (2)$$

Then, as CO_2 is concentrated in the gas phase, the separated solution increases in pH. In other cases, the variation in pH of a solution can take place in several different ways within hydrothermal systems (Fournier 1985). Moreover, because high-temperature solutions are very reactive with the surrounding rocks, the nature of the solution will be further modified.

Table 7.2 shows the compositions of hydrothermal fluids from several active geothermal fields. For a comparison, the chemical compositions of some ore-forming fluids are given in Table 7.3. The origins of hydrothermal fluids are varied, i.e., meteoric water, seawater, magmatic fluid, and so on. However, irrespective of the origin, chloride is a major constituent in hydrothermal fluids. Among many cation species, Na and K predominate and the concentration of Mg is usually low. The Na/K and Na/Ca ratios in fluids are variable from place to place. Fournier and Truesdell (1973) demonstrated that the Na/K and Ca/K ratios in hydrothermal fluids were equilibrated with respect to rock-forming feldspars at a given temperature and pressure. The study by Giggenbach (1980, 1981) exemplifies the attainment of mineral–gas equilibria in a geothermal system of New Zealand type. Similar observations have been made in many geothermal fields. Generally speaking, it is likely that neutral

Table 7.2. Composition of geothermal well waters (mg/kg at atmospheric pressure from the discharge) (Ellis and Mahon 1977; Henley and Ellis 1983)

Source Well depth (m)	Temperature (°C)	pH (20°C)	Li	Na	K	Mg	Ca	Cl	SO_4	HCO_3	SiO_2
Hveragerdi Well G3 Iceland (650 m)	216	9.6	0.3	212	27	0.0	1.5	197	61	55	480
Otake Well 7 Japan (350 m)	200	8.4	4.5	846	105	0.02	9.9	1219	214	56	425
Broadlands Well 13 New Zealand (1080 m)	260	8.6	12.6	980	200	0.02	2.4	1668	6.5	117	750
Kizildere Well 1A Turkey (430 m)	200	9.0	4.5	1280	135	0.2	2.5	117	770	1860	325
Cerro Prieto Well 5 Mexico (1285 m)	340	7.7	38	9062	2287	1	520	16045	6	56	1250
Carlsbad Spring Czech Republic	73	7.65	3.3	1718	104	46.5	102.5	617	1662	2100	68
Reykjanes Well H8 Iceland (1750 m)	275	7.1	6.6	12730	1990	9.8	2249	25054	2.4	–	943
Mahiao Well 401 Philippines (1947 m)	324	6.7	40	7800	2110	0.28	219	14370	32	27	995
Matsukawa Well 1 Japan (945 m)	300	4.9	–	264	144	8.7	22.9	12.4	1780	26	635
Salton Sea IID Well 1 California (1600 m)	340	4.7	215	50400	17500	54	28000	155000	5	7100	400
Matsao Well E-205 Taiwan (1500 m)	245	2.4	26	5490	900	131	1470	13400	350	2	639

Table 7.3. Composition of ore-forming fluids. (After Shikazono, 1976)

	Kuroko	Seafloor Massive sulfides	Epithermal Pb-Zn deposit	Porphyry Mo deposit
Temperature (°C)	250 ± 50	260 ~350	330	
pH	4.5 ± 0.5	3.6		
Na (mg/kg)	12500	9936 ~ 11730	11800	64000
K (mg/kg)	4000	907 ~ 1009	4800	13000
Mg (mg/kg)	1 ~10	~ 0	120	2400
Ca (mg/kg)	>500	469 ~ 834	2500	26000
Cl (mg/kg)	33000	17335 ~ 20526	27400	70000
SO_4 (mg/kg)	450	0		
H_2S (mg/kg)	160	228 ~ 296		
Notes	Estimated	East Pacific Rise, 21°N	Fluid inclusion in sphalerite	Fluid inclusion in quartz
References	Shikazono (1976)	Von Damm et al. (1985)	Rye and Haffty (1969)	Hall et al. (1974)

solutions (near pH 7) are closer to an equilibrium with rocks than are acid solutions (Giggenbach 1988). The pH of a hydrothermal solution is a function of its salinity and is ultimately controlled by the interactions of aluminosilicate minerals and water (Arnorsson et al. 1978; Henley and Ellis 1983; Giggenbach 1984).

A solution, which is equilibrated with the surrounding rocks at depth and which moves to a shallower level, will change toward a new equilibrium state. Mineralogical and chemical changes by hydrothermal alteration occur so that a new equilibrium between the rocks and the circulating solution is approached.

7.5 Formation of Alteration Minerals and Their Zoning

As mentioned above, hydrothermal alteration reflects the response of preexisting rock-forming minerals to thermal and/or chemical changes under the influence of exchange with aqueous solution. As a result, new phases consists mainly of hydrous minerals; clay minerals are produced. In order for a new mineral or a mineral assemblage to be stable within a hydrothermal alteration

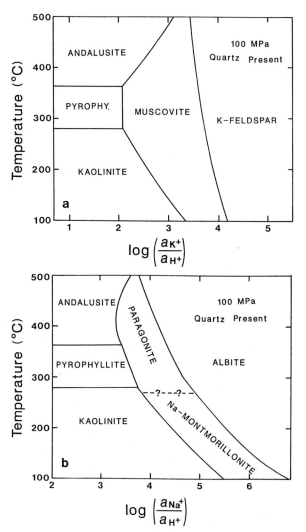

Fig. 7.3. a Temperature–logarithmic cation activity ratio diagram in the system K_2O-Al_2O_3-SiO_2-H_2O at a total pressure of 100 MPa and quartz saturation (redrafted from experiments by Sverjensky et al. 1991). **b** Temperature–logarithmic cation activity ratio diagram in the system Na_2O-Al_2O_3-SiO_2-H_2O at a total pressure of 100 MPa and quartz saturation. (Redrafted from experiments by Montoya and Hemley 1975)

environment, it must not only satisfy requirements regarding thermal stability, but it must also be compatible with the composition of the coexisting solution. The stability relationships of hydrothermal minerals can be usually examined in terms of the two variables: temperature and the activity ratio of the aqueous species in solution. As a first approximation, hydrothermal alteration can be considered to occur at essentially constant pressure.

Representative examples of hydrothermal mineral equilibria can be expressed in a temperature–activity diagram as illustrated in Fig. 7.3. Figure 7.3a expresses the stability relations in potassic systems of kaolinite, pyrophyllite, and muscovite (in reality the potassic minerals are illite or sericite) which are alteration products produced by reactions of rocks with acid to neutral solutions (see Sects. 7.6 and 7.8). Figure 7.3b is concerned with the formation of Na-bearing clay minerals instead of K. The stability of CaAl-silicates is also a

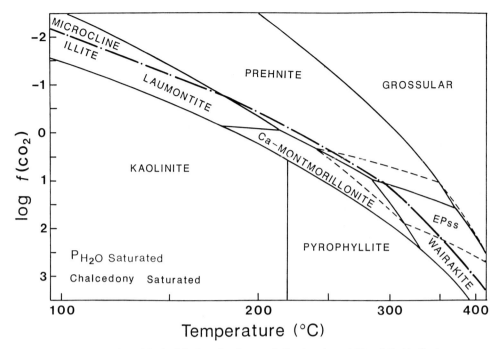

Fig. 7.4. Temperature–logarithmic CO_2-fugacity diagram indicating the stability of CaAl-silicates in equilibrium with chalcedony and calcite under liquid–vapor equilibrium pressures for water (after Giggenbach 1984). *Dash-dotted* cruve represents the locus of microcline-illite coexistence. The stability field of epidote solid solution (*Epss*, delineated by *dash line*) was determined in assuming activity of clinozoisite = –0.5

function of the CO_2-fugacity (f_{CO_2}) of the system (Fig. 7.4). The f_{CO_2} itself may be controlled by CaAl-silicates and calcite (Arnorsson et al. 1978; Giggenbach 1981, 1984).

By combining two different diagrams, as depicted in Fig. 7.3, it is possible to construct isothermal diagrams for some important hydrothermal silicate minerals and mineral assemblages, and to express these stability relations in terms of solution composition (Fig. 7.5). Similar diagrams can be referred to in many related papers and textbooks (Bowers et al. 1984, for a summary). Those diagrams were calculated by using thermodynamic data for minerals and aqueous species which are available today (e.g., Helgeson et al. 1978; Bowers et al. 1984; Henley et al. 1984; Berman 1988). Unfortunately, the thermodynamic data for clay minerals, except for some exceptions, are still erratic. The detailed shape delineating the stability fields of clay minerals depicted in Figs. 7.3–7.5 will be revised in the future by improving the thermodynamic data. For the present purpose, however, Figs. 7.3–7.5 provide a useful understanding of the basic concept of spatial and temporal variations in alteration mineralogy.

Changes of mineral assemblages in an interaction between multi-component solution and rock are actually complex at high temperatures. Let us consider a simple case of reaction sequence in the K_2O–Al_2O_3–SiO_2–H_2O system, according to Beane (1982). To explain the variation in both time and

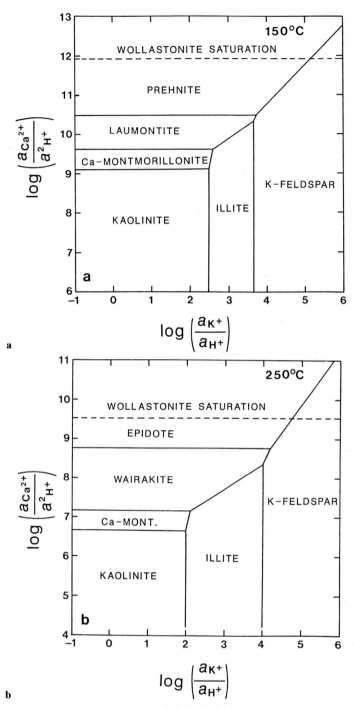

Fig. 7.5. Calculated stability relations among hydrothermal silicate minerals as a function of cation activity ratios in a coexisting solution at constant temperature. **a** The system K_2O–CaO–Al_2O_3–SiO_2–H_2O at 150 °C; **b** the same system at 250 °C; **c** the system K_2O–Na_2O–CaO–Al_2O_3–SiO_2–H_2O at 150 °C and 250 °C; **d** the system K_2O–MgO–Al_2O_3–SiO_2–H_2O at 250 °C. The silica

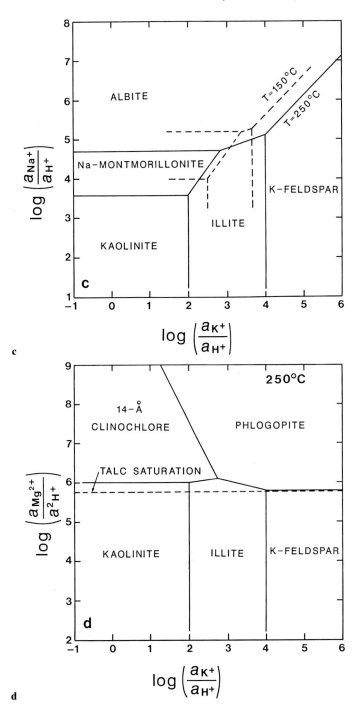

activity in the solution was assumed to be controlled by chalcedony (cryptocrystalline SiO_2) at 150 °C and by quartz at 250 °C. The pressure is controlled by the liquid–vapor curve of water in all systems. Thermodynamic data of minerals were adopted from Bowers et al. (1984), except for montmorillonite and wairakite which were adopted from Lonker et al. (1990)

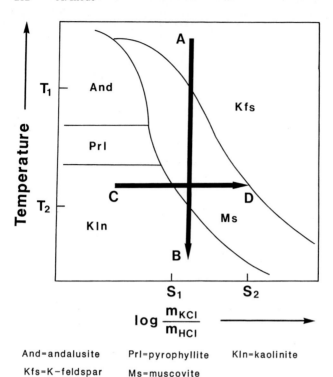

Fig. 7.6. Schematic representation for different reaction paths in hydrothermal alteration (after Beane 1982). The cation activity ratio depicted in Fig. 7.3a is replaced by total concentrations (m) of dissolved KCl and HCl

space in alteration mineralogy, Beane (1982) used the hypothesis of two fundamentally different mechanisms, one was a decrease in temperature and another was an evolving chemical reaction (Fig. 7.6). In the case of temperature-defined alteration assemblages, it is assumed that a fluid initially has a composition in equilibrium with K-feldspar at high temperature. If the fluid has an initial temperature and a [K$^+$]/[H$^+$] activity ratio corresponding to point A in Fig. 7.6, then, upon cooling along the line A–B, the fluid composition moves progressively through the stability fields of K-feldspar, muscovite, and kaolinite. When the fluid moves through a rock mass containing K-feldspar at continually decreasing temperature, it remains in equilibrium with this mineral until it cools to temperature T1. With further temperature decrease, K-feldspar will alter to muscovite from T1 to T2. The reaction is described by:

$$3KAlSi_3O_8 + 2H^+ + 12H_2O = KAl_3Si_3O_{10}(OH)_2 + 2K^+ + 6H_4SiO_4. \tag{3}$$

K-feldspar will alter to kaolinite below T2:

$$2KAlSi_3O_8 + 2H^+ + 9H_2O = Al_2Si_2O_5(OH)_4 + 2K^+ + H_4SiO_4. \tag{4}$$

These reactions result in the formation of an alteration sequence which is K-feldspar → muscovite → kaolinite, in the surrounding rocks. Each of the reactions in the sequence consumes H$^+$ from the aqueous solution. The [K$^+$]/[H$^+$] activity ratio then decreases from the initial value. With decreasing temperature the reaction path continues to a hydrous and low potassium content mineral.

In the case of reaction-defined alteration assemblages, a fluid having a relatively low pH and low [K$^+$]/[H$^+$] activity ratio, reacts with rock containing K-feldspar. Point C in Fig. 7.6, for example, might represent the composition of an original fluid that is in equilibrium with kaolinite. If this fluid interacts with a granite, the K-feldspar in the surrounding rock is out of equilibrium with respect to the hydrothermal fluid and, by consequence, it is converted to kaolinite compatible with the fluid composition, following reaction (4). The resulting consumption of H$^+$ and release of K$^+$ shifts the fluid composition along the line C–D in Fig. 7.6. When the muscovite–kaolinite boundary is reached at S1, kaolinitization of K-feldspar ceases, because the fluid is now in equilibrium with muscovite instead of kaolinite. The formation of muscovite now proceeds by reaction (3) and fluid composition continues to evolve in the direction of the arrow until the [K$^+$]/[H$^+$] activity ratio has increased to a value corresponding to equilibrium between muscovite and K-feldspar at S2. At this stage, interaction between fluid and rock ceases because the fluid composition is compatible with the initial rock-forming K-feldspar. A more detailed reaction path in the case of reaction-defined alteration has been examined in terms of the reaction progress by Helgeson (1979). By either mechanism, temperature or chemistry, a set of mineral assemblages is arranged systematically in terms of space and paleo-time within the surrounding rocks through the agent of moving fluids. A volume where a neo-formed mineral, or mineral assemblage, occurs is defined as a mineral (or an alteration) zone in hydrothermal alteration.

Intensive hydrothermal alteration may be a type of metasomatism. For instance, reactions (3) and (4) represent a hydrogen metasomatic reaction. Korzhinskii (1959) and Thompson (1955) have noted that in metasomatic situations, the mineralogical phase rule should be modified from the usual Gibbs phase rule to read $P = C - M$, where P is the maximum number of phases present in an assemblage under a condition of temperature, pressure, and solution composition chosen arbitrarily, C is the total number of independent chemical variables in the system, and M is the number of mobile components for which the chemical potential is controlled externally to the system rather than by an initial bulk composition. In hydrothermal systems, many of the chemical components are mobile in the thermodynamic sense, and as a result, the number of phases present in an alteration zone is small, in some case unity. This is typical of hydrothermal systems, the new clay-bearing mineralogy contains fewer phases than that of the initial "host" rock.

In active hydrothermal systems, zonal patterns and the directional distribution of alteration minerals are basically related to mass transfer among minerals and hydrothermal solutions as temperature decreases when the hydrothermal solution rises in the Earth's crust (Helgeson 1979). In other cases, the formation of mineral zones can be significantly influenced by many other effects, e.g. the kinetic effect of mineral-fluid interaction, the effect of the flow rate of the solution, the effect of the water to rock ratio, and the boiling effect of a rapidly rising fluid. Recent research seems to be centered on establishing a model of hydrothermal alteration which takes into consideration both the fluid mechanical behavior of solute movement and the kinetic effects of mineral–fluid interaction (e.g., Norton and Cathles 1979; Lasaga 1984; Lichtner 1991). Steefel and Lasaga (1992) noted the importance of the hydrodynamic dispersion with regard to the spatial distribution of alteration zones. As distances increase in space, the ratio of solutions to rock must change in favor of the rocks.

7.6 Classification of Hydrothermal Alteration

It has been recognized that similar clay-mineral assemblages have been recurrently produced in many hydrothermal systems at different times and places. A good example for this can be found in porphyry copper deposit systems (Creasey 1959). In this context, the alteration mineral assemblages or the defined zones in porphyry copper systems have been classified using a specific terminology: potassic, phyllic, argillic, and propylitic types. The classification is unrelated to the genesis and is based on observed minerals in each zone, as listed in Table 7.4. Utada (1980) classified a variety of hydrothermal alteration occurrences found in young orogenic belts according to three types defined in terms of variables of temperature and activity ratio of aqueous cation species in the hydrothermal solution (Fig. 7.7). According to the different activity ratio values of aqueous species in the solution, alteration is first grouped as follows: an acid-type of alteration at low cation/hydrogen ratios in the solution; an intermediate alteration at medium cation/hydrogen ratios; and an alkaline type of alteration at high cation/hydrogen ratios. Each alteration comprises several mineral zones corresponding to different temperature conditions.

The acid types of alteration are characterised mainly by minerals comprising the three components Al_2O_3, SiO_2, and H_2O (Fig. 7.8). The other components are almost all leached out from the rocks during the intensive alteration. The combination of the three components forms silica minerals, kaolin minerals, pyrophyllite, andalusite, corundum, and Al-hydroxides. Under an SiO_2-

Table 7.4. Alteration types and their diagnostic mineral assemblages in porphyry copper systems

Type of alteration	Diagnostic minerals
Potassic	K-feldspar (orthoclase), biotite, sericite, chlorite, quartz
Phyllic	Quartz, sericite, pyrite, chlorite, interstratified illite/smectite
Argillic	Kaolinite, montmorillonite, chlorite
Propylitic	Chlorite, epidote, albite, carbonates

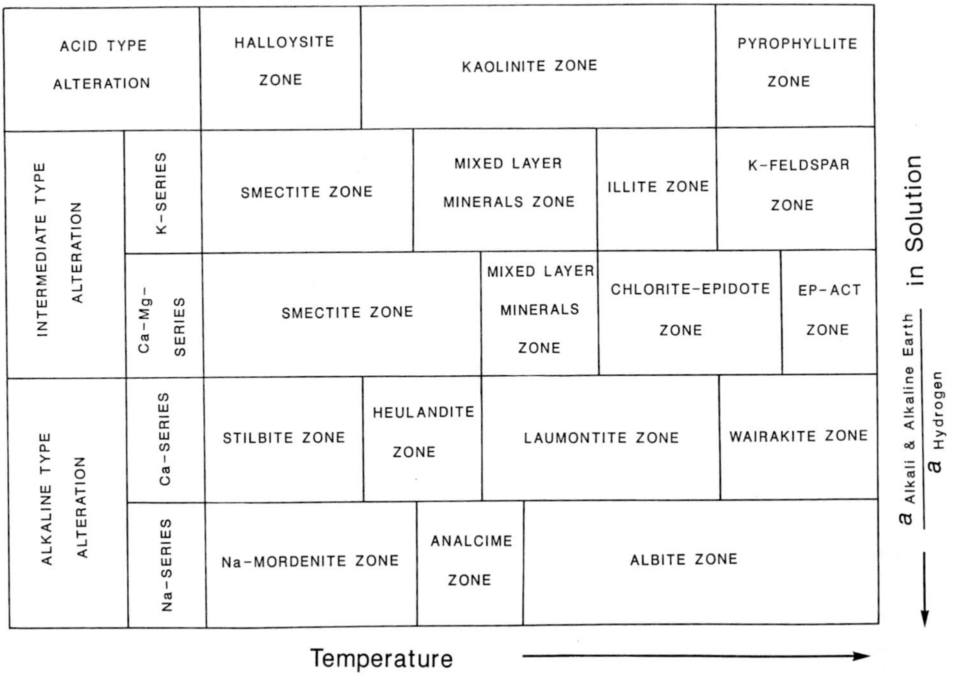

Formation of Clay Minerals in Hydrothermal Environments   285

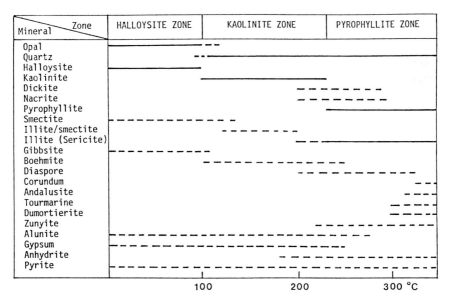

Fig. 7.7. Division of zones in three types of hydrothermal alteration. (After Utada 1980)

Fig. 7.8. Generalized summary of mineral assemblages observed in the acid type of alteration as a function of approximate temperature. *Solid* and *dashed* lines give an indication of the most common and rare occurrences, respectively. The temperature scale attached is approximate

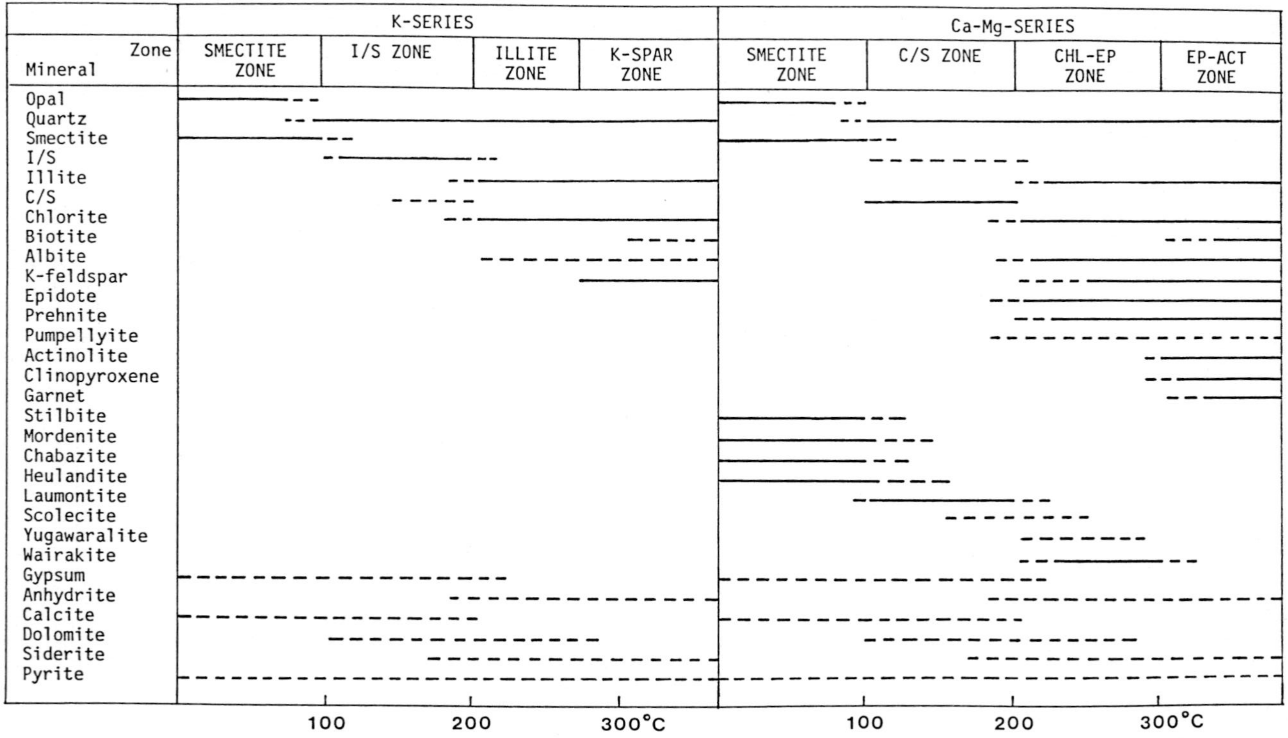

Fig. 7.9. Generalized summary of mineral assemblages observed in intermediate to alkaline types of alteration as a function of approximate temperature. *Solid* and *dashed* lines give an indication of the most common and rare occurrences, respectively. The temperature scale attached is approximate. *CHL-EP* chlorite-epidote zone, *EP-ACT* epidote-actinolite zone, *I/S* illite/smectite mixed layer mineral zone, *C/S* chlorite/smectite mixed layer mineral zone

Table 7.5. Classification of natural zeolites appeared commonly in hydrothermal alteration environments. (After Gottardi and Galli 1985)

Group	Species	Idealized Composition
Fibrous zeolites	Natrolite	$Na_{16}(Al_{16}Si_{24}O_{80}) \cdot 16H_2O$
	Mesolite	$Na_{16}Ca_{16}(Al_{48}Si_{72}O_{240}) \cdot 64H_2O$
	Scolecite	$Ca_8(Al_{16}Si_{24}O_{80}) \cdot 24H_2O$
	Thomsonite	$Na_4Ca_8(Al_{20}Si_{20}O_{80}) \cdot 24H_2O$
Zeolites with singly connected 4-ring chains	Analcime	$Na_{16}(Al_{16}Si_{32}O_{96}) \cdot 16H_2O$
	Wairakite	$Ca_8(Al_{16}Si_{32}O_{96}) \cdot 16H_2O$
	Laumontite	$Ca_4(Al_8Si_{16}O_{48}) \cdot 16H_2O$
	Yugawaralite	$Ca_2Al_4Si_{12}O_{32} \cdot 8H_2O$
Zeolites with doubly connected 4-ring chains	Phillipsite	$K_2(Ca_{0.5},Na)_4(Al_6Si_{10}O_{32}) \cdot 12H_2O$
Zeolites with 6-ring chains	Chabazite	$Ca_2(Al_4Si_8O_{24}) \cdot 12H_2O$
Zeolites of the mordenite group	Mordenite	$Na_3KCa_2(Al_8Si_{40}O_{96}) \cdot 28H_2O$
	Epistilbite	$Ca_3(Al_6Si_{18}O_{48}) \cdot 16H_2O$
	Ferrierite	$(Na,K)Mg_2Ca_{0.5}(Al_6Si_{30}O_{72}) \cdot 20H_2O$
Zeolites of the heulandite group	Heulandite	$(Na,K)Ca_4(Al_9Si_{27}O_{72}) \cdot 24H_2O$
	Clinoptilolite	$(Na,K)_6(Al_6Si_{30}O_{72}) \cdot 20H_2O$
	Stilbite	$NaCa_4(Al_9Si_{27}O_{72}) \cdot 30H_2O$

saturated condition, the zonal patterns in the acid type of alteration are characterized by the sequential appearance of clay minerals which are the same as with increasing temperature; halloysite, kaolinite, and pyrophyllite associated with a stable silica polymorph at respective temperatures (Figs. 7.3, 7.7, and 7.8). In contrast when the system is undersaturated with silica, aluminium hydroxides such as gibbsite, boehmite, and diaspore occur characteristically, and even corundum may occur at temperatures in excess of about 350 °C. When the alteration is caused by a hydrothermal solution containing abundant sulfate ions, alunite (an aluminum sulfate) occurs as an important product with or without the above minerals. Acid-type alteration favors pure oxide or hydrated minerals, either silica or alumina.

In the case of intermediate and alkaline types of alteration, considerable amounts of alkali and alkaline earth elements are present in the solution. Utada (1980) further subdivided these two types of alteration into four series, K, Na, Ca, and Ca-Mg series, which are distinguished by the predominant cation species in the solution (Fig. 7.7). The resulting mineralogical zones are characterized by the mineral assemblages containing these cations in aluminosilicates in addition to secondary mafic (Fe, Mg) minerals. By way of example, diagnostic mineral assemblages in the K and Ca-Mg series of alteration, which are identified in many hydrothermal systems, are summarized in Fig. 7.9.

It is noteworthy that zeolites, as well as clay minerals, characterize the mineral assemblages of alteration zones in the Na, Ca, and Ca-Mg alteration series. The zeolites which are commonly present in the hydrothermal systems discussed here are listed in Table 7.5. Na-mordenite and analcime are common in the Na-series of the alkaline type of alteration. In the Ca and Ca-Mg series of alteration, stilbite, mordenite, chabazite, and heulandite occur in most rock types at low temperatures: they are replaced at moderate temperatures by laumontite, epistilbite, scolecite, and yugawaralite, and finally wairakite is a stable zeolite at temperatures above 230–250 °C. The formation of fibrous zeolites such as natrolite, mesolite, scolecite, and thomsonite is favored by low-silica activity conditions and they are recognized sporadically as vug fillings in propylitically altered basaltic rocks (see Sect. 7.8). Ferrielite follows mordenite when Mg activity in the solution is high. Many other zeolites have been reported in natural hydrothermal environments (Gottardi and Galli 1985) but their occurrence is, in general, rare.

7.7 Distribution and Morphology of Alteration Zones

Fracture distribution, porosity, and permeability of the rocks basically influence the distribution and morphology of alteration zones, because the physical variables which control the fluid circulation within the system are those of the rock structure. Concerning the morphology of alteration zones, at least two types of hydrothermal systems can be identified: one is a fracture-controlled, fluid-dominated system and the other is a heat-conductive, rock-dominated system (Utada 1980; Izawa 1986). Although various morphologies of alteration zones exist in nature, most may be thought of as a derivative or a combination of the two fundamental types.

First, let us consider a hypothetical case of the formation of hydrothermal alteration zones around a single vertical fracture at various depths, as depicted in Fig. 7.10. Based on the reaction mechanisms in Fig. 7.6, alteration zones should be distributed symmetrically from the center of the fractuce to the periphery, following the gradients in temperature and/or chemical potential of the fluid in contact with the rock. The size of each zone which has developed is a function of the mass and flow rate of the solutions and the reactivity of the rock (this variable is related to its crystallinity and porosity); it usually ranges from millimetric to plurimetric. Generally speaking, it is expected that the conditions such as slow flow rate of solution, porous rocks, and glassy materials, facilitate the growth in size of alteration zones around the fracture. The mineral assemblages of the alteration zones are different from level to level of depth because of differences in the temperatures; it is obvious that a high-temperature mineral assemblage occurs at greater depth and that a low-temperature formed assemblage zone overlies the high-temperature formed zones. The distance between the high-temperature and low-temperature zones is a factor of the vertical thermal gradient at a given place.

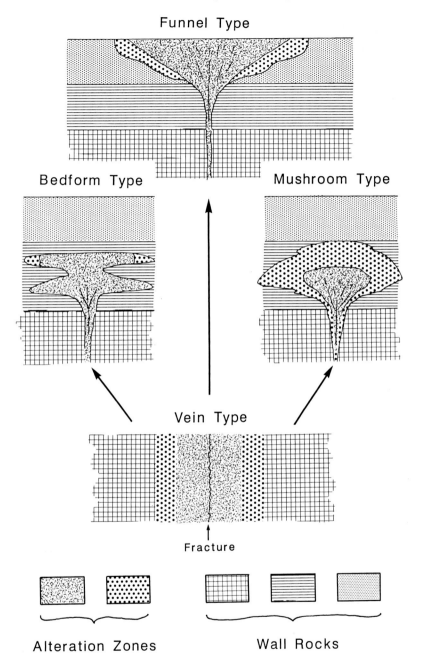

Fig. 7.10. Illustrations of various types of alteration zone morphology which are essentially controlled by the distribution of fractures in the system

If a fluid rising along a vertical fracture encounters a formation which is more porous and permeable than the deeper rocks, alteration proceeds selectively through this permeable formation, because of the easier movement of fluid. As a result, the alteration zone exhibits a bed-form type of morphology (Fig. 7.10). In general, a situation is diagnostic of hydrothermal alteration where field evidence shows that the boundary between alteration zones intersects a stratigraphic boundary. However, the zone boundary in this type of alteration is often apparently parallel to the stratigraphic boundary of the beds. This feature is especially visible in the marginal portion of the bedform type of alteration.

As seen in Fig. 7.10, if there is an overburden or a caprock, the fluid movement will be restrained in the vertical direction, while it will still be possible in the lateral direction. This results in the formation of a mushroom-shaped type of alteration. This morphology may also be formed by the subsurface boiling of a rising fluid, or by mixing with meteoric water at depth (Hayba et al. 1985). If the rising fluid comes into contact with meteoric water near the ground surface, the acidity of the fluid increases following reaction (1) and the oxidation of H_2S at the surface. Such acidic fluids form an intensive zone of acid-type alteration near the surface. In some case, the subsurface boiling may even bring about a hydrothermal explosion (Hedenquist and Henley 1985; Hulen and Nielson 1988). In this case the overburden will be blown away. This modifies the morphology of the alteration zone to form a funnel-type morphology. The fracture-controlled type of alteration is characterized by the development of widespread veins and veinlets in altered rocks. The distribution of fractures and veins is usually controlled by the regional tectonics at a given area (Watanabe 1986; Otsuki 1989; Merceron and Velde 1991).

Another type of alteration-zone morphology is seen in the outer hydrothermal alteration zones adjacent to the inner contact of metamorphic zones. This type of alteration is essentially created by heat conduction from a heat source such as an intrusive mass. Such zones show a concentric halo of alteration morphology around the heat source. The alteration zones which have developed are usually characterized by a distribution on a regional scale, but are delineated by lines subparallel to the stratigraphic boundaries. The altered rocks exhibit a pervasive or stagnant style of alteration. Regional propylitization demonstrates this type of alteration.

The two types of alteration-zone morphology mentioned above may also be related to the evolutional history of a hydrothermal system associated with a deep magmatic heat source. At an incipient stage of the hydrothermal activity, an alteration halo controlled by heat conduction envelopes the subjacent heat source, which produces an effect similar to contact metamorphic aureoles. This is due to the fact that heat diffuses some 15 to 30 times faster than ions in solution (McNabb and Henley 1979). At this stage, the establishment of a fluid-circulation system may have been incomplete and the alteration would have taken place under low water/rock ratio conditions. At

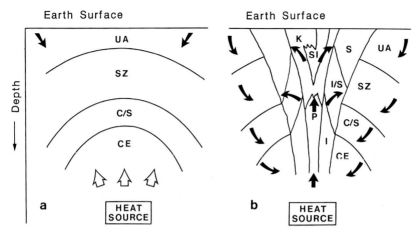

Fig. 7.11. Two types of zoning in hydrothermal alteration, **a** a heat conduction-controlled, rock-dominated type and **b** a fracture-controlled, fluid-dominated type (modified from Izawa 1986). *CE* Chlorite-epidote zone; *C/S* chlorite–smectite mixed-layer mineral zone; *SZ* smectite-zeolite zone; *UA* unaltered rocks; *SI* silicified zone; *K* kaolinite-sericite zone; *P* pyrophyllite zone; *I* illite-chlorite zone; *I/S* illite–smectite mixed-layer mineral zone; *S* smectite zone. *Arrows* indicate flow directions of hydrothermal solutions. Temporal relations between the two types of zonal patterns are discussed in the text

later stages of the activity history, fractures may be developed within the system, a stable fluid-circulation system is then established, and finally the fracture-controlled, fluid-dominated type of alteration overprints the alteration halos which had developed earlier. These conceptual models are schematically illustrated in Fig. 7.11. Such a chronological interpretation for the distribution and morphology of alteration zones has been applied to the genesis of propylitic alteration in porphyry copper systems (Gustafson and Hunt 1975), in epithermal vein systems (Izawa 1986), and in active and fossil geothermal systems (Kimbara 1983; Inoue and Utada 1991a).

7.8 Case Studies of Hydrothermal Alteration

7.8.1 Acid-Type Alteration

This type of alteration is commonly found not only in silica deposits and sulfur fumaroles in volcanic regions, but also in many active and fossil geothermal fields. As mentioned previously, a high-temperature acid solution has the potential to leach most rock-forming elements such as Na, K, Mg, Ca, etc., with the result that relatively simple, homogeneous mineral assemblages are formed, regardless of the different parent rock and mineral types. These very

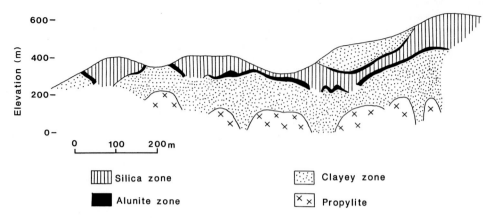

Fig. 7.12. Cross section showing the alteration zoning at the Ugusu silica deposit, Shizuoka, Japan. (After Nagasawa 1978)

aggressive solutions produce kaolin minerals and pyrophyllite, in addition to silica minerals. Native sulfur, pyrite, and alunite, are major constituent minerals characteristic of this type of alteration, as shown in Fig. 7.8.

Figure 7.12 shows a schematic cross section of the zonal mineralogical distribution at the Ugusu silica deposit, Japan (after Iwao 1970). This deposit was formed by acid hydrothermal solutions associated with very recent volcanic activity which is post-Pliocene. The zonal arrangement is well-developed from the center to the marginal parts as follows: silica zone → alunite zone → clayey zone → original rocks, which are Miocene propylites. The clayey zone is composed mainly of kaolin minerals, i.e., dickite, nacrite, and kaolinite. Uno and Takeshi (1982) recognized a local occurrence of pyrophyllite adjacent to the alunite zone. Smectite, illite, their interstratified minerals, and chlorite, in addition to kaolinite, represent the mineral assemblage in the outer portion of the clayey zone.

In general, the component Al_2O_3 is conventionally considered to be conserved in rocks during hydrothermal alteration, because of its low solubility in moderately acid solutions. However, if the pH of solution is extremely low (less than pH 4 at 20 °C and this value is further reduced at high temperatures), Al_2O_3 is more easily dissolved in the solution, whereas SiO_2 is relatively insoluble. In this example, a porous silica zone occurs in the central portion of the alteration zone, which corresponds to a vent of ascending fluid. The central silica zone grades outward into a kaolin mineral zone and then into a smectite zone, as hydrogen ions n solution are consumed by hydrolysis reactions, e.g., reactions (3) and (4), with the rock-forming minerals upon decreasing temperature (see Fig. 7.6), as shown typically in the case of the Ugusu silica deposit. This type of alteration by extremely acid solutions has also been reported in many hydrothermal systems related to sulfur deposits (e.g., Nishiazuma, Japan; Mukaiyama 1970) and acid-sulfate type epithermal gold deposits (e.g., Summitville, Colorado; Stoffregen 1987).

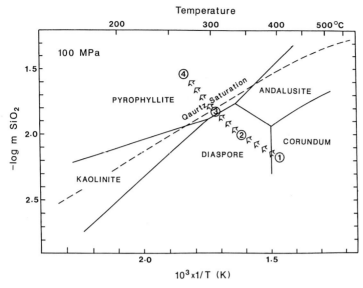

Fig. 7.13. Stability relations of minerals in the system Al_2O_3–SiO_2–H_2O at a total pressure of 100 MPa (after Hemley et al. 1980). The path line from *1* to *4* indicates the conditional variation in the formation of mineral zones shown in Fig. 7.14

In the case where the alteration zone is composed of quartz, alunite, and kaolin, the variation in the pH of reacting solutions through the zone can be estimated by determining the variation in amounts of kaolin, alunite, and quartz in the altered rocks, because the three minerals are interrelated by the reaction:

$$2(K,Na)\text{–alunite} + 6 \text{ quartz} + 3 \text{ water}$$
$$= 3 \text{ kaolin} + 2(K,Na)^+ + 4(SO4)^{2-} + 6H^+. \tag{5}$$

Reaction (5) indicates that a lower pH solution favors precipitation of alunite, and that alunite reacts with quartz to form kaolin in higher pH solutions (Inoue and Utada 1991c).

Examples of the alteration associated with moderately acid solutions can be found in kaolin or pyrophyllite deposits. The major mineral constituent depends on the temperature attained in the system; pyrophyllite occurs at temperatures in excess of 270°C, at a pressure of 100 MPa and quartz saturation, whereas kaolinite replaces pyrophyllite at temperatures below 270°C (Figs. 7.3 and 7.13). This phase-boundary temperature decreases with a reduction in fluid pressure. At temperatures roughly below 100°C, halloysite is an important product in place of kaolinite. Other polymorphs of kaolinite, dickite and nacrite, are often associated with pyrophyllite, which suggests that they are high-temperature phases of kaolinite. In addition to these aluminous clay minerals, andalusite (a polymorph of Al_2SiO_5), corundum (Al_2O_3), diaspore

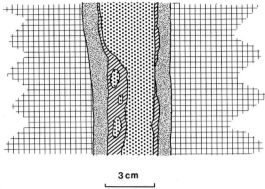

Fig. 7.14. A sketch showing the occurrence of a corundum vein at the Goto pyrophyllite deposit, Japan. The formation process of this zonal arrangement is referred to in Fig. 7.13

(AlOOH), and/or boehmite (AlOOH) are occasionally recognized in the highest-grade (temperature) zone of alteration (Figs. 7.3 and 7.13). Figure 7.14 shows a corundum vein observed in the Goto pyrophyllite deposit, Japan. The outward zonal sequence from a central corundum vein reflects both a decrease in temperature and an increase in silica activity when the fluids react, following approximately the pathway shown in Fig. 7.13.

Two different origins for the acid fluids can be considered, one is meteoric and the other magmatic. In the acid type of alteration formed by reactions where the solution was significantly dominated by magmatic fluids, topaz, zunyite, dumortierite, and/or phosphate-bearing alunite usually coexist with kaolinite or pyrophyllite (Rye et al. 1992). Marumo et al. (1982) distinguished the genetic origin (supergene or hydrothermal) of kaolin deposits in Japan by measuring their ^{18}O and D isotope compositions (Fig. 7.15). Figure 7.15b suggests that kaolinites in some deposits formed at about 100 °C, if the related waters were of meteoric origin and equilibrated with the kaolinites.

7.8.2 Intermediate to Alkaline Types of Alteration

Examples of the K-series type of alteration, following Utada's classification (Fig. 7.7), are found in porphyry copper deposits (Beane 1982), adularia (potassium feldspar)–sericite type epithermal gold vein deposits (Hayba et al. 1985; Heald et al. 1987;), and many sericite deposits (e.g., Miyaji and Tsuzuki 1988). Figure 7.16 shows a schematic illustration for the alteration zoning (for mineral assemblages see Table 7.4) developed within a porphyry copper deposit system which was proposed by Lowell and Guilbert (1970). Concerning the K(potassic)-series alteration, a zonal sequence from potassic to phyllic to

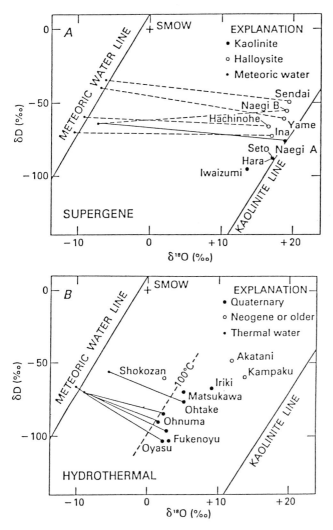

Fig. 7.15. A $\delta D - \delta^{18}O$ diagram for kaolinite and halloysite formed from supergene alteration of some Japanese deposits (after Hayba et al. 1985). *Dashed* lines connect mineral analyses with their corresponding surface water compositions. **B** $\delta D - \delta^{18}O$ diagram for hydrothermal kaolinites and of thermal waters near mineral localities (after Hayba et al. 1985). The *dashed line* indicates the $\delta D - \delta^{18}O$ relation of kaolinite in equilibrium with meteoric water at 100 °C

argillic zones, is well developed from the center to the margin of the alteration. These zones are entirely enclosed by an outer propylitic zone. Gustafson and Hunt (1975) showed the propylitic alteration to be formed prior to the development of the main K-series alteration related to copper ore mineralization at El Salvador. The zonal sequence of K-series alteration in porphyry copper systems is basically interpreted by a temperature-defined mineral paragenesis

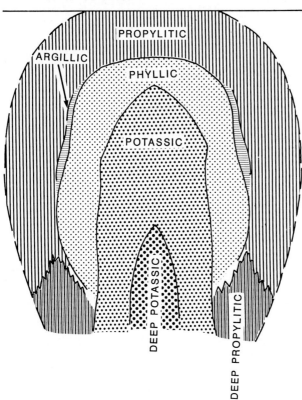

Fig. 7.16. Schematic cross section showing the alteration zoning in a porphyry copper deposit. (After Beane 1982)

as shown in Table 7.4. Taylor (1974) pointed out, using stable isotopic studies of alteration minerals in porphyry copper deposits, that early alteration minerals such as biotite in the potassic zone have $\delta^{18}O$ and δD compositions diagnostic of magmatic water, whereas alteration minerals in the phyllic and argillic zones indicate significant contribution from meteoric waters. Accordingly, the origin of waters related to the formation of alteration zones in porphyry copper deposits, may have varied with time during the development of entire alteration zones. The K-series alteration in adularia–sericite type of epithermal gold vein systems may be a low-temperature analogue, which occurred at relatively shallow levels (lower temperatures) compared to those in porphyry copper systems.

The alteration aureoles enveloping a Kuroko-type ore deposit in Japan may be thought of as another example for the K-series type of alteration (Fig. 7.17). The zonal arrangement is well-developed from the central to the marginal parts: K-feldspar zone → illite-chlorite zone → illite/smectite mixed-layer mineral zone → smectite zone. The sequence of alteration zones corresponds to the variations in temperature and activity of aqueous cation

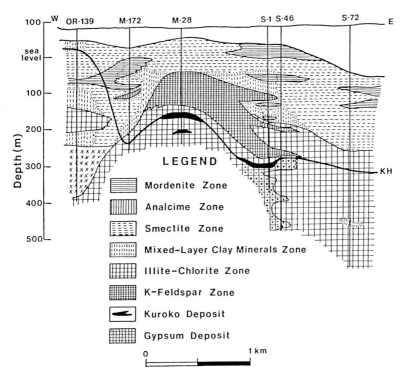

Fig. 7.17. Cross section showing the alteration zoning at the Matsumine Kuroko ore deposit, Akita, Japan (after Inoue 1987). *KH* line indicates the horizon of Kuroko ore deposits

species in solution from the center to the periphery, as deduced from the stability of minerals depicted in Figs. 7.3 and 7.5. However, closer observation of Fig. 7.17 reveals that the K-series alteration zones in Kuroko-type hydrothermal alteration tend to grade outward to the Ca-Mg and/or Na-series alteration zones, which are characterized by the presence of zeolite minerals and in some places interfinger with each other in the marginal portion of the alteration. Thus the K-series type of alteration in Kuroko deposits is considered to be a mixed system with different types of alteration series. This is due to the contribution of seawater in the fluid which produces the alteration zones. The fact that Kuroko deposits occurred on the seafloor at a few thousand meters depth (Ohmoto and Skinner 1983) suggests that the seawater acted as a cover for discharging hydrothermal fluids from a vent, and this resulted in the formation of alteration zones with a mushroom or bed-form type of morphology as shown in Fig. 7.17.

Typical examples of the Ca and Ca-Mg-series alteration are found in the hydrothermal alteration of basalts on the sea floor (e.g., Alt et al. 1986) and at the mid-ocean ridges (e.g., Tomasson and Kristmannsdottir 1972), and in many subaerial geothermal systems formed in andesitic to basaltic rocks as well as by initially calcareous sedimentary rocks (Schiffman et al. 1985). This

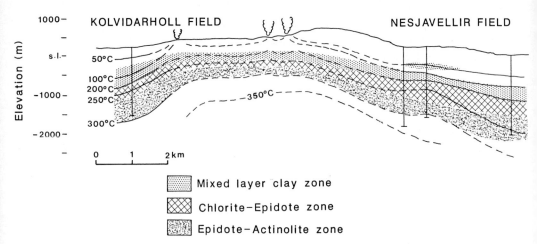

Fig. 7.18. Cross section showing the alteration zoning at the Nesjavellir geothermal field, Iceland. (After Schiffman and Fridleifsson 1991)

type of alteration is customarily called propylitic alteration. Figuer 7.18 shows the zonal distribution examined in a geothermal field of Iceland (Schiffman and Fridleifsson 1991). Chlorite is found with epidote, albite, and/or calcite, and sporadically prehnite represents the mineral assemblage of propylitic alteration at high temperatures, in excess of about 200°C (see Figs. 7.5 and 7.9). At even higher temperatures (>300°C), actinolite, clinopyroxene, biotite, and/or garnet coexist with chlorite (Bird et al. 1984). At lower temperatures, chlorite is replaced by interstratified chlorite/smectite in an intermediate zone and grades toward smectite in the outer zone. In this type of alteration, smectite, chlorite/ smectite, and chlorite exhibit a range of textural occurrence in basaltic and andesitic rocks. Smectite occurs primarily as infilling of vugs, replacement of glass or groundmass, and sporadically as replacement of mafic phenocrysts. Plagioclase and clinopyroxene phenocrysts are usually unaltered in the smectite zone. The occurrence of chlorite/smectite is essentially identical to that of smectite. It also occurs as inclusions within plagioclase phenocrysts. Chlorite occurs not only as replacements of mafic phenocrysts and glassy groundmass but also as an inclusion of albitized plagioclase phenocrysts. It seems that smectite is transformed to chlorite/smectite or chlorite, more progressively in initially permeable rocks, characterized by fractured and brecciated textures, than it is in massive rocks at a similar temperature. A great variety of zeolites are found associated with the clays in this type of alteration. In particular, Ca-zeolites show a characteristic zoning which overprints that of clay minerals: from the center to the margin, one finds a wairakite zone → laumontite zone → stilbite zone. This zonal distribution of zeolites corresponds basically to the decrease in temperature as shown in Figs. 7.5 and 7.9. Some specific zeolites, e.g., thomsonite, yugawaralite, and scolecite, etc., are found sporadically as infillings of vesicles in basalts.

As the propylitic alteration noted above is characterized by the assemblages of calc-silicate minerals, the stability of a mineral or a mineral assemblage is strongly controlled, not only by temperature, but also by fugacity of carbon dioxide and pH of the solution (Figs. 7.4 and 7.5). Under high f_{CO_2} conditions at a given temperature, calc-silicates alter to an assemblage of calcite and smectite, chlorite/smectite, or chlorite. If the fugacity of carbon dioxide is significantly low, pumpellyite forms in addition to the above calc-silicates, even in hydrothermal environments (Inoue and Utada 1991d), despite the fact that pumpellyite has been considered to be stable under relatively higher-pressure metamorphic conditions. Fugacities of oxygen and sulfur in the system are also important factors which affect the stability of minerals in propylitic alteration, because many exhibit a solid solution between the Mg- and Fe-members. In general propylitic alteration conserves original rock composition, excepting water and gases. Accordingly, the mineral assemblage and their compositions are controlled by original bulk rock composition.

The Na-series alteration rarely occurs separately. One example may be found at the marginal zones in some Kuroko-type alterations which have been described by Utada and Ishikawa (1973). The alteration is characterized by extensive occurrences of analcime and Na-smectite.

In some hydrothermal systems, acid-type alteration coexists with intermediate to alkaline types of alteration. The formation of this complex alteration may be interpreted in part by the reaction-defined mineral paragenetic mechanism described in Section 7.5. However, the field occurrence and the K-Ar dating of the associated micaceous minerals indicate that part of such acid-type alteration occurred at different periods from the major intermediate to alkaline alterations (e.g., Gustafson and Hunt 1975; Utada 1988; Inoue and Utada 1991a).

7.8.3 Deep Sea Hydrothermal Alteration

It is worthwhile to describe in a separate section the deep sea hydrothermal alteration of basalt, because this is a special case in which the nature of both the solution and the rock participating in the alteration are well documented by field and laboratory data.

Deep sea hydrothermal alteration can be classified into two groups, low-temperature alteration and high-temperature alteration. Dredging and drilling of the oceanic crust has shown the ubiquitous presence of basalts that have undergone some reaction with seawater. They have interacted with seawater at temperatures close to those of bottom waters, and thus the low-temperature alteration of basalts is generally referred to as weathering on the sea floor. According to Honnorez (1981) and Thompson (1983), the alteration of the glassy part of basalts, e.g., the rim of a lava pillow, has a different style of alteration from that of the crystalline part. The deep sea alteration of basaltic glass at low temperatures is characterized by palagonitization, which includes the formation of phillipsite, smectite, and Fe-Mn oxides. The entire palagonitization process is classified into three stages, taking the chemical balance into account. The initial stage of palagonitization is characterized by the crystallization of an intergranular Na > K phillipsite associating with K-Mg-smectite. The bulk rock is enriched in K, Na, and Mg and depleted in Ca. The glass is entirely hydrated and oxidized. The mature stage of palagonitization gives similar mineralogy but the phillipsite mineral now contains K > Na, and the smectite is a potassic nontronite or, in any case, a Mg-smectite with a high K concentration. At the final stage, the palagonitized glass is completely replaced by an intimate mixture of Ca-poor K > Na phillipsite, with potassic-nontronite and Fe-Mn oxides. In contrast, the original igneous phenocrysts are generally unaffected by the low-temperature alteration. Olivine is only partially altered. The formation of zeolites and celadonite is sometimes found in basalts sampled from greater depth by drilling. The low-temperature alteration may include the weathering at bottom-water temperatures (0–3 °C) up to interactions that took place at around 50–100 °C (Thompson 1983). Seyfried and Bischoff (1979) showed experimentally that the style of seawater–basalt interaction differed markedly between the reactions at 70 and 150 °C.

Geophysical evidence suggests that deep circulation of seawater takes place in the oceanic crust (Lister 1972). This results in high temperature alteration of oceanic crust at depths, as confirmed by rocks recovered, by drilling, from sea floor and ridge areas. These rocks show a range of mineral assemblages, including clays, zeolites, and CaAl-silicates, characteristic of respective grades of the alteration (Humphris and Thompson 1978). The mode of occurrence and the mineral assemblages are essentially identical to those in the Ca and Ca-Mg-series of alteration described above. Saponite and celadonite characterize the mineral assemblages of relatively low-grade alteration; celadonite is entirely restricted in occurrence to the oxidative environment. The saponite is generally a Mg-rich Fe-poor type under oxidation environments, and a Fe-rich, Mg-poor type under non-oxidation ones (Andrews 1980). It is relatively enriched in K. Saponites are replaced by interstratified chlorite/smectite associated with Ca- and Na-zeolites, at intermediate-grade alteration, and finally transform to the mineral assemblages of chlorite, albite, epidote, and actinolite, characteristic of those in propylitic alteration, at high-grade (temperature) alteration. Original igneous minerals are extensively altered at higher grades of alteration. Dredged rocks often show the mineral assemblages hornblende, plagioclase, chlorite, epidote, and biotite, which are similar to those of amphibolite facies in regional metamorphism (Humphris and Thompson 1978). This suggests the occurrence of even higher-temperature alteration, above 400 °C, in the deeper part of the oceanic crust.

The variation of mineral assemblages observed in drilled and dredged

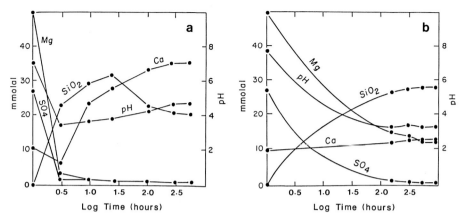

Fig. 7.19. Concentrations of Ca, Mg, SiO$_2$, SO$_4$, and pH (25 °C) in seawater as a function of time during reaction with basaltic glass **a** at 360 °C, 70 MPa, and water/rock ratio of 3 and **b** at 300 °C, 50 MPa, and water/rock ratio of 62. (After Rosenbauer and Bischoff 1983)

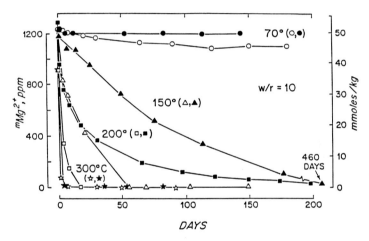

Fig. 7.20. Concentrations of Mg in seawater as a function of time during reaction with basaltic glass (*open symbols*) or the crystalline equivalent (*closed symbols*) at 70 °C, 101 kPa, and 150–300 °C, 50 MPa, at water/rock ratio of 10. (After Mottl 1983)

rocks from the ocean bottom, has been successfully reproduced in laboratory experiments of seawater–basalt interaction. Figures 7.19–7.21 summarize the results of seawater–rock interaction experiments which were performed under a wide range of conditions: temperatures from 20–500 °C, pressures from 101 kPa (1 bar) to 100 MPa, basalt crystallinity from glass to holocrystalline, seawater/rock mass ratio from 1–125, and duration from 14–602 days. Based on these hydrothermal experiments, Mottl (1983) pointed out the important effect of seawater/rock mass ratio on the mineral assemblage produced by this type of alteration at high temperatures. In the course of the reaction between

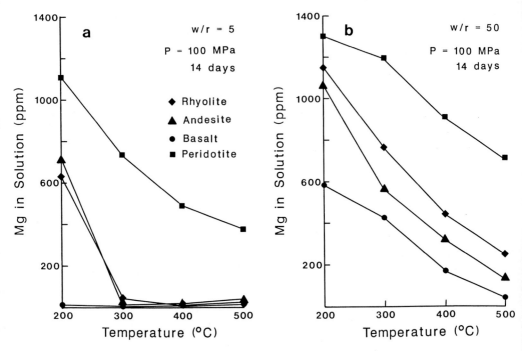

Fig. 7.21. Concentrations of Mg in seawater as a function of temperature in reaction with rhyolite, andesite, basalt, or peridotite at pressure of 100 MPa and duration of 14 days: **a** at water/rock ratio of 5; **b** at water/rock ratio of 50. (After Hajash and Chandler 1981)

seawater and basalt, Mg^{2+} is removed from solution and incorporated as chlorite into rock. Removal of Mg^{2+} from solution occurs more rapidly under low water/rock ratio conditions (Fig. 7.19). On the other hand, removal of Mg^{2+} brings about a drop in the pH of the solution by the following reaction:

$$Mg^{2+} + 2H_2O = Mg(OH)_2 + 2H^+. \qquad (6)$$

Accordingly, as long as the Mg^{2+} concentration in solution is appreciable, as in the experiments at high water/rock ratios (Fig. 7.19b), the solution remains quite acid. However, when the Mg^{2+} concentration drops to a low value, as in the experiments at low water/rock ratios (Fig. 19a), H^+ in solution is rapidly consumed by hydrolysis reactions of rock-forming silicate minerals, and so the pH of the solution comes back up to near neutrality. Epidote and albite are produced from plagioclase by this hydrolysis reaction and Ca in solution is consumed to form epidote. Similar mechanisms can be applied to the formation of the other Ca-bearing minerals such as actinolite, zeolites, and titanite characteristic of the assemblage of high-temperature alteration. Consequently, one can conclude that the greater the water/rock ratio, the more the chlorite-bearing assemblage will be predominant relative to the epidote-bearing assemblage (Mottl 1983). The boundary value of the water/rock ratio was experimentally determined to be about 50.

Figure 7.21 shows that removal of Mg from solution and concomitant release of Ca, Na, and K occur more progressively in mafic rocks than in felsic rocks, except for the case of peridotite. This suggests that the formation of chlorite and Ca,Al-silicates mentioned above is favored in mafic rock alteration. In the reaction of peridotite with seawater, olivine and pyroxene alter to serpentine which does not contain any Ca, Al, Fe, Na, or K. This reaction may be distinctive as a deuteric alteration (serpentinization) of oceanic crust (Velde 1985).

7.9 A Brief Summary on the Effect of Rock Type

In the preceding section, the case studies of hydrothermal alteration are discussed principally in terms of fluid chemistry and physical variables controlling the mineralogy. The type of rock also influences the mineralogy. This may be particular true where the system is nearly closed for solute movement.

Acidic alteration is associated with the formation of kaolin and pyrophyllite in felsic rocks, as in the cases of Ugusu silica deposit, Goto pyrophyllite deposit, and epithermal gold deposits of the acid-sulfate type (Sect. 7.8). This alteration is also often found in mafic rocks. The predominance in felsic rocks may be related to the genetic origin of the acid solution. Felsic volcanism occurs abundantly in orogenic belt zones and often brings about explosive eruption on land. The resulting caldera structure acts as an efficient catchment of meteoric water with the fractures generated by the explosion providing pathways for water circulation. The felsic magma is itself rich in volatile components which contribute to the generation of acidity in solution. These factors favor the generation of acid hydrothermal solutions and acid-type alteration.

In the intermediate to alkaline types of alteration, dioctahedral clays such as montmorillonite, interstratified illite/smectite, and illite are dominant in rocks which were originally felsic. This is typically seen in the cases of Kuroko deposits, porphyry copper deposits, epithermal deposits of adularia–sericite type, and sericite deposits. This preferential occurrence may be in part due to fact that felsic rocks contain more Na and K, and less Mg and Fe than do mafic rocks. Na-rich plagioclase and glass are replaced by dioctahedral clays in these types of alteration. K-feldspar is usually unaltered, or only partially altered, to clays. Mafic minerals which are of low abundance in felsic rocks are altered to saponite, chlorite/smectite, or chlorite. The trioctahedral clays including saponite, chlorite/smectite, and chlorite are dominant in mafic rocks originally rich in Mg, Fe, and Ca. They usually occur as replacements of mafic phenocrysts and groundmass, and are often found as inclusions in plagioclase. These features are typically observed in the hydrothermal alteration of ocean floor basalts, as well as in geothermal systems developed in andesitic to basaltic rocks (see Sects. 7.8 and 7.10). When both saponite and montmorillonite are identified in a rock by X-ray powder diffraction, a situation often encountered in hydrothermal alteration of pyroclastic rocks, when observed under the optical microscope each mineral is observed to replace only one kind of rock fragment. In ultramafic rocks, serpentine and talc are formed in the intermediate to alkaline types of alteration. Sepiolite and palygorskite are often found as veins in hydrothermally altered serpentinite and calcareous rocks (e.g., Imai and Otsuka 1984). This suggests that the wall rocks contribute significant amounts of calcium and magnesium to the circulating solution and that the minerals have formed under considerably saline conditions at low temperatures, below 100 °C.

7.10 Detailed Mineralogy of Selected Clay Mineral Types

As mentioned in the previous sections, a mineral zone is often defined by a diagnostic mineral assemblage of silicates, carbonates, and/or sulfates. Recent detailed mineralogical studies in hydrothermal systems indicate that a single mineral species can display several variations in structure and composition, even within a designated mineral zone. This variation is especially remarkable in interstratified (mixed-layer) minerals. Understanding the structural and compositional variations of clay minerals provides more detailed information about the processes of hydrothermal alteration.

7.10.1 Interstratified Illite/Smectite

Interstratified illite/smectite (I/S) minerals, which are characteristic minerals of neutral to alkaline types of alteration in felsic rocks, occur as intermediate products in the smectite to illite sequence of mineral transformation which is the result of increasing temperature. Of course, it is possible to consider the reverse case in which I/S is produced retrogressively from illite or muscovite by the action of superimposed alteration. However, I/S formed in this manner occurs typically in weathering and is rarely reported from hydrothermal systems.

From another viewpoint, mudstones and sandstones usually contain smectite in the matrix. It is reasonable to assume that in the diagenetic development of these rocks there is a precursor to smectite at the beginning of the smectite-to-illite conversion, and that the smectite will convert to illite through I/S with increasing temperature due to burial. General features of smectite-to-illite conversion observed in both diagenetic and hydrothermal environments will be described in detail below. Nevertheless, Bethke et al. (1986) questioned the direct application of the diagenetic reaction series of smectite to illite to the series found in hydrothermal systems. They pointed out the possibility that the formation of I/S minerals in hydrothermal systems is a simple result of direct precipitation from solution, controlled by the ambient temperature, pressure, and solution composition at arbitrary conditions. In contrast to this view, Inoue et al. (1992) indicated that the formation of I/S minerals in hydrothermal systems is basically identical to that which occurs in diagenetic environments, i.e., that it is due to a consecutive reaction from an early-formed smectite to illite through I/S. Evidence for this process was obtained from a comparison of structural variations of I/S in many geothermal fields (see Sect. 7.10.4, mineral stability).

7.10.1.1 Structural Variation

With the illitization of smectite, the proportion of smectite layers (%S) and the ordering (termed *reichweite*) of these layers, that is a layer arrangement in a vertical stacking sequence along the c^* axis, change systematically. Thus the characteristics of interstratification of minerals with compositions between smectite and illite are expressed by the two variables, %S (or %I) and *reichweite*. To understand the interstratified structure in more detail, the information about the transition (or junction) probabilities between the component layers must be considered (see Reynolds 1980).

Several convenient techniques determining %S value in I/S have been proposed by many researchers, for example, a saddle/001 peak intensity method (Weir et al. 1975; Rettke 1981; Inoue et al. 1989), a Srodon method (Srodon 1980, 1981), and a $\Delta 2\theta_1 - \Delta 2\theta_2$ method (Watanabe 1981, 1988). Figure 7.22 shows a $\Delta 2\theta_1 - \Delta 2\theta_2$ diagram of Watanabe (1981), where $\Delta 2\theta_1$ is defined by an angular difference between a peak (l_2) at 8.9–10.2 °2θ (Cu Kα radiation) and a peak (l_1) at 5.1–7.6 °2θ and $\Delta 2\theta_2$ is an angular difference between a peak (l_3) at 16.1–17.2 °2θ and l_2. These $\Delta 2\theta_1$ and $\Delta 2\theta_2$ values show systematic changes with expandability (%S) at constant *reichweite* (given by each curve in Fig. 7.22). Plotting in this diagram gives the benefit of showing both %S and *reichweite* at the same time. Tomita et al. (1988) presented a modified Δ2θ diagram by drawing isopleths of transition probabilities. By using the diagram, more detailed information on the interstratified structure is available.

Plots in Fig. 7.22 show an example for the structural variation of I/S from a Kuroko-type hydrothermal alteration at Shinzan, Japan (Inoue et al. 1978; Inoue and Utada 1983). K-Ar dating of these minerals indicated that the formation of I/S at Shinzan in fact resulted from the effect of a later episode of hydrothermal activity associated with an intrusion of a daci-andesite dike related to the Kuroko mineralization (Utada et al. 1988; Inoue et al. 1992). The structural variation of I/S at Shinzan is characterized as follows. (1) Plots are rarely found between *reichweite* 0 and 1 and possibly between *reichweite* 1 and 2. (2) Occurrences of perfectly ordered I/S with 55–45%S, *reichweite* 1 are

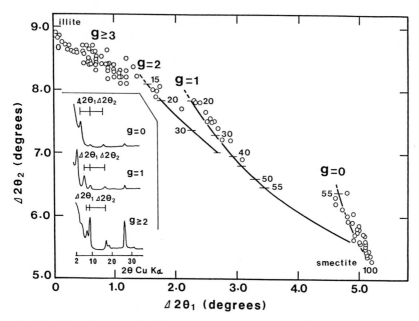

Fig. 7.22. Plots of interstratified illite/smectite minerals from the Shinzan hydrothermal alteration field on the $\Delta 2\theta_1$-$\Delta 2\theta_2$ diagram (after Inoue and Utada 1983). *Reichweite* is abbreviated by g. *Numerical values* indicate the percentages of smectite layers

rare. (3) I/S with *reichweite* 2 is rare. These characteristics are commonly recognized in I/S samples from many geothermal systems hosted by felsic volcanic rocks.

Point (1) has been correctly established by refining the structure parameters of interstratification involving the transition probabilities, in comparing them with the structure parameters of I/S from diagenetic shales and bentonites (Bethke et al. 1986; Inoue et al. 1990). Concerning point (2), perfectly ordered I/S with 55–45%S is usually called rectorite (Bailey et al. 1982). However, rectorite shows quite different characteristics in structure, chemistry, and thermal stability from usual I/S minerals, as will be discussed below. As for point (3), the paucity of occurence of *reichweite* 2 I/S minerals is the same in diagenetic environments as it is in hydrothermal systems.

On the other hand, the occurrence of the continuous reaction series from smectite to illite through I/S has been rarely documented in the outer zones surrounding kaolinite, pyrophyllite, and sericite deposits. In these deposits I/S minerals occur as local veinlets and spots and their %S is usually less than 50% smectite. Whether a map-scale mixed-layer mineral zone is formed depends on the prevailing thermal gradient. If the thermal gradient of the system is great, the zone is compressed to within a narrow space and an illite or sericite zone is apparently in direct contact with a smectite zone or unaltered rocks at the margin of alteration. Many other factors can of course be related to the lack of mixed-layer minerals in the specific hydrothermal systems. The details are unknown.

7.10.1.2 Chemical Variation

In the X-ray diffraction (XRD) model of interstratification, I/S is thought to be composed of smectite and illite component layers. Therefore, if each of the component layers has a constant chemical composition the chemical composition of an intermediate I/S product can be determined definitively by the proportion of each component layer. In fact, smectite is a group name including montmorillonite, beidellite, and nontronite as the end members. They make a solid solution with each other and the layer charge varies from 0.25 to 0.60 per $O_{10}(OH)_2$ (Weaver and Pollard 1973; Newman and Brown 1987). Illite is also a solid solution mineral.

Velde and Brusewitz (1986) and Meunier and Velde (1990) demonstrated that the chemical composition of I/S changes continuously from one end point of a montmorillonite smectite to another end point of illite, but the path line is different for various geologic origins because the precursor smectites have different layer charges. Smectite of hydrothermal origin tends to show higher layer charge than the average value of 0.33 in smectite. Despite this, the smectite layer in I/S appears to be almost identical despite different origins and different stages of the reaction series; it is montmorillonitic in composition. The composition of illite may be relatively constant regardless of different origins. Hower and Mowatt (1966) showed, on the basis of a statistical analysis of many published examples that the illite layer charge is about 0.75/$O_{10}(OH)_2$. Analyses of illite by means of recent new techniques have revised the illite layer charge to be about $0.9/O_{10}(OH)_2$ (Meunier and Velde 1989; Srodon et al. 1992).

Figure 7.23 illustrates variation of chemical composition of I/S with origin. It is apparent that as a whole there are few differences in composition between minerals of hydrothermal and diagenetic origins. The marked characteristics in compositional variation are an increase in the tetrahedral substitution of Al for Si, and a concomitant increase in the interlayer K with decreasing %S. The octahedral Fe and Mg contents are nearly constant for the first half of the reaction series and there is a later decrease toward an illite composition. Sato et al. (1990) indicated, by using the so-called Greene-Kelly test of I/S for diagenetic shales and sandstones, that smectite illitization did not proceed directly from precursor montmorillonite to ordered I/S, until the precursor montmorillonite was altered to a beidellitic composition. Then the beidellite component was converted to ordered I/S. No chemical and XRD evidence supporting the presence of beidellite was detected in I/S from hydrothermal systems hosted by felsic volcanic rocks (Inoue, unpubl. data).

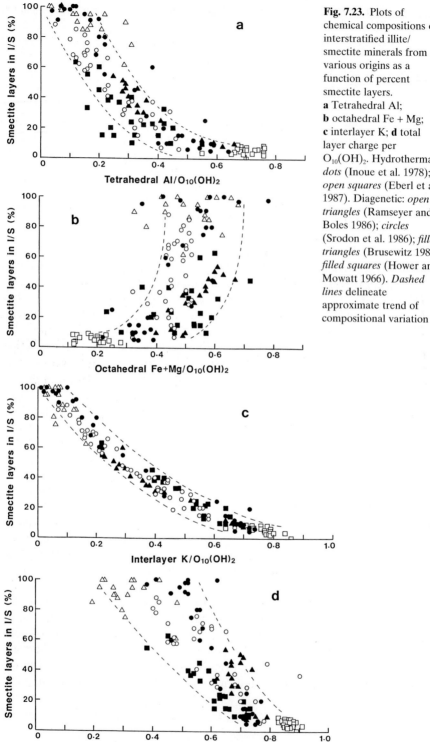

Fig. 7.23. Plots of chemical compositions of interstratified illite/smectite minerals from various origins as a function of percent smectite layers. **a** Tetrahedral Al; **b** octahedral Fe + Mg; **c** interlayer K; **d** total layer charge per $O_{10}(OH)_2$. Hydrothermal: *dots* (Inoue et al. 1978); *open squares* (Eberl et al. 1987). Diagenetic: *open triangles* (Ramseyer and Boles 1986); *circles* (Srodon et al. 1986); *filled triangles* (Brusewitz 1986); *filled squares* (Hower and Mowatt 1966). *Dashed lines* delineate approximate trend of compositional variation

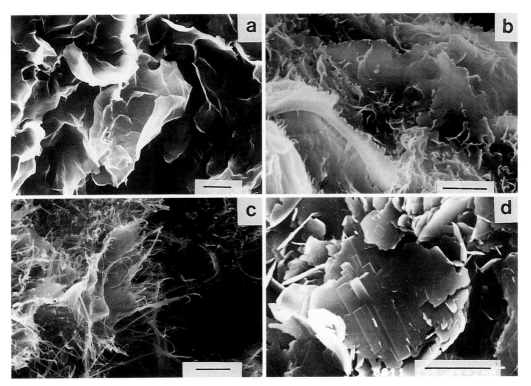

Fig. 7.24. Scanning electron micrographs of critical-point dried illite/smectite samples from the Shinzan hydrothermal alteration field (after Inoue 1986). **a** 100% Smectite sample; **b** 70% smectite sample; **c** 23% smectite sample; **d** 0% smectite sample. Scale *bars* 4mm

7.10.1.3 Morphology Variation

The grain morphology of I/S is observed under scanning electron microscope (SEM) (Fig. 7.24), and under transmission electron microscope (TEM). Randomly interstratified I/S (*reichweite* 0) displays a flaky form (Fig. 7.25b), basically similar to the morphology of pure smectite (Fig. 7.25a). Ordered I/S exhibits a lath form (Fig. 7.25c); the lath becomes larger in size with decreasing %S. Pure illite is usually a mixture of large lath and hexagonal plate (Fig. 7.25d). These laths have a 1M polytype of mica structure elongating to the *a* axis direction. The hexagonal plate usually has a $2M_1$ polytype structure. These morphological variations of I/S under SEM and TEM were commonly observed in several hydrothermal systems hosted by felsic volcanic rocks (Inoue et al. 1992).

In reference to the reaction mechanisms, the morphological variation observed suggests that dissolution of precursor material and recrystallisation of new material occurred during illitization of smectite (Inoue et al. 1987, 1988). Partitioning of ^{18}O isotopes between solid and solution phases is likely

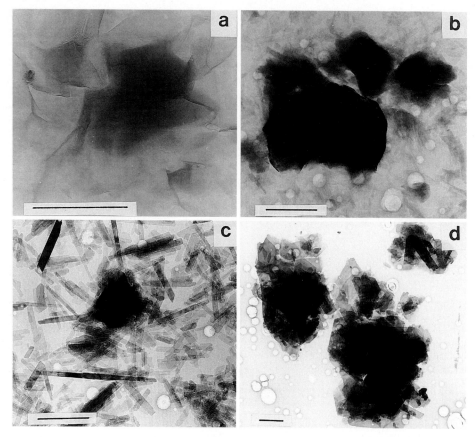

Fig. 7.25. Transmission electron micrographs of illite–smectite samples from the Shinzan hydrothermal alteration field (after Inoue 1986). **a** 100% Smectite sample; **b** 70% smectite sample; **c** 23% smectite sample; **d** 0% smectite sample. Scale *bars* = 1 mm

to occur through these dissolution and recrystallization reactions (Whitney and Northrop 1988). The growth of lath shaped I/S crystals, which occurs during the later half of smectite illitization, is controlled by Ostwald ripening through spiral growth mechanisms, as deduced from quantitative measurements of the grain-size distribution by means of Pt shadowing techniques (Inoue et al. 1988; Inoue and Kitagawa 1994) and from direct observations of the spiral patterns of the Au decorated crystal surfaces of the I/S and illite crystallites under TEM observation (Inoue and Kitagawa 1994).

Figure 7.26 shows a relationship between the mean equivalent diameter (D) and the mean aspect ratio (L/W) of I/S and illite grains of different origins, locations, and rock types. Here, the equivalent diameter of a grain is defined as $D = 2(LW/\pi)^{1/2}$, where L and W are the greatest distance (length) and the smallest distance (width) of a grain, respectively (Inoue and Kitagawa 1994). It is apparent that plots of I/S from hydrothermal systems give a unique trend

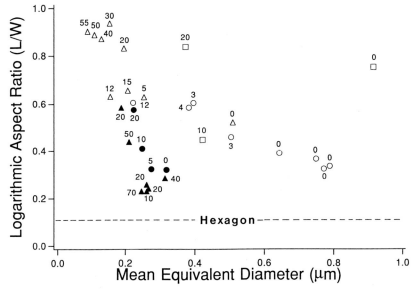

Fig. 7.26. Plots of mean equivalent diameter and logarithmic mean aspect ratio of illitic clay grains from various origins and rock types (after Inoue and Kitagawa 1994). *Open circles* Kamikita hydrothermal system (Inoue and Kitagawa 1994); *open triangles* Shinzan hydrothermal system (Inoue et al. 1988); *filled triangles* diagenetic bentonites (Nadeau 1985); *dots* diagenetic shales (Lanson and Champion 1991); *squares* diagenetic sandstones (Nadeau 1985). *Numerical values* indicate the percentages of smectite layers

different from the cluster of plots for diagenetic I/S. Grains of 50%S I/S from hydrothermal systems form a thin lath which has a mean aspect ratio of about 8, whereas those from bentonites and diagenetically altered shales give a smaller value. The aspect ratio decreases with illitization, up to about 2 at 0%S, corresponding to the morphological variation from lath to hexagon observed under TEM, noted above. Huang (1990) and Small et al. (1992) noted, from their experimental studies on illite morphology, that fibrous illite is favored by high silica activity in the solution. The fact that host rocks in the hydrothermal systems studied are predominantly felsic in composition is concordant with the above experimental observations. On the other hand, values for I/S grains from diagenetic sandstones seem to scatter along the hydrothermal trend (Fig. 7.26), although there are still large deviations owing to the scarcity of data. These suggest that the two distinct trends might be related to the difference in porosity of rocks rather than the difference in origins, such as hydrothermal and diagenesis (Inoue et al. 1990; Whitney 1990; Inoue and Kitagawa 1994).

7.10.1.4 Stability

Smectite illitization is basically a temperature-dependent reaction. It is generally accepted that the reaction series in diagenetic environments is completed between approximately 50–230 °C (Srodon and Eberl 1984). This temperature range seems to be applicable to the I/S reaction series in hydrothermal environments (McDowell and Elders 1980; Horton 1985; Jennings and Thompson 1986). However, it has often been documented from many active geothermal fields that smectite, and more expandable I/S, occur at temperatures higher than the above values. For instance, smectite appears at about 200 °C at Otake, Japan (Hayashi 1973) and Cerro Prieto, Mexico (Elders et al. 1979), and at 160 °C at Yellowstone, Wyoming (Bargar and Beeson 1981) and Wairakei, New Zealand (Steiner 1968). A detailed clay-mineral study in the Sumikawa active geothermal field, Japan (Inoue et al. 1992) indicated that when smectite appeared at bottom-hole temperatures higher than 200 °C, it coexisted with more than one I/S phase, having different %S in the same rock specimen. This feature was characteristically visible in the high-temperature area of the Sumikawa field. In the low-temperature area, on the other hand, the assemblage of I/S was rather homogeneous, similar to those in many fossil geothermal fields. This may be due to the fact that the I/S in the low-temperature area has been kept for a longer time under the same temperature conditions and thereby the expandability of I/S has become more homogeneous. Consequently one must consider that smectite and more expandable I/S in hydrothermal systems, can form metastably at temperatures higher than those deduced from diagenetic environments, because in a hydrothermal system increases in temperature can occur abruptly for a relatively short time period. The early formed smectite has been converted to I/S or illite, having a constant expandability with time according to the ambient temperature.

Thermodynamic status of interstratified I/S has been controversial since a first inquiry by Zen (1962, 1967). Garrels (1984) and Aja et al. (1991) considered I/S to be a two-phase mixture in a thermodynamic sense. Aagaard and Helgeson (1983) and Giggenbach (1985) regarded I/S as a single solid-solution phase. Inoue et al. (1987) divided an entire I/S series into two independent solid solutions, one is I/S from 100 to 50%S and the other is I/S from 50 to 0%S. Each series of I/S was regarded as a solid solution of smectite and a solid solution of illite, respectively. From another viewpoint, the importance of the kinetic aspect of the smectite to illite reaction has been stressed by many researchers (e.g. Eberl and Hower 1976; Velde 1992; Velde and Vasseur 1992 among others).

The smectite illitization in hydrothermal systems is generally completed during a short time period (probably within one million years), although there are a few exceptions, as determined by K-Ar dating of I/S minerals (Inoue et al. 1992). This time scale is almost equivalent to the lifetime of a single hydrothermal event (Wohletz and Heiken 1992). K-Ar dating of micaceous materials including I/S and illite is beneficial in allowing one to distinguish each stage of the different hydrothermal alterations superimposed at a site (Sawai et al. 1989; WoldeGabriel and Goff 1989, 1992; Inoue and Utada 1991a).

7.10.2 Rectorite

This mineral is often found in kaolinite, pyrophyllite, and sericite deposits. This suggests that the formation of rectorite is favored under acid to neutral conditions of solution–rock interaction. In the Kamikita fossil geothermal field, Japan, rectorite occurs as replacement of plagioclase phenocrysts and groundmass in volcanics of the outer zone adjacent to the central pyrophyllite-diaspore zone (Inoue and Utada 1989, 1991a). Rectorite is thought to be formed at temperatures of about 200 °C or slightly more. This is still hotter than the formation temperature of a usual I/S of equivalent expandability.

The chemical composition of rectorite is characterized by high values in Al content and total layer charge, with low Mg and Fe contents, compared to those of usual I/S equivalents of the same smectite content (Inoue and Utada 1983, 1989; Matsuda 1984). Kodama (1966) determined precisely the structure and chemistry of rectorite from Baluchistan, whose component layers are paragonite and beidellite. Since his study, many occurrences of rectorite-like minerals have been documented from pyrophyllite deposits. They are often characterized by interlayer cations other than Na (Nishiyama and Shimoda 1981; Matsuda 1984); they are termed Na, K, or Ca-rectorite according to the dominant interlayer cation species (Bailey et al. 1982). Even in K-rectorite, however, the interlayer composition tends to be rich in Na and Ca relative to those in common in I/S of the same smectite content. Inoue and Utada (1989) showed that in rectorite there is a negative correlation between the interlayer (Na + Ca) content and the octahedral (Fe + Mg) content. This behavior is similar to that observed in paragonite and margarite (Guidotti 1984). Based on these observations, rectorite may be regarded as a single phase of a low-temperature equivalent to paragonite or margarite.

It is well-known that allevardite, which is a perfectly ordered mica/smectite with 50%S (Brindley 1956), exhibits a long ribbon form under TEM (Weir et al. 1962). Baluchistan Na-rectorite shows a similar morphology to the allevardite (Sudo and Takahashi 1971). In contrast, many rectorite-like minerals reported from kaolinite, pyrophyllite, and sericite deposits usually exhibit an irregular flake form different from the ribbon-like form. The reason why they are different in morphology is uncertain.

7.10.3 Dioctahedral Smectite

Dioctahedral smectite is widespread in the lowest grade alteration zone, i.e. at lowest temperatures. In active geothermal fields, smectite is formed by the reaction of rocks with solutions from neutral to alkaline pH. In some cases, this mineral itself concentrates in a restricted space and forms a so-called bentonite deposit. There are many bentonite deposits of this hydrothermal origin, in Algeria, Spain, Greece, Hungary and Japan, among other countries. They are mostly formed in felsic rocks (Grim and Güven 1978). In these deposits, smectite occurs as a monomineralic layer, but it is often associated with halloysite, allophane, zeolites, and/or cristobalite. This mineral association suggests a low temperature of formation.

The interlayer cation of smectite can be easily exchanged with various species of cations in the solution. The selectivity sequence in smectite is generally $Na < K=NH_4 < Rb < Cs$ and $Mg < Ca < Sr < Ba$, but this sequence can change with temperature. Shirozu and Iwasaki(1980) and Inoue et al. (1984a) demonstrated that the Ca/Na ratio of smectite in Kuroko-type hydrothermal alteration, decreased systematically from the inner to the outer portion within the marginal smectite zone. It is known that the higher the temperature, the more the selectivity for Ca in smectite increases. The outward decrease in the Ca/Na ratio in smectite probably reflects a decrease in temperature and concomitant change in solution composition brought about by movement of the solution. In many drill core samples from geothermal fields, smectite occurring near the surface, tends to have increased Mg in the interlayer (Inoue et al. 1984a). Uno and Takeshi (1979) showed that smectite from weathered bentonite deposits is characterized by the presence of Mg, as well as H, in the interlayer position. Th enrichment of Mg in smectite may be attributed to the circulation of ground water.

7.10.4 Sericite

There still remains some confusion in terminology for the terms illite and sericite. Sericite is a petrographic term used to indicate highly birefringent, fine-grained, micaceous material. Recently, the term illite has been preferably used for these micaceous materials (Srodon and Eberl 1984). On the other hand, some researchers insist that the identity of sericite, in its structure and composition, is different from illite. Higashi (1980) and Shirozu (1985) examined the structure and composition of sericite from many Kuroko ore, sericite, and pyrophyllite deposits in Japan, and characterized it as having a large grain size, high K and Al contents, and low Fe, Mg, and Si contents compared to those in illite. The polytype structure is predominantly $2M_1$ and 1M in sericite, but 1M and 1Md in illite. However these differences seem to be related to the differences in environmental conditions of the formation of such micaceous materials. Inoue and Utada (1991a) showed that in the Kamikita fossil geothermal field there are distinct differences in the chemical compositions of micaceous materials produced in the acid type of alteration and the neutral to alkaline types of alteration. In the former, the composition is close to that of ideal muscovite, whereas in the latter, including Kuroko-type and propylitic alterations, the composition is more phengitic. These phengitic micas grade toward I/S minerals in the periphery. As mentioned previously, the I/S minerals change in expandability in a continuous manner. Such a continuous gradation, on the other hand, is not clear in micaceous materials from the acid-type alteration, and only one type of interstratified mica/smectite mineral with 35%S occurs discretely in the outer zone (Inoue and Utada 1989). Yoder and Eugster (1955) and Velde (1965) demonstrated experimentally that the polytype structure of white mica changes from $1Md \to 1M \to 2M_1$, with increasing temperature.

Alt and Jiang (1991) indicated that micaceous materials which were precipitated on the seafloor, associated with massive sulfide deposits, are characterized by relatively high Al and low Fe and Mg contents. They related these compositions the those of sericite associated with Kuroko ore and its ferruginous chert (Higashi 1980; Kalgeropoulos and Scott 1983). The formation of Fe, Mg-poor mica may be caused by the selective removal of Fe and Mg from the solution due to the co-precipitation of Fe-sulfides, Fe-oxides, and Mg-chlorite.

Yamamoto (1967) and Higashi et al. (1991) showed the appreciable presence of NH_4 as an interlayer cation of micaceous materials accompanied by sericite and pyrophyllite deposits. If the proportion of NH_4 in the interlayer position exceeds 50%, the mineral is called tobelite (Higashi 1982). The origin of the NH_4 is unknown.

7.10.5 Interstratified Trioctahedral Chlorite/Smectite (C/S)

Trioctahedral chlorite/smectite (C/S) is a diagnostic mineral for the hydrothermal alteration of andesitic to basaltic composition rocks under neutral to alkaline alteration-fluid conditions. This alteration shows various modes of occurrences such as vein and vug fillings, and replacement of mafic minerals and volcanic glasses in the outer zone of propylitic alteration.

7.10.5.1 Structural Variation

The structural variation of C/S, determined by the XRD method of Reynolds (1980), has been examined in the outer zone of propylitic alteration in various geologic situations, e.g., thermally altered volcanic rocks (Inoue and Utada 1991b), stagnant hydrothermal alteration of caldera-filled deposits (Inoue et al. 1984b), hydrothermal alteration of mid-ocean ridge basalts (Tomasson and Kristmannsdottir 1972; Alt et al. 1986; Schiffman and Fridleifsson 1991), and ophiolite (Bettisson and Schiffman 1989; Shau et al. 1990; Bettisson-Varga et al. 1991).

Figure 7.27 shows an example of structural variation in C/S as a function of temperature (actually the distance from an intrusive mass) which was observed in the Kamikita fossil geothermal field (Inoue and Utada 1991b). The structural variation is characterized by the fact that %S in C/S decreases discontinuously with a rise in temperature, with a step at 100–80%, 50–40%, and 10–0%; the C/S often exhibits a bimodal or trimodal frequency in

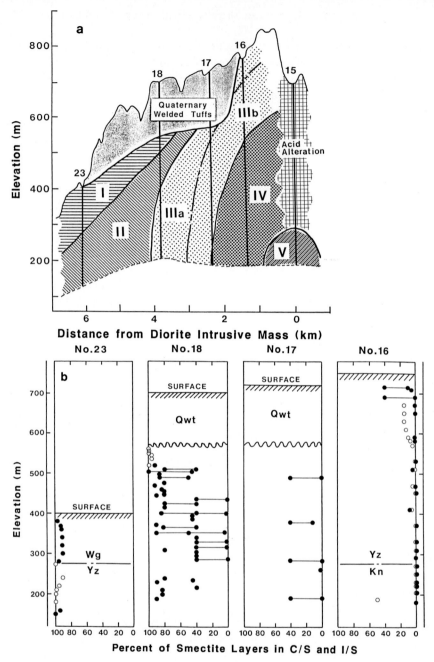

Fig. 7.27. a Cross section showing the alteration zoning at the Kamikita hydrothermal alteration field, Aomori, Japan after Inoue and Utada (1991). *I* Smectite zone; *II* smectite–zeolite zone; *III* corrensite–laumontite zone without epidote (IIIa) and with epidote (IIIb); *IV* chlorite-epidote zone; *V* biotite–actinolite zone. **b** Variations of precentages of smectite layers in chlorite/smectite (*solid cricles*) and illite/smectite (*open circles*) as a function of depth of drill holes in the Kamikita field (after Inoue and Utada 1991). *Tie-lines* indicate the presence of more than two chlorite/smectite samples having different smectite layer percentages in a rock specimen. *Qwt* Quaternary welded tuffs; *Wg* Wadagawa Fromation; *Yz* Yotsuzawa Formation; *Kn* Kanegasawa Formation

expandability in a single rock specimen. The C/S having 50–40%S, which has a superlattice reflection with d = 31 Å after ethylene glycol saturation, has a *reichweite* value of 1. The minerals should be termed corrensite (a perfectly ordered trioctahedral C/S), because the coefficients of variation of the d(001) values, defined by Bailey et al. (1982), gave more than 0.75. The C/S minerals that are more or less expandable than corrensite have *reichweite* values of 0. These characteristics in the structural variation had been known in hydrothermal alteration of caldera-filled deposits as demonstrated by Inoue et al. (1984b). Schiffman and Fridleifsson (1991) also confirmed this in the Nesjavellir geothermal field, Iceland. In contrast, some researchers indicated the continuous variation of %S in C/S in diagenetic environments, similar to that found in I/S (Schultz 1963; Helmold and van der Kamp 1984; Chang et al. 1986). Bettison-Varga et al. (1991) showed the occurrence of randomly interstratified C/S with more than 50%S by means of high-resolution transmission electron microscope (HRTEM).

7.10.5.2 Chemical Variation

The end member smectite layer in C/S is usually saponite which is a trioctahedral smectite whose interlayer cation is predominantly calcium. The saponite tends to have a greater layer charge than the average value of $0.33/O_{10}(OH)_2$ as usually cited for smectite.

Saponite shows an extensive solid solution between the Mg and Fe. Mg-saponite is described in several hydrothermal systems where it is for the most part associated with gypsum, anhydrite, or carbonates (Shirozu et al. 1975). Fe-saponite is rare under terrestrial surface conditions. Kohyama et al. (1973) and Kohyama and Sudo (1975) demonstrated that Fe-rich saponite easily alters to the mineral hisingerite by oxidation of iron under surface weathering conditions. Published data indicate that the Fe/Mg + Fe ratio in saponite from hydrothermal systems ranges from 0 to 0.7. The occupancy of octahedral Al is usually less than 10% in atomic proportion (Brigatti and Poppi 1984; Inoue 1985).

Corrensite exhibits a unique range of chemical composition. The octahedral Al occupancy ranges from 10 to 30% (Inoue 1985). The Fe/Mg + Fe ratio is influenced by many factors, e.g., bulk rock composition, fugacities of oxygen and sulfur in the system, etc. The maximum ratio according to the published data is about 0.7; this apparent limit may be due to the effects of surface weathering. Inoue and Utada (1991b) noted that the C/S, which provides a superlattice reflection at 31 Å after ethylene glycol saturation, contains approximately $Si_{5.4}Al_{2.6}$–$Si_{6.4}Al_{1.6}$ per $O_{20}(OH)_{10}$ in the tetrahedral sheet and 8.8–9.8 cations in the octahedral sheet. Electron microprobe analyses of C/S minerals in hydrothermal systems often show a continuous variation in compositions from saponite to chlorite (Inoue 1985, 1987; Meunier et al. 1991; Schiffman and Fridleifssnn 1991). On the other hand, the XRD structural data indicate that they are an intimate mixture of more than two phases of saponite, corrensite and/or chlorite.

Previous electron microprobe analyses of C/S minerals has indicated that the Fe/Mg + Fe ratios of smectite, C/S, and chlorite are nearly constant through the reaction series smectite-to-chlorite, within a given geothermal field (Inoue and Utada 1991b; Schiffman and Fridleifsson 1991). This means that the Fe/Mg + Fe ratios of these minerals are essentially unrelated to temperature under propylitic alteration environments. The chemical compositions of these minerals are rather strongly related to the bulk rock composition and the mode of occurrence, e.g., a replacement of mafic minerals, a replacement of glass, a vug-filling, or a vein-filling.

7.10.5.3 Stability

Kimbara and Ohkubo (1978) stated that corrensite was formed at more than 110 °C coexisting with mordenite in the Satsunan active geothermal field, Japan. Corrensite appears at 150–200 °C in Icelandic geothermal fields (Tomasson and Kristmannsdottir 1972). The mineral usually coexists with laumontite in andesitic to basaltic rocks. Deducing its stability from that of laumontite (Liou 1971), the temperature at which corrensite appears is estimated to be about 100–150 °C. The reported upper limit of thermal stability of corrensite is 170 °C at Satsunan (Kimbara and Ohkubo 1978), 220 °C in Iceland (Tomasson and Kristmannsdottir 1972). At these temperatures, wairakite is seen instead of laumontite. The upper limit of corrensite stability is approximately 230–250 °C. Mg-corrensite can be formed at lower temperatures under brine conditions (Weaver and Pollard 1973). Thus, corrensite shows a limited thermal stability field in addition to the unique characteristics in structure and composition as mentioned above. This suggests that corrensite is a single phase in the thermodynamic and crystallochemical sense (Inoue et al. 1984b; Shau et al. 1990; Inoue and Utada, 1991b).

7.10.6 Trioctahedral Chlorite in Hydrothermal Deposits

Trioctahedral chlorite appears ubiquitously in the higher-grade zones of alteration where the temperature exceeded more than 200 °C under neutral to alkaline solution conditions. Observations in geothermal fields show that chlorite reacts with illite and/or K-feldspar to form biotite when temperature exceeds 300 °C (Fig. 7.9).

The chemical composition of chlorite is influenced by the modes of occurrence and the bulk rock composition as well as temperature and solution composition (Shikazono and Kawahata 1987). Mottl (1983) showed that in deep sea hydrothermal alteration of basalt, chlorite occurring as vein-fillings, shows higher values in Fe/Fe + Mg ratio than chlorite which occurs in massive rocks. In other words, a higher water/rock ratio favors the precipitation of Fe-rich chlorite. Most chlorite compositions in propylitic alteration range from ripidolite to brunsvigite or diabantite, following the classification of Foster (1962). This suggests that the composition of chlorite in propylitic alteration is more strongly controlled by the bulk composition of original rocks, because the propylitic alteration occurred under a nearly isochemical condition. In chlorite from vein-type ore deposits and their adjacent alteration, the Fe/Mg + Fe ratio is greater than that of chlorites in propylitic alteration (Nagasawa et al. 1976; Shikazono 1985), indicating strong influence by temperature, f_{O_2}, f_S, pH, and salt composition of the solution. Mg-chlorite appears to form under slightly oxidizing and low pH conditions. In this case, it is sometimes associated with Al-chlorite (sudoite). Fe-chlorite is favored by reducing conditions. This mineral is likely to precipitate during boiling of solution (Reed and Spycher 1985; Lonker et al. 1990). Under high f_S conditions, the existence of pyrite results in Fe-poor chlorite.

The occurrence of Mg-chlorite is also reported from Kuroko-type alteration, where it is typically associated with sulfides and/or gypsum ores (Shirozu 1974; Shirozu et al. 1975). Alt and Jiang (1991) reported the presence of Mg-chlorite in sea floor hydrothermal deposits. In these environments, the mineral is often associated with talc (see Fig. 7.5d). The origin of Mg may be attributed to seawater. In the outer zone of Kuroko-type alteration, Fe, Mg-chlorite is predominant. This increase in Fe content in the exterior parts of the deposit is thought to reflect the decrease in temperature from the center to the margin of alteration (Tamura 1982; Urabe et al. 1983).

There are several types of polytype structure in chlorite (Bailey and Brown 1962). In burial diagenesis, chlorite changes polytype from Ib ($\beta = 97°$) to IIb through Ib ($\beta = 90°$) with increasing temperature (Hayes 1970). The polytype structure is also related to the composition; Fe-chlorite is usually Ib ($\beta=90°, 97°$), whereas Fe,Mg-chlorite and Mg-chlorite are predominantly IIb structural types (Nagasawa et al. 1976).

Cathelineau (Cathelineau and Nieva 1985; Cathelineau 1988) proposed an empirical geothermometer of chlorite, by utilizing the tetrahedral Al content or the octahedral vacancy. The tetrahedral Al content increases with a rise in temperature: T (°C) = −61.92 + 321.98X, where X is the tetrahedral Al content per $O_{10}(OH)_8$. The empirical equation was successfully applied to the evaluation of crystallization temperature of chlorite from 150–300 °C in Los Azufres and Salton Sea geothermal fields. When the formation temperature of chlorite was correlated with the homogenization temperature of fluid inclusions in the coexisting minerals, chlorite compositions from epithermal vein-type ore deposits (Yoneda 1989) and those from propylitic alteration of basalts (Yagi 1992) all scatter along the Cathelineau curve. However, plots of chlorite formed at relatively low temperatures tend to deviate from the calibration curve. These chlorites probably contain a small amount of expandable layer.

7.10.7 Aluminous Chlorite and Chlorite/Smectite

Aluminous chlorites discussed here include donbassite, cookeite, and sudoite. These interstratified minerals with smectite are called tosudite, which is a perfectly ordered dioctahedral C/S with 50%S, *reichweite* 1. The formation of these minerals is favored under acid to neutral solution conditions, for instance in kaolinite, pyrophyllite, and sericite deposits (Shimoda 1978). At times sudoite and its tosudite occur with Mg-chlorite in Kuroko-type deposits associating with a lot of gypsum and/or anhydrite (Shirozu 1978) and with boehmite in sea floor basalt (Howard and Fisk 1988). They are also common in acid–type alteration in geothermal fields. Tosudite forms at lower temperatures than sudoite. In the Kamikita geothermal field (Inoue and Utada 1989), tosudite is intimately associated with rectorite. It was probably formed at about 200 °C, as noted previously. Donbassite and its interstratified mineral are reported from acid alteration of granitic rocks (e.g., Merceron et al. 1988, 1992). Cookeite is rare in geothermal systems.

The Li content of these minerals is noteworthy, and is in fact often the reason for their formation. The Li content of sudoite is negligible, whereas donbassite usually contains a significant amount of Li by substituting for Al in the interlayer sheet (Newman and Brown 1987; Merceron et al. 1988). Dioctahedral chlorite and tosudite from Kuroko-type alteration and many geothermal fields, do not contain Li at all, whereas those from kaolinite, pyrophyllite, and sericite deposits contain Li (Shimoda 1980; Higashi 1990). The origin of Li in such acid-type alteration is unknown.

7.11 Concluding Remarks

Clays formed under hydrothermal environments merit some special attention. Hydrothermal alteration occurs within a wide range of physico-chemical conditions, rock types, and geologic settings and thereby a great variety of clay minerals are formed. The clay formed often occurs as a monomineralic unit and is relatively homogeneous in its mineralogical properties, thus such a material is suitable for closer crystallochemical investigation. Much mineralogical knowledge has therefore been accumulated to date. Nevertheless, it is difficult to sufficiently summarize the conditions and processes of clay formation in hydrothermal environments. For example, thermodynamic data of clay minerals are insufficient to predict the stability relationships in the great variety of hydrothermal systems. This may be related in part to a poor understanding of thermodynamic status of interstratified minerals, for example. A more serious deficiency in clay mineralogy is the poor understanding of kinetics and the mechanisms of clay formation. It seems that these points remain a subject for future research in clay mineralogy.

On the other hand, clays are commonly accompanied by hydrothermal ore deposits and active geothermal systems. Studies of clay minerals in hydrothermal environments contribute to the evaluation of the processes of ore deposition and their exploration, and assist in our understanding of the internal structure and evolution of geothermal systems as an indicator of temperature and solution chemistry. The accumulation of knowledge regarding the kinetics and mechanism of clay formation, will lead us to understand comprehensively the dynamic process of hydrothermal alteration.

Acknowledgements. This chapter benefited from discussions over the years with Professor M. Utada, University of Tokyo to whom I owe my greatest thanks.

References

Aagaard P, Helgeson HC (1983) Activity/composition relations among silicates and aqueous solutions. II. Chemical and thermodynamic consequences of ideal mixing of atoms on homological sites in montmorillonites, illites, and mixed-layer clays. Clays Clay Min 31: 207–217

Aja SU, Rosenberg PE, Kittrick JA (1991) Illite equilibria in solutions. I. Phase relationships in the system $K_2O-Al_2O_3-SiO_2-H_2O$ between 25° and 250°C. Geochim Cosmochim Acta 55: 1353–1364

Alt JC, Jiang WT (1991) Hydrothermally precipitated mixed-layer illite-smectite in recent massive sulfide deposits from the sea floor. Geology 19: 570–573

Alt JC, Honnorez J, Laverne C, Emmermann R (1986) Hydrothermal alteration of a 1 km section through the upper oceanic crust. Deep Sea Drilling Project Hole 504B: mineralogy, chemistry, and evolution of seawater-basalt interactions. J Geophys Res 91: 10309–10335

Anderson AT, Newman S, Williams S, Druitt T, Skirius C, Stolper E (1989) H_2O, CO_2, Cl and gas in the Plinian and ash-flow Bishop rhyolite. Geology 17: 221–225

Andrews AJ (1980) Saponite and celadonite in layer 2 basalts, DSDP Leg 37. Contrib Mineral Petrol 73: 323–340

Arnorsson S, Gronvold K, Sigurdsson S (1978) Aquifer chemistry of four high-temperature geothermal systems of Iceland. Geochim Cosmochim Acta 42: 523–536

Bacon CR, Newman S, Stolper E (1992) Water, CO_2, Cl, and F in melt inclusions in phenocrysts from three Holocene explosive eruptions, Crater Lake, Oregon. Am Mineral 77: 1021–1030

Bailey SW, Brown BE (1962) Chlorite polytypism. I. Regular and semi-random one-layer structures. Am Mineral 47: 819–850

Bailey SW, Brindley GW, Kodama H, Martin RT (1982) Nomenclature for regular interstratifications. Clays Clay Min 30: 76–78

Bargar KE, Beeson MH (1981) Hydrothermal alteration in research drill hole Y-2, Lower Geyser Basin, Yellowstone National Park, Wyoming. Am Mineral 66: 473–490

Beane RE (1982) Hydrothermal alteration in silicate rocks, 117–137. In: Titley SR (ed) Advances in geology of the porphyry copper deposits, southwestern North America. University of Arizona Press, Tucson, pp 117–137

Berman RG (1988) Internally-consistent thermodynamic data for minerals in the system $Na_2O-K_2O-CaO-FeO-Fe_2O_3-Al_2O_3-SiO_2-TiO_2-H_2O-CO_2$. J Petrol 29: 442–522

Bethke CM, Vergo N, Altaner SP (1986) Pathways of smectite illitization. Clays Clay Min 34: 125–135

Bettison LA, Schiffman P (1989) Compositional and structural variations of phyllosilicates from the Point Sal ophiolite, California. Am Miner 73: 62–76

Bettison-Varga L, Mackinnon IDR, Schiffman P (1991) Integrated TEM, XRD, and electron microprobe investigation of mixed layered chlorite/smectite from Point Sal ophiolite, California. J Metamorph Geol 9: 697–710

Bird DK, Schiffman P, Elders WA, Williams AE, McDowell SD (1984) Calc-silicate mineralization in active geothermasl systems. Econ Geol 79: 671–695

Bischoff JL, Rosenbauer RJ (1985) An empirical equation of state for hydrothermal seawater (3.2 percent NaCl). Am J Sci 285: 725–763

Bowers TS, Jackson KJ, Helgeson HC (1984) Equilibrium activity diagrams for coexisting minerals and aqueous solutions at pressures and temperatures to 5 kb and 600 °C. Springer, Berlin Heidelberg New York

Brigatti MF, Poppi L (1984) Crystal chemistry of corrensite: a review. Clays Clay Min 32: 391–399

Brindley GW (1956) Allevardite. Am Mineral 41: 91–103

Browne PRL (1978) Hydrothermal alteration in active geothermal fields. Annu Rev Earth Planet Sci 6: 229–250

Brusewitz AM (1986) Chemical and physical properties of Paleozoic potassium bentonites from Kinnekulle, Sweden. Clays Clay Min 34: 442–454

Cathelineau M (1988) Cation site occupancy in chlorites and illites as a function of temperature. Clay Min 23: 471–485

Cathelineau M, Nieva D (1985) A chlorite solid solution geothermometer, the Los Azufres (Mexico) geothermal system. Contrib Mineral Petrol 91: 235–244

Chang HK, Mackenzie FT, Schoonmaker J (1986) Comparisons between the diagenesis of dioctahedral and trioctahedral smectite, Brazilian offshore basins. Clays Clay Min 34: 407–423

Craig H (1961) Isotopic variations in meteoric waters. Science 133: 1702–1703

Creasey SC (1959) Some phase relations in hydrothermally altered rock of porphyry copper deposits. Econ Geol 54: 351–373

Devine JD, Sigurdsson H, Davis AN, Self S (1984) Estimates of sulfur and chlorine yield to the atmosphere from volcanic eruptions and potential climate effects. J Geophys Res 89: 6309–6325

Eberl DD, Hower J (1976) Kinetics of illite formation. Geol Soc Am Bull 87: 161–172

Eberl DD, Srodon J, Lee M, Nadeau PH, Northrop HR (1987) Sericite from the Silverton caldera, Colorado: correlation among structure, composition, origin, and particle thickness. Am Mineral 72: 914–934

Elders WA, Hoagland JR McDowell SD, Cobo JM (1979) Hydrothermal mineral zones in the Cerro Prieto geothermal field of Baja California, Mexico. Geothermics 8: 201–209

Ellis AJ, Mahon WAJ (1977) Chemistry and geothermal systems. Academic Press, New York

Foster MD (1962) Interpretation of the composition and a classification of the chlorites. US Geol Surv Prof Pap 414-A, Washington, DC, 22 pp

Fournier RO (1985) The behavior of silica in hydrothermal solutions. In: Berger BR, Bethke PM (eds) Geology and geochemistry of epithermal systems. Rev Econ Geol 2: 45–61

Fournier RO (1987) Conceptual models of brine evolution in magmatic hydrothermal systems. US Geol Surv Prof Pap 1350: 1487–1506

Fournier RO, Truesdell AH (1973) An empirical Na-K-Ca geothermometer for natural waters. Geochim Cosmochim Acta 37: 1255–1275

Garrels RM (1984) Montmorillonite/illite stability diagrams. Clays Clay Min 32: 161–166

Giggenbach WF (1980) Geothermal gas equilibria. Geochim Cosmochim Acta 44: 2021–2032

Giggenbach WF (1981) Geothermal mineral equilibria. Geochim Cosmochim Acta 45: 393–410

Giggenbach WF (1984) Mass transfer in hydrothermal alteration systems–a conceptual approach. Geochim Cosmochim Acta 48: 2639–2711

Giggenbach WF (1985) Construction of thermodynamic stability diagrams involving dioctahedral clayminerals. Chem Geol 49: 231–242

Giggenbach WF (1988) geothermal solute equilibria: derivation of Na–K–Mg–Ca geo-indicators. Geochim Cosmochim Acta 52: 2749–2765

Gottardi G, Galli E (1985) Natural zeolites. Springer, Berlin Heidelberg New York

Grim RE, Guven N (1978) Bentonites: geology, mineralogy, properties and uses. Elsevier, Amsterdam

Guidotti CV (1984) Micas in metamorphic rocks, 357–456, In: Bailey SW (ed) Micas. Reviews in mineralogy 13. Mineralogical Society of America, Washington

Gustafson LB, Hunt JP (1975) The porphyry copper deposits at El Salvador, Chile. Econ Geol 70: 857–912

Hajash A, Chandler GW (1981) An experimental investigation of high-temperature interactions between seawater and rhyolite, andesite, basalt and peridotite. Contrib Mineral Petrol 78: 240–254

Hall WE, Friedman I, Nash JT (1974) Fluid inclusion and light stable isotope study of the Climax molybdenum deposits, Colorado. Econ Geol 69: 884–901

Hayashi M (1973) Hydrothermal alteration in the Otake geothermal area, Kyushu. J Jpn Geotherm Energy Assoc 10: 9–46

Hayba DO, Bethke PM, Heald P, Foley NK (1985) Geologic, mineralogic, and geochemical characteristics of volcanic-hosted epithermal precious-metal deposits. In: Berger BR, Bethke PM (eds) Geology and geochemistry of epithermal systems. Rev Econ Geol 2: 129–167

Hayes JB (1970) Polytypism of chlorite in sedimentary rocks. Clays Clay Min 18: 285–306

Heald P, Foley NK, Hayba DO (1987) Comparative anatomy of volcanic-hosted epithermal deposits: acid-sulfate and adularia-sericite types. Econ Geol 82: 1–26

Hedenquist JW (1987) Volcanic-related hydrothermal systems in the circum-Pacific Basin and their potential for mineralization. Mining Geol Jpn 37: 347–364

Hedenquist JW, Henley RW (1985) Hydrothermal eruptions in the Waiotapu geothermal system, New Zealand: their origin, associated breccias, and relation to precious metal mineralization. Econ Geol 80: 1640–1668

Helgeson HC (1979) Mass transfer among minerlas and hydrothermal solutions. In: Barnes HL (ed) Geochemistry of hydrothermal solutions. John Wiley Sons, New York, pp 568–610

Helgeson HC, Delany JM, Nesbitt HW, Bird DK (1978) Summary and critique of the thermodynamic properties of rock-forming minerals. Am J Sci 278-A

Helmold RP, van der Kamp PC (1984) Diagenetic mineralogy and controls on albitization and laumontite formation in Paleogene arkoses, Santa Ynez Mountains, California. Am Assoc Petrol Geol Mem 37: 239–276

Hemley JJ, Montoya JW, Marinenko JW, Luce RW (1980) Equilibria in the system Al_2O–SiO_2–H_2O and some general implications for alteration/mineralization processes. Econ Geol 75: 210–228

Henley RW (1985) The geothermal framework of epithermal deposits. In: Berger BR, Bethke PM (eds) Geology and geochemistry of epithermal systems. Rev Econ Geol 2: 1–24

Henley RW, Ellis AJ (1983) Geothermal systems ancient and modern: a geochemical review. Earth Sci Rev 19: 1–50

Henley RW, Truesdell AH, Barton PB Jr (1984) Fluid-mineral equilibria in hydrothermal systems. Rev Econ Geol 1 Soc Econ Geol, El Paso

Higashi S (1980) Mineralogical studies of hydrothermal dioctahedral mica minerals. Mem Facult Sci Kochi Univ E-1: 1–39

Higashi S (1982) Tobelite, a new ammonium dioctahedral mica. Mineral J 11: 138–146

Higashi S (1990) Li-tosudite in Tobe pottery stone. J Mineral Soc Jpn Spec Issue 19: 3–9

Higashi S, Okazake K, Makamura T, Masaki K (1991) Mineralogical features of sericite comprising Roseki clay deposits in the Chugoku province of western Japan, with special reference to interlayer NH_4 composition, mica/smectite interstratification, $2M_2$ polytype and thermal behavior. Mining Geol Jpn 41: 351–366

Honnorez J (1981) The ageing of the oceanic crust at low temperature. In: Emiliani C (ed) The oceanic lithosphere. The sea, vol 7. John Wiley, New York, pp 525–587

Horton DG (1985) Mixed-layer illite/smectite as a paleotemperature indicator in the Ametheyst vein system, Creede district, Colorado, USA. Contrib Mineral Petrol 91: 171–179

Howard KJ, Fisk MR (1988) Hydrothermal alumina-rich clays and boehmite on the Gorda Ridge. Geochim Cosmochim Acta 53: 2269–2279

Hower J, Mowatt TC (1966) The mineralogy of illite and mixed-layer illite-montmorillonites. Am Mineral 51: 825–854

Huang WL (1990) Illitic clay formation during experimental diagenesis of arkoses. Abstr Annu Meet 27th Clay Minerals Conf, Columbia MO, 62 pp

Hulen JB, Nielson DL (1988) Hydrothermal brecciation in the Jemez fault zone, Valles calldera, New Mexico: results from continental scientific drilling program core hole VC-1. J Geophys Res 93: 6077–6090

Humphris SE, Thompson G (1978) Hydrothermal alteration of oceanic basalts by seawater. Geochim Cosmochim Acta 42: 107–125

Imai N, Otsuka R (1984) Sepiolite and palygorskite in Japan, 211–232. In: Singer A, Galan E (eds) Palygorskite-sepiolite: occurrences, genesis and uses. Develop Sediment 37. Elsevier, Amsterdam

Inoue A (1985) Chemistry of corrensite: a trend in composition of trioctahedral chlorite/smectite during diagenesis. J Coll Arts Sci Chiba Univ B-18: 69–82

Inoue A (1986) Morphological change in a continuous smectite-to-illite conversion series by scanning and transmission electron microscope. J Coll Arts Sci Chiba Univ B-19: 23–33

Inoue A (1987) Conversion of smectite to chlorite by hydrothermal and diagenetic alterations in the Hokuroku Kuroko mineralization area, northest Japan. Proc 8th Int Clay Conf, Denver, pp 158–164

Inoue A, Kitagawa R (1994) Morphological characteristics of illitic clay minerals from a hydrothermal system. Am Mineral 79: 700–711

Inoue A, Utada M (1983) Further investigations of a conversion series of dioctahedral mica/smectites in the Shinzan hydrothermal alteration area, northeast Japan. Clays Clay Min 31: 401–412

Inoue A, Utada M (1989) Mineralogy and genesis of hydrothermal aluminous clays containing sudoite, tosudite, and rectorite in a drill hole near the Kamikita Kuroko ore deposit, northern Honshu, Japan. Clay Sci 7: 193–217

Inoue A, Utada M (1991a) Hydrothermal alteration in the Kamikita Kuroko mineralization area, northern Honshu, Japan. Mining Geol Jpn 41: 203–218

Inoue A, Utada M (1991b) Smectite-to-chlorite transformation in thermally metamorphosed volcanoclastic rocks in the Kamikita area, northern Honshu, Japan. Am Mineral 76: 628–640

Inoue A, Utada M (1991c) Hudrothermal alteration related to Kuroko mineralization in the Kamikita area, northern Honshu, Japan, with special reference to the acid-sulfate alteration. Geol Surv Jpn Rep 277: 39–48

Inoue A, Utada M (1991d) Pumpellyite and related minerals from hydrothermally altered rocks at the Kamikita area, northern Honshu, Japan. Can Mineral 29: 255–270

Inoue A, Minato H, Utada M (1978) Mineralogical properties and occurrence of illite/montmorillonite mixed layer minerals formed from Miocene volcanic glass in Waga Omono district. Clay Sci 5: 123–136

Inoue A, Utada M, Kusakabe H (1984a) Clay mineral composition and their exchangeable interlayer cation composition from altered rocks around the Kuroko deposits in the Matsumine-Shakanai-Matsuki area of the Hokuroku district, Japan. J Clay Sci Soc Jpn 24: 69–77

Inoue A, Utada M, Nagata H, Watanabe T (1984b) Conversion of trioctahedral smectite to interstratified chlorite/smectite in Pliocene acidic pyroclastic sediments of the Ohyu district, Akita Prefecture, Japan. Clay Sci 6: 103–116

Inoue A, Kohyama N, Kitagawa R, Watanabe T (1987) Chemical and morphological evidence for the conversion of smectite to illite. Clays Clay Min 35: 111–120

Inoue A, Velde B, Meunier A, Touchard G (1988) Mechanism of illite formation during smectite-to-illite conversion in a hydrothermal system. Am Mineral 73: 1325–2334

Inoue A, Bouchet A, Velde B, Meunier A (1989) Convenient technique for estimating smectite layer percentage in randomly interstratified illite/smectite minerals. Clays Clay Min 37: 227–234
Inoue A, Watanabe T, Kohyama N, Brusewitz AM (1990) Characterization of illitization of smectite in bentonite beds at Kinnekulle, Sweden. Clays Clay Min 34: 241–249
Inoue A, Utada M, Wakita K (1992) Smectite-to-illite conversion in natural hydrothermal systems. Appl Clay Sci 7: 131–145
Iwao S (1970) Clays and silica deposits of volcanic affinity in Japan, 267–283. In: Tatsumi T (ed) Volcanism and ore genesis. University of Tokyo Press, Tokyo
Izawa S (1986) Clays minerals in epithermal deposits. J Mineral Soc Jpn Spec Issue 17: 17–24
Jennings S, Thompson GR (1986) Diagenesis in Plio-Pleistocene sediments in the Colorado River delta, southern California. J Sediment Petrol 56: 89–98
Kalgeropoulos SI, Scott SD (1983) Mineralogy and geochemistry of tuffaceous exhalites (Tetsusekiei) of the Fukazawa mine, Hokuroku district, Japan. Econ Geol Monogr 5: 412–432
Keller WD, Reynolds RC, Inoue A (1986) Morphology of clay minerals in the smectite-to-illite conversion by scanning electron microscopy. Clays Clay Min 34: 187–197
Kimbara K (1983) Hydrothermal rock alteration and geothermal systems in the eastern Hachimantai geothermal area, Iwate Prefecture, northern Japan. J Jpn Assoc Miner Petrol Econ Geol 78: 479–490
Kimbara K, Ohkubo T (1978) Hydrothermal altered rocks found in an exploration bore-hole (No. SA-1), Satsunan geothermal area, Japan. J Jpn Assoc Miner Petrol Econ Geol 73: 125–136
Kodama H (1966) The nature of the component layers of rectorite. Am Mineral 51: 1035–1055
Kohyama N, Sudo T (1975) Hisingerite occurring as a weathering product of iron-rich saponite. Clays Clay Min 23: 215–218
Kohyama N, Shimoda S, Sudo T (1973) Iron-rich saponite (ferrous and ferric forms). Clays Clay Min 21: 229–237
Korzhinskii DS (1959) Physico-chemical bases of the analysis of the paragenesis of minerals. Consultants Bureau, New York (translated in English from Russian)
Krauskopf KB (1979) Introduction to geochemistry, 2nd edn. McGraw-Hill, New York
Lanson B, Champion D (1991) The I/S-to-illite reaction in the late stage diagenesis. Am J Sci 291: 473–506
Lasaga AC (1984) Chemical kinetics of water-rock interactions. J Geophys Res 89: 4009–4025
Lichtner PC (1991) The quasi-stationary state approximation to fluid/rock reaction: local equilibrium revisited. In: Ganguly J (ed) Diffusion, atomic ordering, and mass transport. Adv Phys Geochem 8. Springer Berlin Heidelberg, New York, pp 452–560
Liou JG (1971) Stilbite-laumontite equilibrium. Contrib Mineral Petrol 31: 171–177
Lister CRB (1972) On the thermal balance of a mid-ocean ridge. Geophys J R Ast Soc 39: 465–509
Lonker SW, FitzGerald JD, Hedenquist JW, Walshe JL (1990) Mineral-fluid interactions in the Broadlands-Ohaki geothermal system, New Zealand. Am J Sci 290: 995–1068
Lowell JD, Guilbert JM (1970) Lateral and vertical alteration-mineralization zoning in porphyry ore deposits. Econ Geol 65: 373–408
Marumo K, Matsuhisa Y, Nagasawa K (1982) Hydrogen and oxygen isotopic compositions of kaolin minerals. In: Jpn Proc Int Clay Conf, Development in Sedimentology 35. Elsevier, Amsterdam, pp 315–320
Matsuda T (1984) Mineralogical study on regularly interstratified dioctahedral mica-smectites. Clay Sci 6: 117–148
McDowell SD, Elders WA (1980) Authigenic layer silicate minerals in borehole Elmore 1, Salton Sea geothermal field, California, USA. Contrib Mineral Petrol 74: 293–310
McNabb A, Henley RW (1979) Water-rock interaction during reinjection of geothermal waste water. Proc NZ Geothermal Worksh, Univ Auckland, pp 68–71
Merceron T, Velde B (1991) Application of Cantor's dust method for fractal analysis of fractures in the Toyoha mine. Jpn J Geophys Res 96: 16641–16650

Merceron T, Inoue A, Bouchet A, Meunier A (1988) Lithium-bearing donbassite and tosudite from Echassieres, Massif Central France. Clays Clay Min 36: 31–39

Merceron T, Vieillard P, Fouillac AM, Meunier A (1992) Hydrothermal alterations in the Echassieres granitic cupola (Massif Central France). Contrib Mineral Petrol 112: 279–292

Meunier A, Velde B (1990) Solid solutions in I/S mixed-layer minerals and illite. Am Mineral 74: 1106–1112

Meunier A, Inoue A, Beaufort D (1991) Chemiographic analysis of trioctahedral smectite-to-chlorite conversion series from the Ohyu caldera, Japan. Clays Clay Min 39: 409–415

Meyer C, Hemley JJ (1969) Wall rock alteration. In: Barnes HL (ed) Geochemistry of hydrothermal ore deposits. Holt, Rinehart and Winston, New York, pp 166–235

Miyaji K, Tsuzuki Y (1988) Hydrothermal alteration genetically related to the Mannen and Uebi pottery stone deposits in Tobe district, Ehime Prefecture. J Clay Sci Soc Jpn 28: 183–199

Mizawa E (1986) Clay minerals in epithermal deposits. J Mineral Soc Jpn Spec Issue 17: 17–24

Montoya JW, Hemley JJ (1975) Activity relations and stabilities in alkali feldspar and mica alteration reactions. Econ Geol 70: 577–594

Mottl MJ (1983) Metabasalts, axial hot springs, and the structure of hydrothermal systems at mid-oceanic ridges. Bull Geol Soc Am 94: 161–180

Mukaiyama H (1970) Volcanic sulphur deposits in Japan. In: Tatsumi T (ed) Volcanism and ore genesis. University of Tokyo Press, Tokyo, pp 285–294

Nadeau PH (1985) The physical dimensions of fundamental clay particles. Clay Min 20: 499–514

Nagasawa K (1978) Kaolin minerals. In: Sudo T, Shimoda S (eds) Clays and clay minerals in Japan. Development in Sedimentology 26. Elsevier, Amsterdam, pp 189–219

Nagasawa K, Shirozu H, Nakamura T (1976) Clay minerals as constituents of hydrothermal metallic vein-type deposits. Mining Geol Jpn Spec Issue 7: 75–84

Newman ACD, Brown G (1987) The chemical constitution of clays. In: Newman ACD (ed) Chemistry of clays and clay minerals. Mineralogical Society, London, pp 1–128

Nishiyama T, Shimoda S (1981) Ca-bearing rectorite from Toho mine, Japan. Clays Clay Min 29: 236–240

Norton D, Cathles L (1979) Thermal aspects of ore deposition. In: Barnes HL (ed) Geochemistry of hydrothermal ore deposits, 2nd edn. John Wiley, New York, pp 611–631

Ohmoto H, Skinner RJ (1983) The Kuroko and related volcanogenic massive sulfide deposits: introduction and summary of new findings. Econ Geol Monogr 5: 1–8

Otsuki K (1989) Reconstruction of Neogene tectonic stress fields of northeast Honshu Arc from metalliferous veins. Mem Geol Soc Jpn 32: 281–304

Ramseyer K, Boles JR (1986) Mixed-layer illite/smectite minerals in Tertiary sandstones and shales, San Joaquin basin, California. Clays Clay Mineral 34: 115–124

Reed MH, Spycher NF (1985) Boiling, cooling, and oxidation in epithermal systems: a numerical modelling approach. In: Berger BR, Bethke PM (eds) Geology and geochemistry of epithermal systems. Rev Econ Geol 2, Society Economic Geologists, El Paso, pp 249–272

Rettke RC (1981) Probable burial diagenesis and provenance effects on Dakota Group clay mineralogy, Denver basin. J Sediment Petrol 51: 541–551

Reynolds RC (1980) Interstratified clay minerals. In: Brindley GW, Brown G (eds) Crystal Structures of Clay Minerals and their X-ray identification. Mineralogical Society, London, pp 249–303

Roedder E (1984) Fluid Inclusions. Reviews in Mineralogy 12. Mineralogical Society of America, Washington, DC

Rona PA, Bostrom K, Laubier L, Smith KL Jr (1983) Hydrothermal processes at seafloor spreading centers. Plenum Press, New York

Rose AW, Burt DM (1979) Hydrothermal alteration. In: Barnes HL (ed) Geochemistry of hydrothermal ore deposits, 2nd edn. John Wiley, New York, pp 173–235

Rosenbauer RJ, Bischoff JL (1983) Uptake and transport of heavy metals by heated seawater: a summary of the experimental results. In: Rona PA, Bostrom K, Laubier L, Smith KL Jr (eds) Hydrothermal processes at seafloor spreading centers. Plenum Press, New York, pp 177–197

Rye RO, Haffty J (1969) Chemical composition of the hydrothermal fluids responsible for the lead-zinc deposits at Providencia, Zacatecas, Mexico. Econ Geol 64: 629–643

Rye RO, Bethke PM, Wasserman MD (1992) The stable isotope geochemistry of acid sulfate alteration. Econ Geol 87: 225–262

Sato T, Fujii M, Watanabe T, Otsuka R (1990) Properties of expandable layer in illitization of smectite. J Mineral Soc Jpn Spec Issue 19: 17–22

Sawai O, Okada T, Itaya T (1989) K-Ar ages of sericite in hydrothermally altered rocks around the Toyoha deposits, Hokkaido, Japan. Mining Geol Jpn 39: 191–204

Schiffman P, Fridleifsson GO (1991) The smectite-chlorite transition in drill hole NJ-15, Nesjavellir geothermal field, Iceland: XRD, BSE and electron microprobe investigations. J Metamorphic Geol 9: 679–696

Schiffman P, Bird DK, Elders WA (1985) Hydrothermal mineralogy of calcareous sandstones fro the Colorado River delta in the Cerro Prieto geothermal system, Baja California, Mexico. Min Mag 49: 435–449

Schultz LG (1963) Clay minerals in the Triassic rocks of the Colorado Plateau. US Geol Surv Bull 1147-C: 1–71

Seyfried WE, Bischoff JL (1979) Low temperature basalt alteration by seawater: an experimental study at 70 °C and 150 °C. Geochim Cosmochim Acta 43: 1937–1947

Shau YH, Peacor DR, Essene EJ (1990) Corrensite and mixed layer chlorite/corrensite in metabasalt from northern Taiwan: TEM/AEM. EMPA, XRD, and optical studies. Contrib Mineral Petrol 105: 123–142

Shikazono N (1976) Chemical composition of Kuroko ore forming solution. J Jpn Assoc Miner Petrol Econ Geol 71: 201–215

Shikazono N (1985) Gangue minerals from Neogene vein-type deposits in Japan and an estimate of their CO_2 fugacity. Econ Geol 80: 754–768

Shikazono N (1988) Hydrothermal alteration associated with epithermal veintype deposits in Japan: a review. Mining Geol Jpn Spec Issue 12: 47–55

Shikazono N, Kawahata H (1987) Compositional differences in chlorite from hydrothermally altered rocks and hydrothermal ore deposits. Can Mineral 25: 465–474

Shimoda S (1978) Interstratified minerals. In: Sudo T, Shimoda S (eds) Clays and clay minerals of Japan, Development in Sedimentology 26. Elsevier, Amsterdam, pp 265–322

Shimoda S (1980) A Li-bearing tosudite and some mineralogical problems of tosudite found in Japan. Sci Rep Inst Geosci Univ Tsukuba B-1: 97–105

Shinohara H (1991) Geochemistry of magmatic hydrothermal system. J Jpn Geochem Soc 25: 27–38

Shirozu H (1974) Clay minerals in altered wall rocks of the Kuroko-type deposits. Mining Geol Jpn Spec Issue 6: 303–310

Shirozu H (1978) Chlorite minerals. In: Sudo T, Shimoda S (eds) Clays and clay minerals of Japan. Development in Sedimentology 26. Elsevier, Amsterdam, pp 243–264

Shirozu H (1985) Formation of clay minerals by hydrothermal action and their mineralogical properties. J Clay Sci Soc Jpn 25: 113–118

Shirozu H, Iwasaki T (1980) Clay minerals in alteration zones of Kuroko deposits, with special reference to montmorillonite. J Jpn Assoc Mineral Petrol Econ Geol Spec Issue 2: 115–121

Shirozu H, Sakasegawa T, Katsumoto N, Ozaki M (1975) Mg-chlorite and interstratified Mg-chlorite/saponite associated with Kuroko deposits. Clay Sci 4: 305–321

Small JS, Hamilton DL, Habesch S (1992) Experimental simulation of clay precipitation within reservoir sandstones 2: mechanism of illite formation and controls on morphology. J Sediment Petrol 62: 520–529

Srodon J (1980) Precise identification of illite/smectite interstratification by X-ray powder diffraction. Clays Clay Min 28: 401–411

Srodon J (1981) X-ray identification of randomly interstratified illite-smectites in mixtures with discrete illite. Clay Min 16: 297–304

Srodon J, Eberl DD (1984) Illite. In: Bailey SW (ed) Micas. Reviews in Mineralogy 13, Mineralogical Society of America, Washington, DC, pp 495–544

Srodon J, Morgan DJ, Eslinger EV, Eberl DD, Karlinger MR (1986) Chemistry of illite/smectite and end-member illite. Clays Clay Min 34: 368–378

Srodon J, Elsass F, McHardy WJ, Morgan DJ (1992) Chemistry of illite-smectite inferred from TEM measurements of fundamental particles. Clay Min 27: 137–158

Steefel CI, Lasaga AC (1992) Putting transport into water-rock interaction models. Geology 20: 680–684

Steiner A (1968) Clay minerals in hydrothermally altered rocks in Wairakei, New Zealand. Clays Clay Min 16: 193–213

Stoffregen R (1987) Genesis of acid sulfate alteration and Au–Cu–Ag mineralization at Summitville, Colorado. Econ Geol 82: 1575–1591

Sudo T, Takahashi H (1971) The chlorites and interstratified minerals. In: Gard JA (ed) Electron-Optical Investigation of Clays. Mineralogical Society London, pp 277–300

Sverjensky DA, Hemley JJ, D'Angelo WM (1991) Thermodynamic assessment of hydrothermal alkali feldspar-mica-aluminosilicate equilibria. Geochim Cosmochim Acta 55: 989–1004

Tamura M (1982) Alteration minerals and mineralization in the Shakanai Kuroko deposit, Akita Prefecture. Mining Geol Jpn 32: 379–390

Taylor HP Jr (1974) The application of oxygen and hydrogen isotope studies to problems of hydrothermal alteration and ore deposition. Econ Geol 69: 843–883

Thompson G (1983) Basalt–seawater interaction. In: Rona PA, Bostrom K, Laubier L, Smith KJ Jr (eds) Hydrothermal processes at seafloor spreading centers. Plenum Press, New York, pp 225–278

Thompson JB Jr (1955) The thermodynamic basis for the mineral facies concept. Am J Sci 253: 65–103

Tomasson J, Kristmannsdottir H (1972) High temperature alteration minerals and thermal brines, Reykjanes, Iceland, Contrib. Mineral Petrol 36: 123–134

Tomita K, Takahashi H, Watanabe T (1988) Quantification curves for mica/smectite interstratifications by X-ray powder diffraction. Clays Clay Min 36: 258–262

Uno Y, Takeshi H (1979) Exchange cations and structural formulae of montmorillonites in the Nakajo acid clay deposit, Niigata Prefecture. J Mineral Soc Jpn Spec Issue 14: 90–103

Uno Y, Takeshi H (1982) Rock alteration and formation of clay minerals in the Ugusu silica deposit, Izu Peninsula, Japan. Clay Sci 6: 9–42

Urabe T, Scott SD, Hattori K (1983) A comparison of footwall-rock alteration and geothermal systems beneath some Japanese and Canadian volcanogenic massive sulfide deposits. Econ Geol Monogr 5: 345–364

Utada M (1980) Hydrothermal alteration related to igneous acidity in Cretaceous and Neogene formations of Japan. Mining Geol Jpn Spec Issue 8: 67–83

Utada M (1988) Hydrothermal alteration envelop relating to Kuroko-type mineralization: a review. Mining Geol Jpn Spec Issue 12: 79–92

Utada M, Ishikawa T (1973) Alteration zones surrounding "Kuroko-type" ore deposits in Nishiaizu district–Especially the analcime zone for an indicator of exploration of the ore deposits. Mining Geol Jpn 23: 213–226

Utada M, Aoki M, Inoue A, Kusakabe H (1988) A hydrothermal alteration envelope and alteration minerals in the Shinzan mineralized area, Akita Prefecture, northeast Japan. Mining Geol Jpn Spec Issue 12: 67–77

Velde B (1965) Experimental determination of muscovite polymorph stabilities. Am Mineral 50: 436–449

Velde B (1985) Clay Minerals: A physico-chemical explanation of their occurrence. Development in Sedimentology 40, Elsevier, Amsterdam

Velde B (1992) The stability of clays. In: Price GD, Ross NL (eds) The Stability of Minerals. Chapman & Hall, London, pp 329–351

Velde B, Brusewitz AM (1986) Compositional variation in component layers in natural illite/smectite. Clays Clay Min 34: 651–657

Velde B, Vasseur G (1992) Estimation of the diagenetic smectite to illite transformation in time-temperature space. Am Mineral 77: 967–976

Von Damm KL, Edmond JM, Grant M, Walden B, Weiss RF (1985) Chemistry of submarine hydrothermal solutions at 21 °N, East Pacific Rise. Geochim Cosmochim Acta 49: 2197–2220

Watanabe T (1981) Identification of illite/montmorillonite interstratifications by X-ray powder diffraction. J Mineral Soc Jpn Spec Issue 15: 32–41

Watanabe T (1988) The structural model of illite/smectite interstratified minerals and the diagram for its identification. Clay Sci 7: 97–114

Watanabe Y (1986) Neogene regional stress field inferred from the pattern of ore veins in Hokkaido, Japan. Mining Geol Jpn 36: 209–218

Weaver CE, Pollard LD (1973) The chemistry of clay minerals. Development in Sedimentology 15, Elsevier, Amsterdam

Weir AH, Nixon, HL, Wood RD (1962) The thickness of dispersed clay flakes. Clays Clay Min 9: 419–423

Weir AH, Ormerod EG, ElMansey IMI (1975) Clay mineralogy of sediments of the western Nile delta. Clay Min 10: 369–386

White DE (1957) Thermal waters of volcanic origin. Bull Geol Soc Am 68: 1637–1658

Whitney G (1990) Role of water in the smectite-to-illite reaction. Clays Clay Min 38: 343–350

Whitney G, Northrop HR (1988) Experimental investigation of the smectite to illite reaction: Dual reaction mechanisms and oxygen-isotope systematics. Am Mineral 73: 77–90

Wohletz K, Heiken G (1992) Volcanology and Geothermal Energy. University of California Press, Berkeley

WoldeGabriel G, Goff F (1989) Temporal relations of volcanism and hydrothermal systems in two areas of the Jemez volcanic field, New Mexico. Geology 17: 986–989

WoldeGabriel G, Goff F (1992) K/Ar dates of hydrothermal clays from core hole VC-2B, Valles Caldera, New Mexico and their relation to alteration in a large hydrothermal system. J Volcano Geotherm Res 50: 207–230

Yagi M (1992) Characteristics of hydrothermal alteration and its effects on oil reservoirs related to Miocene volcanism in the Yurihara oil and gas field, northern Honshu, Japan. Res Rep JAPEX Res Cent 8: 27–79

Yamamoto T (1967) Mineralogical studies of sericite associated with Poseki ores in the western part of Japan. Mineral J 5: 77–97

Yoder HS, Eugster HP (1955) Synthetic and natural muscovites. Geochim Cosmochim Acta 8: 225–280

Yoneda T (1989) Chemical composition of chlorite with special reference to the iron vs. manganses variation, from some hydrothermal vein deposits, Japan. Mining Geol Jpn 39: 393–401

Zen E-An (1962) Problem of the thermodynamic status of the mixed-layer minerals. Geochim Cosmochim Acta 26: 1055–1067

Zen E-An (1967) Mixed-layer minerals as one-dimensional crystals. Am Mineral 52: 635–659

Subject Index

allavardite 36
allophane 145
alteration (hydrothermal) 2, 193, 202, 268, 269
  acid 284, 291
  alkaline 284, 294
  assemblages, reaction defined 283
  assemblages, temperature defined 282
  calcium-magnesium 297
  classification 284
  deep sea 299
  mechanisms 253
  mineralogy 304
  minerals, zoning type 277, 283
  propylitic 253, 298
  sodic 299
  zones, morphology 288
altered wall rocks 248, 250
alteroplasma 91
alunite 293
Amazon basin 166, 170
amphibolite 115
analcime 40, 237
Andosols 145
anion exchange 146
authigenic 163

basalt weathering 126
  alteration (hydrothermal) 297
basic cations 152
beidellite 31, 90, 109, 307
  high charge 153
  iron 116, 122, 128
bentonite 194
berthérine 24, 29, 199
biotite 37, 114
Brownian motion 177
brucite 34, 189
burial history 230

calcareous crust 154
Cambisol 102

carbonate compensation depth 176
cation selectivity 16
CEC (cation exchange capacity) 16, 107, 142, 188
celadonite 30, 37, 128, 194, 253, 300
chabazite 41
chamosite 33, 200
chemical equilibrium 232
  potential 81
  potential-chemical potential diagram 86, 99, 280, 281
  potential-composition diagram 85, 95, 100
  substitutions 23
Chernobyl 190
Chernozems 60
chlorite 33, 107, 112, 125, 136, 165, 193, 199, 205, 251, 263, 303, 320
  compositions 34
  geothermometer 319
  hydrothermal alteration 318
  soil 126
chlorite/smectite mixed layer clay 247, 303, 315, 317
chlorite/vermiculite mixed layer clay 117
clay cycle 3
  detrital 204
  expanding 12
  illuviation 58
  minerals, structure 17, 20
  -organic complex 113
  origin 5
  size fractions 104
  stability 3
  swelling 12
  transport 14
clay-water interaction 13
climate 47, 51, 212
  paleo 210
  tropical 148
clinoptilolite 40
colloid 190
compaction 220, 224

composition diagram   93
convection systems   272
corrensite   36, 125, 193, 201, 251, 263, 317
   stability   318
critical point   271
crystalline water   22

deep sea, hyrothermal alteration   297, 299
   sediments   182
dehydration   6
deposition   169
   ice   185
   wind   182
detrital minerals   163, 235, 240
diagenesis   2, 211, 221
diaspore   294
dioctahedral   18, 26, 31
dissolution, incongruent   77
dissolution-recrystallization   73
dissolution voids   255
dombassite   252
double layer attraction   179
drainage   49

epistilbite   41
epithermal deposits   268, 296
equilibrium, chemical   232
equipotential   84
erionite   40
erosion   46, 162, 169
estuaries   173, 177
evaporation   49
evapo-transporation   49
exchange ions (see also CEC)   15
expanding minerals (see smectites)
extensive variable   79
extractable ion   103

Ferralsols   60
Ferrisols   148
field water capacity   49
fissures   92
flocculation   177
   pH   182
   salt   178
fluid compositions   273
   regime   260
fractures   288
fugacity, CO_2 279

gabbro, weathering profile   115, 120
geothermal systems, fossil   271
gibbsite   20, 34, 101

gismondine   41
glacial deposition, transport   185
glauconite   30, 37, 129, 195
global flux   168
goethite   39, 148
Goto deposit, pyrophyllite   294
grain size   8, 14
granite weathering   87
gravitational potential   69

halloysite   29, 117, 128, 146, 285, 295
halmyrolysis   194
heat conduction   290
hectorite   191
hematite   39, 148
hemiplagic   176
heulandite   40
humic compounds   53
hydration reactions   5
   states   12
hydraulic conductivity   72
hydrobiotite   36
hydrocarbons   237
hydrogen exchange   5
hydrothermal alteration (see alteration)
hydroxy complex   34, 110, 143

ice rafting   186
iddingsite   127
illite   30, 88, 94, 101, 206, 213, 227, 229, 235,
   265, 286, 303
   ferric   198
illite/smectite mixed layer   37, 132, 247, 255,
   286, 303, 304
   chemical composition   307
   grain morphology   309
   hydrothermal   306, 316
   layer charge   307
   ordering   305
   structure   305
illite/vermiculite   134
illuviation   140
imogolite   145, 240
intensive variable   79
intergrade mineral   110, 113, 139
interlayer cations   28, 314
interstratified minerals (see mixed layered
   minerals)
ion exchange   187
ions absorbed   15
iron hydroxide   103, 143
iron oxide   38, 91, 122, 127, 131
I/S (see illite/smectite)

K-Ar dating 299, 312
kaolinite 20, 90, 94, 101, 122, 131, 147, 171, 205, 209, 234, 244, 250, 252, 285, 295
kinetics, clay transformations 226
  hydrothermal alteration 273
  reaction 312
$K_2O-Al_2O_3-SiO_2-H_2O$ system 279
Kuroko deposit 268, 296

lawsonite 41
lepidocrosite 39
loess 140, 183
Los Angeles basin 229
Luvisols 59

maghemite 39
magmatic fluids 294
mass balance 260
matrix potential 69
mica 107, 137
mica-like 30
mica/vermiculite 107, 114
microaggregates 150
microsites 78, 90
microsystem, fissural 80
microsystem, plasmic 78
mineral families 28
mineral species 27
mineral transformations, transport and deposition 187
mixed layered minerals 35, 247
  random 36, 243
  regular (ordered) 36, 305
montmorillonite (see also smectite) 30, 31, 90, 109, 131, 143, 250, 303
mordenite 285
muscovite 252

natrolite 40
nepheloid layers 174
network, silicon-oxygen 18
nodules 149
nontronite 31, 119, 122, 124, 131, 191, 253, 300

octahedra 19
oolites, iron 200
organic flocculation 180
organic matter 52, 58, 134, 138, 237
osmotic potential 69
Ostwald ripening 310

palagonitization 193

palygorskite 24, 33, 37, 154, 193, 201, 212, 303
Paris basin 228
particle shapes 9, 310
particle size 14
pedolasma 91
pelagic settling 174
peridotite 115
pH 96, 147, 292, 302
phase rule 83, 283
phengite 254
phillipsite 40
phosphorus 190
phyllic alteration 284
phyllosilicates 17, 20
plagonitization 300
Podsols 60
podzolisation 135
polymorphs 22
polymorph, silica 235
polytype, chlorite 319
  mica 314
Porhyry copper deposits 268, 294
porosity 222, 311
potassic alteration 284, 294
pressure gradient 248
  potential 69
propylitic alteration 284
proton-cation exchange 76
provenance 207
pyrophyllite 20, 30, 252, 285

reaction mechanisms 309
  clay mineral 234
  progress 283
  sequence 256
rectorite 36, 313
red clays 176
reichweite 305
repeat distances 20

saline soil 154
Salton Sea 229
saponite 30, 32, 116, 119, 124, 191, 251, 253, 300, 303
saprock 91
saprolite 91
scavenging 177
scoelite 41
sedimentation 2, 14, 162, 222
  history 230
  rate 14, 226
sepiolite 33, 37, 193, 201, 243, 303

sericite 296, 314
serpentine 29, 303
serpentinite 115
  weathering 124
shapes, clay particles 11
silica polymorph 235
siliceous ooze 176
smectite (expanding mineral) 28, 31, 33, 37, 101, 107, 135, 152, 155, 194, 183, 206, 213
  dioctahedral 31
  dioctahedral, hydrothermal alteration 227, 235, 285, 313
  high charge 264
  low charge 264
  -organic complex 153
  trioctahedral 32, 234
soils 44, 205
  aggregates 63
  clays 102
  climates 52
  density 68
  development 55
  horizons 45
  lateral transfer 57
  map 60
  mechanical resistance 68
  -moisture retention curve 70
  organic matter 52
  pores 63, 66
  profiles 141
  saturation 71, 72
  structure 62, 151
  temperature 47
  water regime 50
solid solution 24, 28
solution transport 231
sorption 189
spacing, 7Å 20, 29
  10Å 20, 29
  14Å 20, 33
  060 21, 33
stability, chemical 311
stable isotopes 275, 294, 295
stevensite 32, 191, 193
stilbite 40
submarine weathering 198
substitutions, in crystal structures 24
sudoite 36
surface area of minerals 10
suspension 169
swelling mineral (see also smectite) 12

talc 30, 117, 119, 123, 303

temperature (hydrothermal solutions) 271
  -activity diagram 278
tetrahedra 18
thermal gradient 273
tie-line 83, 95
topography 57
tosudite 36, 252
transformation 165
transport 14
  rivers 169, 171
trioctahedral 19, 26, 303
turbidity currents 174, 176

Usgusu silica deposit 292

vein 248
  crosscutting 262
  drainage 249
  infiltration 249
  injection 248
  zonation 249
verdine facies 199
vermiculite 30, 32, 90, 107, 117, 119, 123, 135, 136, 189
vermiculite/smectite mixed layer mineral 114
Vertisols 60, 152
volcanic ash 194
volcanic gas 275

water suction 73
  tension 73
  to rock ratio 284
weathering 2, 44, 164, 188, 211
  basalt 126
  carbonate 131
  climate 134
  coherent rock 88
  glauconite 130
  mineral reactions 118
  profile 89
  profile, gabbro 120
  profile, serpentinite 124

X-ray diffraction, soil clays 104, 111, 139

zeolite 39, 191, 285
  alkali 40
  chemical series 287
  facies diagenesis 241
  mineralogy 236
zonation, alteration 254
  kinetics 256

Springer-Verlag and the Environment

We at Springer-Verlag firmly believe that an international science publisher has a special obligation to the environment, and our corporate policies consistently reflect this conviction.

We also expect our business partners – paper mills, printers, packaging manufacturers, etc. – to commit themselves to using environmentally friendly materials and production processes.

The paper in this book is made from low- or no-chlorine pulp and is acid free, in conformance with international standards for paper permanency.

Printing: Saladruck, Berlin
Binding: Buchbinderei Lüderitz & Bauer, Berlin